Nanotechnology in Civil Infrastructure

Kasthurirangan Gopalakrishnan,
Bjorn Birgisson, Peter Taylor,
and Nii O. Attoh-Okine (Eds.)

Nanotechnology in Civil Infrastructure

A Paradigm Shift

Springer

Editors

Prof. Dr. Kasthurirangan Gopalakrishnan
Department of Civil, Construction and Environmental Engineering
Iowa State University
353 Town
Ames, IA 50011
USA
E-mail: rangan@iastate.edu

Prof. Bjorn Birgisson
Swedish Royal Institute of Technology (KTH), Kungl Tekniska Högskolan
SE-100 44 Stockholm
Sweden
E-mail: bjornbir@kth.se

Dr. Peter C. Taylor
National Center for Concrete Pavement Technology (CP Tech Center)
Iowa State University
353 Town, Ames, IA 50011
USA
E-mail: ptaylor@iastate.edu

Dr. Nii O. Attoh-Okine
Associate Professor
Department of Civil and Environmental Engineering, University of Delaware
Newark DE 19716
USA
E-mail: okine@ce.udel.edu

ISBN 978-3-642-16656-3 e-ISBN 978-3-642-16657-0

DOI 10.1007/978-3-642-16657-0

Library of Congress Control Number: 2011922618

Typesetting: Data supplied by the authors

Cover Design: Scientific Publishing Services Pvt. Ltd., Chennai, India

Printed on acid-free paper

9 8 7 6 5 4 3 2 1

springer.com

Preface

Nanotechnology was conceived with a vision to advance our understanding and control of matter at nanoscale toward national economic benefit, national and homeland security, and improved quality of life. It enables changes at the molecular level to optimize material behavior and performance of civil infrastructure systems such as buildings and highways at the macro-functional level. The resulting nano-modified high-performance construction materials and systems are characterized by higher strength, greater durability, increased speed of construction, and reduced environmental impact.

Recent advances made in the development and characterization of nanotechnology based high-performance, sustainable, "designer" civil engineering materials, smart structures, and systems forms the motivation for bringing out this timely book. Each chapter in this book has been peer-reviewed by at least two anonymous referees to assure the highest quality. A brief description of each chapter follows.

The chapter entitled "Multifunctional and Smart Carbon Nanotube Reinforced Cement-based Materials" reviews recent progress and advances of Carbon Nano-Tubes (CNTs) reinforced cement-based materials, with attention to their fabrication methods, mechanical properties, electrical and piezoresistive properties, thermal conductive and damping properties, and potential structural applications.

The chapter entitled "Applications of Nanotechnology in Road Pavement Engineering" focuses on providing the required background and information on the status of nanotechnology applications in road pavements. It starts with background on pavements and nanotechnology, and then focuses on materials, characterization and concerns and issues.

The chapter entitled "Application of Nanoscience Modeling to Understand the Atomic Structure of C-S-H" surveys the computational tools available and their limitations for state-of-the-art nanoscience modeling of C-S-H. Application of nanoscience in improving the cement paste by studying the interaction of salt, water etc. using the crystal structure of Tobermorite and Jennite are discussed first followed by a methodology to find the amorphous structure of C-S-H from the crystal structure of Tobermorite.

The chapter entitled "The Effect of SWCNT and Other Nanomaterials on Cement Hydration and Reinforcement" discusses the effects of nano-titania, nano-calcium carbonate and nano-alumina dispersed by sonication with Ordinary Portland Cement (OPC) on hydration and compares it to the effect of Single

Walled CNT (SWCNT) dispersed by the same method. The experimental results are used to draw conclusions about the nucleation mechanism in SWCNT composites and on the impact of SWCNT dispersion method on the performance of those composites.

The chapter entitled "Nanomaterials-enabled Multifunctional Concrete and Structures" reports on the experimental investigations of the effect of various loading states on the self-sensing ability of nanoconcrete and the development of a theoretical model to predict the strain gauge factor of nanoconcrete under various loading or environmental conditions.

The chapter entitled "Nano-Optimized Construction Materials by Nano-Seeding and Crystallization Control" summarizes existing technologies for analyzing and changing the nano-structure of cement and gypsum construction materials. It also shows first results in homogeneous seeding the precipitation of calcium silicate hydrates within a real Portland cement composition.

The chapter entitled "Next-Generation Nano-based Concrete Construction Products: A Review" covers the primary areas of nano-engineering and nano-modifications of cementitious systems, applications in building materials, nanotechnology based devices etc.

The chapter entitled "Optimization of Clay Addition for the Enhancement of Pozzolanic Reaction in Nano-modified Cement Paste" presents a process for obtaining nano-size silicate platelets for use in cement paste from exfoliating Cloisite Na+ clay particles.

The chapter entitled "Characterization of Asphalt Materials for Moisture Damage Using Atomic Force Microscopy and Nanoindentation" describes the use of Atomic Force Microscopy (AFM) and nanoindentation techniques to gain accurate insight into moisture damage performance of asphalt materials at nanoscale.

Finally, the chapter entitled "Nanoclay-modified Asphalt Binder Systems" provides insight on nanomodification of asphalt binder systems to enhance their mechanical properties.

Researchers and practitioners interested in the design and state-of-the-art characterization of nano-engineered civil infrastructure systems will find this book very useful. This book will also serve as an excellent state-of-the-art reference material for undergraduate, graduate and postgraduate students in the field of nanotechnology.

November 9, 2010 Kasthurirangan (Rangan) Gopalakrishnan
Ames, Iowa

Contents

Multifunctional and Smart Carbon Nanotube Reinforced Cement-Based Materials

Baoguo Han, Xun Yu, and Jinping Ou

Abstract. Nanotechnology has changed and will continue to change our vision, expectations and abilities to control the material world. These developments will definitely affect the field of construction and construction materials. Carbon nanotubes (CNTs) are considered to be one of the most beneficial nano-reinforcement materials. The combination of high aspect ratio, small size, low density, and unique physical and chemical properties make them perfect candidates as reinforcements in multifunctional and smart cement-based materials. Here, we review recent progress and advances of CNTs reinforced cement-based materials, with attention to their fabrication methods, mechanical properties, electrical and piezoresistive properties, thermal conductive and damping properties, and potential structural applications. Future challenges for the development and deployment of multifunctional and smart CNTs reinforced cement-based materials and structures are also discussed.

1 Introduction

Many civil infrastructures around the world are in a state of utter disrepair, and significant efforts are needed to render the failing infrastructures back to a serviceable and safe state. The root of the problem is at the apparent lack of durability in construction materials, inability to provide timely maintenance, absence of

Baoguo Han
School of Civil Engineering, Harbin Institute of Technology, Harbin, 150090 China
e-mail: hithanbaoguo@china.com.cn, baohan@d.umn.edu

Baoguo Han · Xun Yu
Department of Mechanical and Industrial Engineering, University of Minnesota, Duluth, MN 55812, USA
e-mail: xunyu@d.umn.edu

Jinping Ou
School of Civil and Hydraulic Engineering, Dalian University of Technology, Dalian, 116024 China
e-mail: oujinping@dlut.edu.cn

K. Gopalakrishnan et al. (Eds.): Nanotechnology in Civil Infrastructure, pp. 1–47.
springerlink.com

advanced condition assessment tools and lack of long-lasting effective repair materials and technologies. The multifunctional and smart composites are ideal materials for achieving these goals [Banthia 2009, Saafi 2009].

Cement-based materials (including cement paste, cement mortar and concrete) have been widely used for many years as a composite material for various types of structures. Compared to other construction materials such as metals and polymers, cement-based materials are significantly more brittle and exhibit a poor tensile strength. Cement-based materials carry flaws and micro-cracks both in the material and at the interfaces even before an external load is applied. These defects and micro-cracks emanate from excess water, bleeding, plastic settlement, thermal and shrinkage strains and stress concentrations imposed by external restraints, etc. Under an applied load, distributed micro-cracks propagate, coalesce and align themselves to produce macro-cracks. When loads are further increased, conditions of critical crack growth are attained at the tips of the macro-cracks and unstable and catastrophic failure is precipitated. Under fatigue loads, cement-based materials crack easily, and cracks create easy access routes for deleterious agents. This will lead to early saturation, freeze-thaw damage and scaling, discoloration and so on [Banthia 2009, Aitcin 2000, Mather 2004]. There has been a huge demand to monitor structures cracking and prevent cracks from propagating further. These efforts are very important for ensuring timely repair, safety and long-term durability of critical structures [Hou and Lynch 2005, Chung 2002, Ou and Han 2009, Azhari 2008]. Non-destructive evaluations, such as attaching or embedding foreign sensors (e.g. resistance strain gauges, optic sensors, piezoelectric ceramic, shape memory alloy and fiber reinforced polymer bar etc. [De Backer et al. 2003, Leng et al. 2006, Park et al. 2006, Li et al. 2004, Hiroshi et al. 2004] onto or into structures, have been used in many ways to accommodate the demand, yet these sensors have some drawbacks including poor durability, low sensitivity, high cost, low survival rate and/or unfavorable compatibility with structures (i.e. loss of structural mechanical properties). It would be desired that the structural material itself has the sensing capability (i.e. structural materials are multifunctional or smart). Self-sensing cement-based materials not only have potential in the field of structural health monitoring and condition evaluation for concrete structures, but also can be used for highway traffic monitoring, border security, structural vibration control and so on [Han and Ou 2007, Han et al. 2008].

Self-sensing (piezoresistive) cement-based materials are reinforced by electrically conductive fillers to increase their ability to sense the strain, stress or crack in themselves while maintaining good mechanical properties[Chung 2002]. The electrically conductive fillers can be classified as fibrous and particle fillers. Examples of suitable fibrous fillers include short carbon fibers (CFs), surface modification CFs, steel fibers, carbon coated nylon fibers and etc, and those of effective particle fillers include carbon black, steel slag, nickel powder and etc[Han and Ou 2007, Han et al. 2008]. As the piezoresistive cement-based materials are deformed or stressed, the contact state between the fillers and the matrix is changed, which affects the electrical resistance of the cement-based materials. Strain, stress, crack and damage can therefore be detected through measurement of the electrical resistance[Ou and Han 2009, Hou and Lynch 2005, Chung 2002, Azhari 2008]. The electrical resistance of the piezoresistive cement-based materials decreases maximally by 79.82% under once compressive loading to destruction. The sensitivities of the materials to compressive stress and strain can reach 0.17/MPa and 1336.5 (extraordinarily high compared to the values of around 2 for conventional resistive strain gauges) respectively [Han et al. 2008]. Therefore, the piezoresistive cement-based materials have the ability to sense tiny structural flaws before they become significant, which could be used in monitoring the internal condition of structures[Ou and Han 2009, Hou and Lynch 2005, Chung 2002, Azhari 2008]. In addition, the fibrous fillers can introduce ductility and toughness without sacrificing existing cement-based materials properties. The presence of the fibrous fillers also controls the cracking so that the cracks do not initiate and propagate catastrophically as in the case of conventional cement-based materials [Banthia 2009].

The field of nanotechnology is rapidly maturing into a fertile and interdisciplinary research area from which new multifunctional and smart materials can be developed. Since carbon nanotubes (CNTs) were discovered by S. Iijima [1991], they have been widely used for a variety of application due to their excellent physical properties: high strength and Young's modulus (the tensile strength and Young's modulus of CNTs are 20 times and 10 times that of CFs respectively), high bonding force with matrix (the interlaminar shear strength of CNTs reinforced epoxy materials is 10 times that of CFs reinforced epoxy materials), large deformation and high ductility (the elongation at break of CNTs is 18%, and is 18 times that of CFs), high aspect ratio (>500), and excellent electrical conductivity. In particular, the extremely high aspect ratio and low density of CNTs makes them easy to form a conductive and mechanical reinforcement networks with a CNT doping level as low as 0.05 wt. %. CNTs also have interesting piezoresistive properties. When CNTs are subjected to stress/strain, their electrical properties change with the level of stress/strain, expressing a linear and reversible piezoresistive response even for a huge strain of 3.4% [Meyyappan 2005, Grunlan et al. 2004, Blanchet et al. 2004, Kim et al. 2005, Tombler et al. 2000, Cao et al. 2003, Yu and Kwon 2009]. Because the properties of CNTs are superior to that of CFs (the most effective fibrous fillers), and the use of CNTs as reinforcements is intended to move the reinforcement behavior from the macroscopic to the nanoscopic level, it is promising to develop multifunctional or smart cement-based materials without adding any additional weight and sacrificing existing cement-based materials properties. In recent years, more and more researchers have started investigations on utilizing CNTs in multifunctional and smart cement-based materials.

Here, we review recent progress and advances that have been made in fabrication, mechanical, electrical and piezoresistive properties, and potential structural applications of multifunctional and smart CNTs reinforced cement-based materials. Future challenges for continued development and deployment of multifunctional and smart CNTs reinforced cement-based materials and smart structures are also discussed.

2 Fabrication of CNTs Reinforced Cement-Based Materials

Similar to other CNTs reinforced materials, one major issue in fabricating high-quality CNTs reinforced cement-based materials is how to homogeneously distribute CNTs in cement-based materials.

2.1 Structure and Dispersion of CNTs

Structurally, CNTs can be approximated as ''rolled-up'' sheets of graphite (Fig.1). In this graphite, sp^2 hybridization occurs, where each carbon atom is connected evenly to other three carbon atoms (120°) in the xy plane, and a weak π bond is present in the z axis. The sp^2 set forms the hexagonal (honeycomb) lattice of a sheet of graphite [Dresselhaus et al. 2000, Odom et al. 2000, Terrones 2003].

Varieties of CNTs include single-walled nanotubes (SWNTs) and multi-walled nanotubes (MWNTs). SWNTs only have one wall that constitutes a tube, whereas MWNTs are made up of multiple walls that can slide against each other. The diameters of CNTs range from just a few nanometres in the case of SWNTs to several tens of nanometres for MWNTs. The lengths of the tubes are usually in the micrometer range [Dresselhaus et al. 2000].

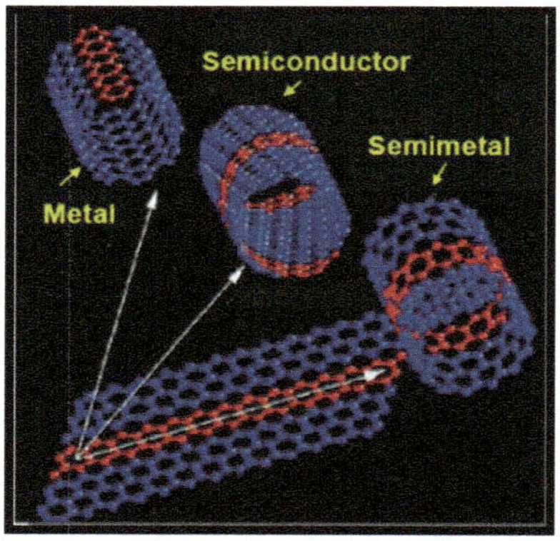

Fig. 1 Structure of CNTs [Images obtained from webpage http://www.tyndall.ie/ research/ electronic-theory-group/nanotube.html].

Either SWNTs or MWNTs presents two main geometrical structural characteristics: nano-scale dimensions and high aspect ratios. Nano-scale dimensions of CNTs turn their dispersion in matrix into a challenge, because as the surface area of a particle increases, so does the attractive forces between the aggregates [Lourie et al. 1998]. High aspect ratios combined with high flexibilities, increase the possibility of nanotube entanglement and close packing [Thess et al. 1996]. The low dispersability stems from the tendency of pristine nanotubes to assemble into bundles or ropes (Fig. 2), which contain hundreds of close-packed CNTs tightly

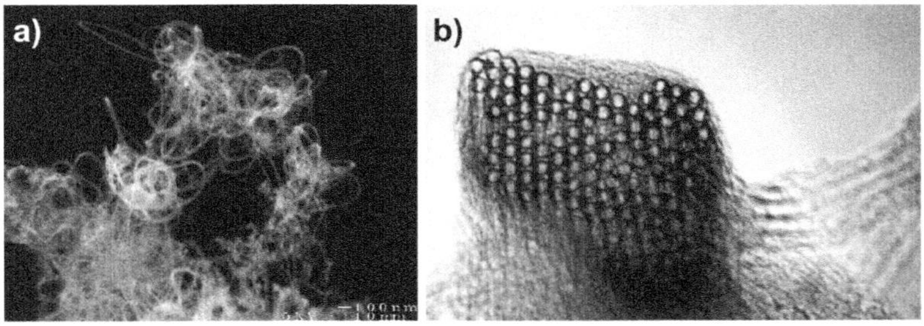

Fig. 2 (a) SEM image of a SWNT bundles (b) TEM image of the cross section of a SWNT bundle [Images obtained from Smalley's webpage http://www.ruf.rice.edu/~smalleyg/image_gallery.htm].

bound by van der Waals attraction energy of 500 eV/μm of tube-tube contact [Girifalco 2000].

In practice, CNTs do not have perfect geometrical structures, and they have very complex surface structures (Fig. 3). The CNTs' surface structures (e.g. surface atom arrangement, roughness, defect and functional group) dominate the surface energy and the wettability of CNTs' surface, which is an important factor affecting solubility and dispersibility of CNTs. For example, a small change of CNTs' surface structures can cause a switch from hydrophobicity to hydrophilicity (Fig. 4) [Kakade and Pillai 2008]. Therefore, the surface modification is commonly used to improve the wettability of CNTs' surfaces, thus enhances the solubility and dispersibility of CNTs[Vaisman et al. 2006].

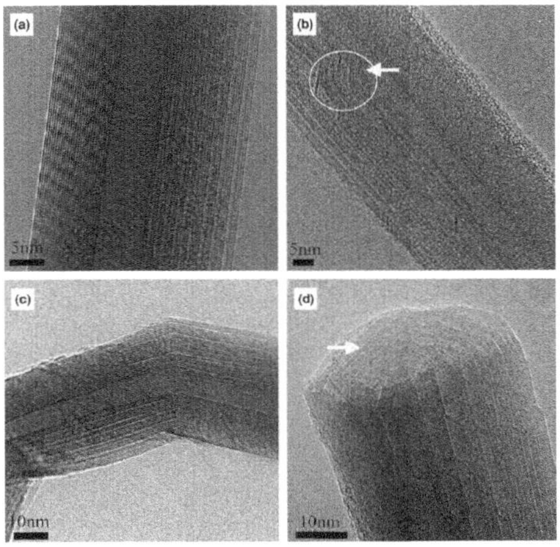

Fig. 3 TEM images of CNTs: (a)-(c) TEM images exhibit highly unevenly spaced lattice fringes on one side of the hollow core and (d) cross-sectional TEM image of nanotubes are not circular, rather show flat regions [Musso et al.2009].

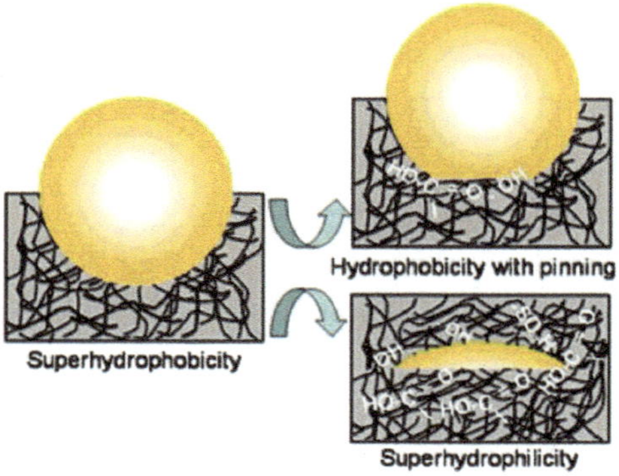

Fig. 4 Schematic of CNTs' hydrophobicity/hydrophilicity.

Since CNTs tend to self-associate into micro-scale aggregates, their disaggregation and uniform dispersion are critical challenges that must be addressed to successfully fabricate multifunctional and smart CNTs reinforced cement-based materials.

2.2 *Current Approaches for Dispersing CNTs in Cement-Based Materials*

Some works have reported disappointing results, showing little or no improvement in cement-based materials' properties by adding CNTs because of insufficient or poor dispersion of CNTs. For example, De Ibarra et. al. [2006] observed that a non-uniform distribution of the CNTs' bundles (Fig. 5) within the matrix provokes a decrease in the mechanical properties of the materials. Because the utilized SWNTs are straighter and more defect-free structures than the utilized MWNTs, SWNTs have a lower dispersion degree in matrix, resulting in a lower Young's modulus and hardness of the SWNTs reinforced cement-based materials. Additionally, a worse dispersion is presented (more agglomeration and less uniform

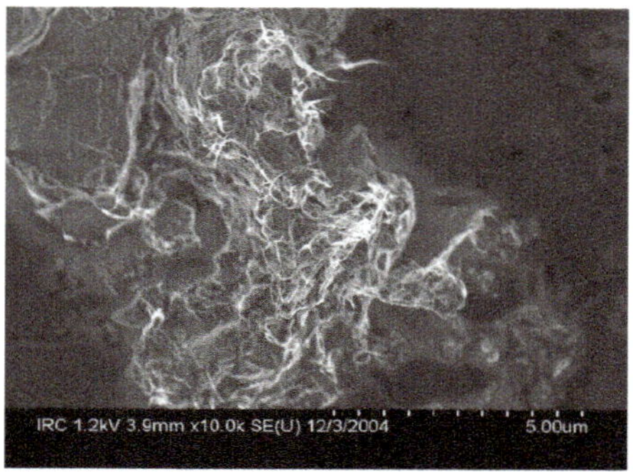

Fig. 5 CNTs bundles distributed on unhydrated cement paste [Makar et al. 2005].

dispersion throughout the matrix) when the nanatube quantity in the matrix is increased, and the hydration of cement is decreased, thus impairing the mechanical performance. As a result, the materials with higher concentration of CNTs present worse mechanical properties in both cases (with SWNTs and MWNTs).

Shah et al. [2009] found that in the CNTs reinforced cement-based materials where dispersion was achieved without the use of surfactant, MWNTs appear poorly dispersed in cement paste, forming large agglomerates and bundles (Fig. 6 a) and b)). This leads to the lowest fracture load of the meterials with no surfactant (Fig. 7).

Fig. 6 Surfactant concentration effect on nanotube dispersion: (a)-(d) represent a dispersant to MWNTs weight ratio of 0, 1.5, 4.0 and 6.25, respectively.

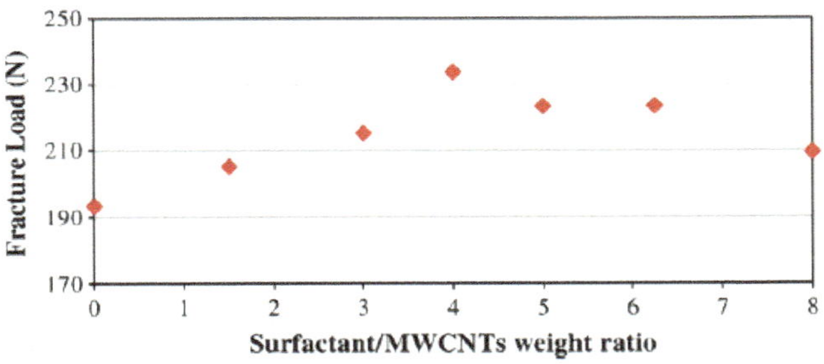

Fig. 7 Fracture load of 28 days w/c=0.5 cement paste reinforced with long MWNTs (0.08% by weight of cement) [Konsta-Gdoutos et al. 2010].

In order to develop high property CNTs reinforced cement-based materials and fully utilize unique properties of the tubes, the thermodynamic drive toward aggregation must be overcome. Makar et. al. [2005] and Sanchez [2009] dispersed SWNTs in ethanol or isopropanol by sonication to aid dispersion of CNTs in cement paste. Musso et al. [2009] dispersed MWNTs in acetone by means of an ultrasonic probe to minimize the size of aggregated MWNTs, thus to enhance dispersion of MWNTs in cement paste.

It is worth noting that a dispersion method, as an important step of the integral production chain, has to be selected in accordance with the processing conditions of the CNTs reinforced cement-based materials. Because water is an indispensable raw material for fabricating cement-based materials, one suitable route is to produce a homogeneous dispersed suspension of CNTs in water or liquor before CNTs are incorporated into cement-based materials. Thus, the fabrication procedure of CNTs reinforced cement-based materials can be divided into two steps: dispersion of CNTs in water or aqueous solution and dispersion of CNTs' suspension in the matrix. The dispersion of CNTs in water or aqueous solution is prerequisite for distributing CNTs in matrix. In addition, the dispersion of CNTs' suspension in matrix can be easily performed by typically mixing method for conventional cement-based materials. Therefore, the intensive research was focused on the dispersion of CNTs in water or aqueous solution for dispersing CNTs in cement-based materials. There are two approaches for dispersing CNTs in water or aqueous solution: the mechanical methods and the methods that are designed to alter the CNTs surface structures, either physically (non-covalent surface modification by surfactants) or chemically (covalent surface modification). Mechanical dispersion methods, such as high shear mixing and ball milling, separate CNTs from each other, but can also fragment the CNTs, decreasing their aspect ratios. Ultrasonication is another common physical method used to disperse CNTs into base fluids. Ultrasonic processors convert line voltage to mechanical vibrations. These mechanical vibrations are transferred into the liquid through creating pressure waves. This action causes the formation and violent collapse of microscopic bubbles. This phenomenon, referred to as cavitation, creates millions of shock waves. The cavitational collapse lasts only a few microseconds. Although the amount of energy released by each individual bubble is small, the cumulative effect causes extremely high levels of energy to be released, resulting in dispersion of objects and surfaces within the cavitation field [Shah et. al. 2009, Konsta-Gdoutos et al. 2010]. Saafi [2009] adopted mixing and ultrasonication method to produce the homogeneous dispersed solution of SWNTs and cement (Fig.8). Chaipanich et al. [2010] dispersed initially CNTs using ultrasonic with some part of mix water and than added the sonicated solution to fly ash cement.

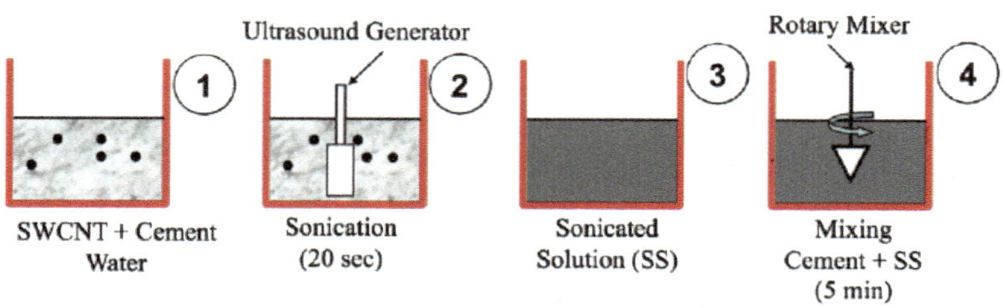

Fig. 8 CNTs reinforced cement paste fabrication process based on mechanical methods.

Ultrasonication alone is hard to separate CNTs from each other because of the CNTs' intrinsic hydrophobicity [De Ibarra el al. 2006], so it is often combined with surface modification method [Saafi 2009]. Li et al. [2005] and Yu et al. [2009] effectively dispersed covalent surface modification CNTs treated with a mixed solution of H_2SO_4 and HNO_3 in water and cement-paste by utilizing ultrasonic energy (Fig.9). Sanchez [2009] improved the dispersion of CNTs in the cement pastes by using covalent surface modification with nitric acid. The CNTs were found as individual fibers throughout the cement pastes and as entangled networks in pockets.

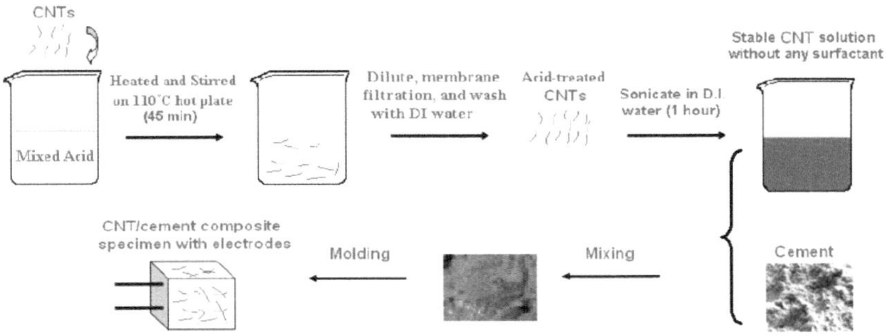

Fig. 9 CNTs reinforced cement paste fabrication process based on the covalent surface modification method combined with mechanical method.

The covalent surface modification method uses surface functionalization of CNTs to improve their chemical compatibility with the target medium, that is to improve their wettability and reduce their tendency to agglomerate. However, aggressive chemical functionalization, such as the use of neat acids at high temperatures, might introduce structural defects resulting in inferior properties for the tubes (e.g. decrease in mechanical property, electrical property and aspect ratio etc) [Vaisman et al. 2006].

The non-covalent surface modification method combined with ultrasonication is particularly attractive because of almost no impairment to pristine CNTs, low energy consuming and controllability. The non-covalent surface modification method is based on adsorption of the hydrophobic part of the surfactant at nanotubes sidewalls through van der Waals, π-π or CH-π, and aqueous solubility is provided by the hydrophilic part of the surfactant. Additionally, the stability of the dispersion of CNTs with adsorbed surfactant on their surface is guaranteed by electrostatic and/or steric repulsion [Hu et al. 2009]. Veedu [2010] and Azhari [2008] dispersed CNTs into water by using methylcellulose as surfactant and by means of ultrasonic energy, and they successfully fabricated cement-based materials with uniformly distributing CNTs. Yu et al. [2009] achieved an effective dispersion of MWNTs in water and cement paste by applying ultrasonication in combination with the use of a surfactant sodium dodecyl sulfate (SDS). Luo et al. [2009] used separately or jointly five surfactants [sodium dodecylbenzene sulfonate (NaDDBS), sodium deoxycholate (NaDC), Triton X-100 (TX100), Arabic gum (AG) and cetyltrimethyl ammonium bromide (CTAB)] to enhance solubilization/dispersion of MWNTs in aqueous solution and cement paste with a tip ultrasonication, and they found that the NaDDBS and TX10 with a mixing ratio of 3:1 by weight of cement and CNTs exhibits the best solubilization/dispersion capability.

Additionally, the covalent surface modification method and the non-covalent surface modification method can be combined to improve CNTs in the cement-based materials. Cwirzen et al. [2008] obtained stable and homogenous dispersions of MWCNTs in water by using non-covalent surface functionalization in combined with modification using polyacrylic acid polymers. They also proved that the surface covalent functionalization combined with modification using polyacrylic acid polymers has more superior dispersion effect compared only using polyacrylic acid polymers or GA. Han et al. [2009] achieved an effective dispersion CNTs in cement paste and cement mortar by using carboxylation MWNTs and NaDDBS.

Although the above-mentioned surfactants are very helpful for dispersing CNTs in cement-based materials, it is deserved to further investigate the effect of surfactants on mechanical, electrical and durable properties of cement based materials and the compatibility issues between surfactants and cement. Han et al. [2009] had found that the excessive doping level of NaDDBS or SDS have bad effect on the hydration of cement and will cause a marked air-entraining effect. Cwirzen et al. [2008] found that the hydration of the cement incorporating GA is slowed down. De Ibarra et al. [2006] also reported that SDS or TX100 impairs the cement hydration.

Considering the nature of the cement-based materials, researches have focused on methods of dispersal that are compatible with cement chemistry. The primary approach is to adopt commonly used admixtures such as water reducing admixtures, plasticizers and superplasticizers as surfactants of CNTs. There is an obvious correspondence with the use of these surfactants to improve the dispersion of cement particles. This can facilitate dispersion process of CNTs and enhance the uniformity and workability of nanocomposites. Initial research at the National Research Council Canada has shown that a small amount of CNTs can be dispersed by sonication in water containing 5% superplasticizer [Makar and Beaudoin 2003]. Shah et al. [2009] and Konsta-Gdoutos et al. [2010] achieved an effective dispersion of MWNTs with different lengths and concentrations in water by applying ultrasonic energy in combination with polycarboxylate-based superplasticizers. They proved that the application of ultrasonic energy is required for proper dispersion. Considering the effect of surfactant concentration and MWNTs' length and concentration on the distribution of CNTs in cement paste, they also suggested the optimum amount of MWNTs and superplasticizer for reinforcement in cement based materials according to the aspect ratio of CNTs (Fig.10).

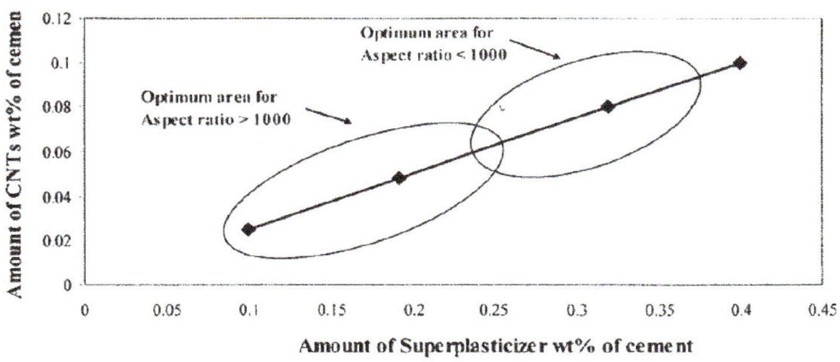

Fig. 10 The optimum amount of MWNTs and superplasticizer for reinforcement in cement-based materials according to the aspect ratio of CNTs.

Recently, researchers proposed a novel approach to solve the dispersion issue of CNTs [Nasibulin et al. 2009, Cwirzen et al. 2009, Ludvig et al. 2009, Dunens et al. 2009]. The CNTs are grown on the surface of cement (or fly ash, a conventional admixtures for fabricating cement-based materials) particles through in-situ synthesis (i.e. a modified Chemical Vapor Deposition (CVD) method) [Fig.11]. The amount and morphology of the grown CNTs depend on the applied temperatures, the chemical component and flow speed of pristine cement, and the type/amount of the introduced gas. It was possible to obtain cement containing up to 20% of nanofibers (including CNTs and CFs) [Cwirzen et al. 2009]. The cement (or fly ash)/CNTs hybrid can be used to fabricate the CNTs reinforced cement-based materials with high dispersion of CNTs and strong bonding of CNTs to matrix.

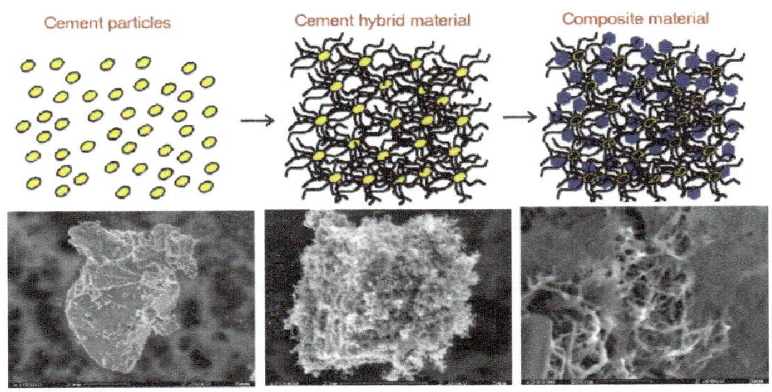

Fig. 11 Schematic representation of the general concept of the incorporation of CNTs into composite material by their direct growth on the surface of matrix particles.

3 Mechanical Properties of CNTs Reinforced Cement-Based Materials

3.1 Mechanical Properties of CNTs

The strength of the sp^2 carbon-carbon bonds endues CNTs amazing mechanical properties (e.g. ultra-high strength and stiffness, and elastic stress-strain behavior). The Young's modulus of the best CNTs can be as high as 1000 GPa which is approximately 5 times higher than steel. The tensile strength or breaking strain of CNTs can be up to 63 GPa or 10%, around 50 times higher than steel. In addition to their high strength and elastic constant, CNTs have extremely high aspect ratios, with values typically higher than 1000:1 and reaching as high as 2,500,000:1. As a result of these properties coupled with the lightness, large surface area (typically 200-300m2/g), and excellent chemical and thermal stability, CNTs reinforcements are expected to produce significantly stronger and tougher cement-based materials than traditional reinforcing materials (e.g. glass fibers or carbon fibers) [Makar et al. 2005].

3.2 Mechanical Properties of Nanocomposites

The CNTs' enhancement capability to the nanocomposites (i.e. CNTs reinforced cement-based materials) depends on many factors, among them the quality of CNTs' dispersion, the final aspect ratio (whether the nanotubes were shortened as a result of disaggregative treatment), CNTs' content level, CNTs' intrinsic structure and properties, composition and structure of matrix, and the interfacial bonding condition between CNTs and matrix. Therefore, very different results have been reported on the mechanical behavior of CNTs reinforced cement-based materials.

Makar et al. [2005] found that the improvements of CNTs are dependent both on matrix composition and hydration time. The nanocomposites produced at 0.8 w/c ratio (i.e water to cement ratio), show reductions in strength at all time periods. In contrast, low w/c samples with a 2% SWNTs content show improvements of up to 600% in hardness at early ages, but essentially no improvement after 14 days of hydration (Fig. 12).

De Ibarra et al. [2006] studied the impact of the addition of 0.05% and 0.1% SWNTs and 0.1% and 0.2% MWNTs to cement paste, and reported modest gains in the Young's modulus and hardness of this kind of nanocomposites. Yakovlev et al. [2006] found that the use of CNTs (0.05% by mass) as the reinforcement for production of form non-autoclave concrete, allows increasing its compressive strength up to 70%. Shah et al. [2009] and Konsta-Gdoutos et al. [2010] studied the effect of MWCNs' length (short versus long) and the concentration on the fracture properties of the nanocomposites for a constant weight ratio of surfactant to MWNTs. They also explored the effect of the surfactant concentration on the fracture properties of the nanocomposites reinforced with 0.08 wt.% of cement long MWNTs. The fracture mechanics test results indicate that the flexural

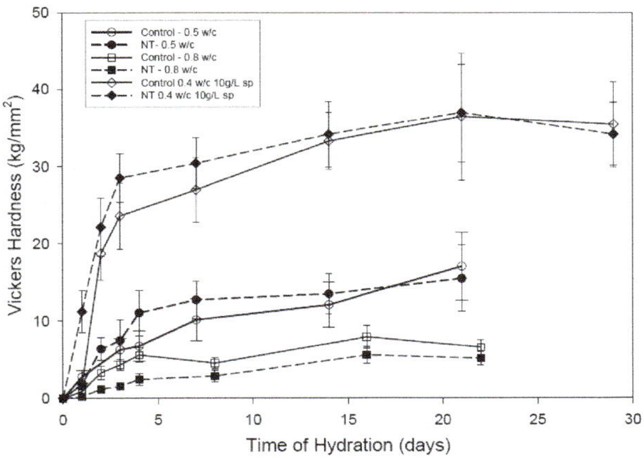

Fig. 12 Hardness measurements on cement paset made at different w/c ratios with (NT) and without (control) 2% SWNTs addtions.

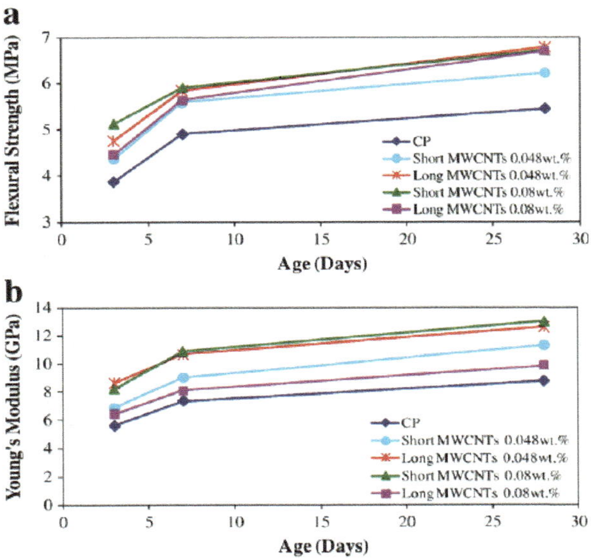

Fig. 13 Effect of different lengths (short and long) of MWNTs and concentration on (a) the flexural strength and (b) the Young's modulus of cement paste (w/c=0.5).

strength and the Young's modulus of cement matrix significantly increase through use of small amounts of MWNTs (0.048 wt.% and 0.08 wt.%) (Fig. 13). As compared to the same material without CNTs, the nanocomposites exhibit an increase of the Young's modulus of at least 15% up to about 55% and an increase in flexural strength of at least of 8% up to about 40%. An over 45% increase in the 28 day flexural strength is realized at the 0.08 wt% MWNTs loading versus the unmodified control. In particular, higher concentrations of short MWNTs are required to achieve effective reinforcement, while lower amounts of longer MWNTs are needed to achieve the same level of mechanical performance.

Matthew et al. [2008] fabricated CNTs filled concrete with different CNTs' content levels. They found that the use of 0.1% CNTs by weight can increase the tensile strength of the nanocomposites, and reduce compressive strength. When at least about 0.2% CNTs were added, there is a small amount of increase in both compressive and tensile strength of the nanocomposites. Luo et al. [2009] prepared the plain cement paste, and nanocomposites with 0.2% MWNTs using the different surfactants. Compared to the plain cement paste, the flexural and compressive strengths of the nanocomposites are enhanced. The maximum increase of flexural strength and compressive strength is 35.45% and 29.5% respectively. Luo [2009] also fabricated the nanocomposites with different concentrations of MWNTs and different concentrations and combinations of surfactants. He found that the fracture toughness and critical opening displacement of the nanocomposites with 0.5 % of MWNTs can be enhanced by 175.21% and 54.77%, relative to the plain cement paste, respectively. Musso et al. [2009] prepared CNTs reinforced cement-based materials by adding 0.5% in weight of as-grown and annealed MWNTs to plain cement paste, and they found that both pristine and annealed MWNTs induce an improvement in the flexural and compressive properties of the nanocomposites (Fig. 14).

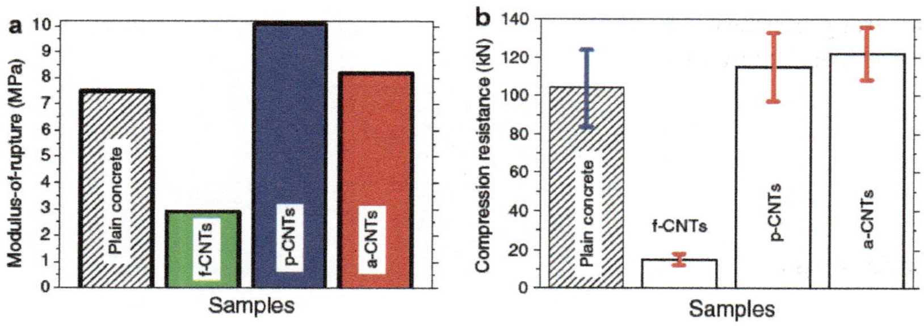

Fig. 14 Modulus-of-rupure (a) and compression resistance (b) of cement paste. With and without MWNTs (p-CNTs is pristine MWNTs, a-CNTs is annealed MWNTs and f-CNTs is carboxyl functionalized MWNTs)

Chaipanich et al. [2010] prepared CNTs reinforced fly ash cement paste by adding CNTs of 0.5 and 1% by weight into a cement-fly ash (20% by weight of cement) system. They found that the use of CNTs results in higher compressive strength of the nanocomposites. The highest strength obtained was found at 1% CNTs where the compressive strength at 28 days was 54.7 MPa (the almost relative strength 100% to that of Portland cement paste) (Fig. 15).

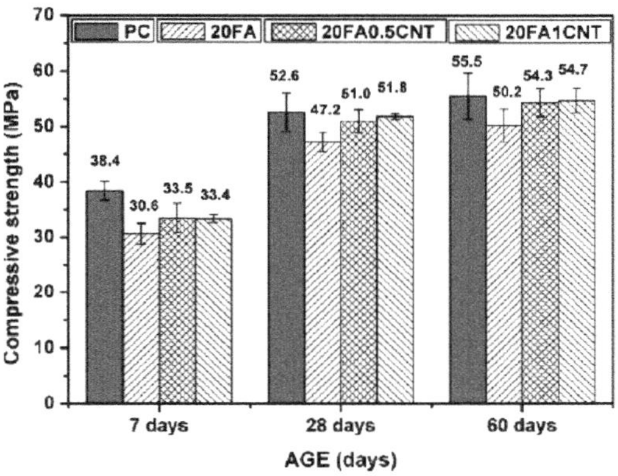

Fig. 15 Compressive strength of CNT-fly ash cement paste at 7, 28 and 60 days.

Veedu (2010) compared the mechanical performances of CFs reinforced cement-based materials and CNTs reinforced cement-based materials, and observed that the flexural and compressive strength of CNTs reinforced cement-based materials are 30% and 100% greater, respectively, than that of CFs reinforced cement-based materials (Table 1).

Table 1 Comparison of mechanical properties of CNTs reinforced cement-based materials and CFs reinforced cement-based materials

Type of materials	Flexural strength (MPa)	Compressive strength (MPa)
CFs reinforced cement-based materials	2.42	14
CNTs reinforced cement-based materials	3.2	26.8

In addition, it is well-known that the interfacial bonding condition between the reinforcement and the matrix is a key factor contributing to improving fiber reinforced materials. The covalent surface modification can produce functional groups on the outer walls of CNTs. These superficial chemical groups will originate strong chemical bonds between CNTs and matrix, thus enhancing the reinforcement efficiency even though some mechanical properties of CNTs are impaired. Li et al. [2005] employed carboxyl functionalized MWNTs and obtained modest improvements in compressive and flexural strengths. The compressive strength increases up to 19%, while the flexural strength increases up to 25%. The use of CNTs can also improve the deformation ability of cement mortar, and lead to a significant increase of the failure strain (Fig. 16). The compressive and flexural strengths of cement paste containing treated CNTs are higher about 2.7 MPa and 0.4 MPa than that 69.41MPa and 9.56MPa of cement paste containing 0.5% untreated CNTs by weight of cement respectively [Li et al. 2007].

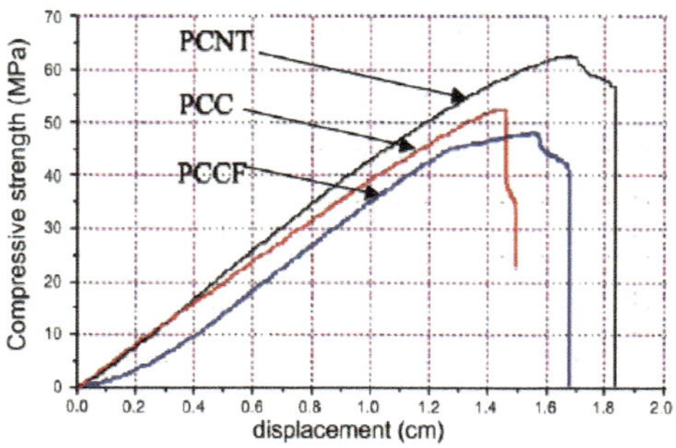

Fig. 16 Typical load-displacement curves of cement-based materials (PCC are control cement mortar, PCCF is cement mortar containing untreated CFs, PCNT is ement mortar containing 0.5% treated CNTs by weight of cement).

Cwirzen et al. [2008] studied the compressive strength of the nanocomposites with fifteen different combinations of hydroxyl functionalized MWNTs, admixtures and w/c ratios. They found that the cement paste incorporating CNTs reveals an increase in the compressive strength of nearly 50% even with only a small addition of the MWNTs, namely 0.045-0.15% of the cement weight. Luo [2009] used 0.5% of carboxyl functionalized MWNTs and obtained enhancements of 149.32% and 34.96% in fracture toughness and critical opening displacement compared to the plain cement paste, respectively. However, Musso et al. [2009] also used carboxyl functionalized MWNTs, but they obtained a different result. Flexural and compressive tests performed on composite containing functionalized CNTs, show a significant reduction of the performances compared to pristine cement (Fig. 13). The possible reason is that the functionalized CNTs are so hydrophilic as to absorb most of the water contained in cement paste. This hampers the proper hydration of cement paste.

Nasibulin et al. [2009], Cwirzen et al. [2009] and Ludvig et al. [2009] utilized cement/CNTs hybrid prepared with different synthesis processes to fabricate CNTs reinforced cement paste. Nasibulin et al. [2009] found that the nanocomposites reveal as high as a 2 times increase in the compressive strength compared plain cement paste (Table 2). Ludvig et al. [2009] achieved a 34.28% increase in the tensile strength by using cement/CNTs hybrid containing 0.3% of CNTs.

Table 2 Compressive strength of cement paste prepared by adding cement/CNTs hybrid

Fraction of cement/CNTs hybrid	Synthesis conditions				Compressive strength (MPa)
	Temp (°C)	Gas flow rate ($cm^3 min^{-1}$)			
		C_2H_2	CO_2	CO	
0	-	-	-	-	25
100	550	860	0	177	22
100	575	660	660	0	55
100	500	500	500	0	40
100	525	660	660	0	56

It can be seen from the above literature reviews that both exciting and disappointing results were obtained in the previous researches. This indicates that the effect of CNTs on the reinforcement of cement based materials is very complex and need further investigation.

3.3 Reinforcement Mechanisms

Why the physical and chemical properties of CNTs can affect the mechanical behavior of composites results from both the traditional reinforcment mechanisms and the nanoscale reinforcement mechanisms, which are summarized below.

(1) Excellent intrinsic properties of CNTs
Sections 2.1 and 3.1 have introduced the excellent intrinsic properties of CNTs, which are not discussed here. It is well-known that the mechanical properties of the nanocomposites depend on many factors affecting CNTs intrinsic properties. For example, the high temperature annealing treatments remove lattice defects from the walls of CNTs, hence obviously increase CNTs' and nanocomposites' mechanical strength [Simone et al. 2009].

(2) Extensive distributing enhancement meshwork of CNTs in the nanocomposites
The diameters and aspect ratios of CNTs (about the same size as the distance between layers in hydrated cement) mean that they can be distributed on a much finer scale than commonly used reinforcing fibers (Fig. 17). As a result, CNTs in the matrix form an extensive distributing enhancement meshwork [Li et al. 2005], which is beneficial for enhancing the integrity of materials [Veedu 2010].

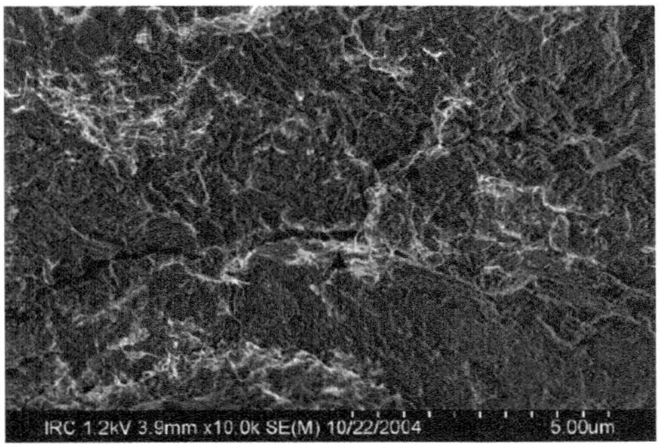

Fig. 17 SEM images of fracture surfaces of the three day hydration sample of CNTs reinforced cement paste [Makar et al. 2005].

(3) Improvement of nanocomposites' microstructures
The nanocomposites exhibit a modified nanostructure so that the average values of stiffness and hardness of C-S-H are higher than that of the same cement-based materials without CNTs (Fig. 18) [Shah et al. 2009].

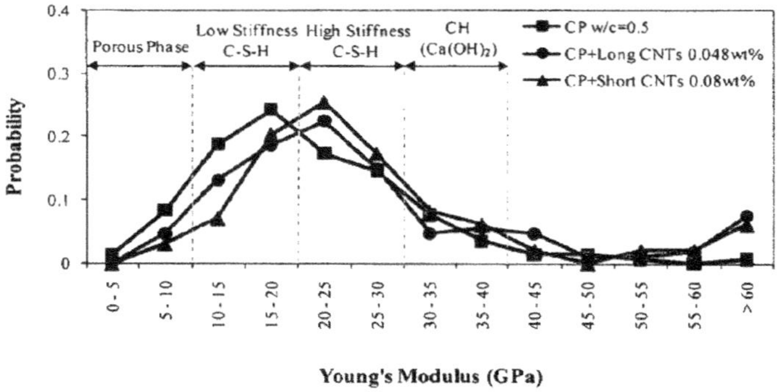

Fig. 18 The probability plot of the calculated Young's modulus of 28 days cement paste (w/c=0.5), cement paste reinforced with 0.048 wt% long CNTs and cement paste reinforced with 0.08 wt% short CNTs.

Moreover, the CNTs' addition inhibits the reduction in nanoscale flaws and fractures and provides a cement matrix essentially "crack free". The autogenous shrinkage in the nanocomposites reduces at least 30% after 96 hours compared to the same cement-based materials without CNTs (Fig. 19).

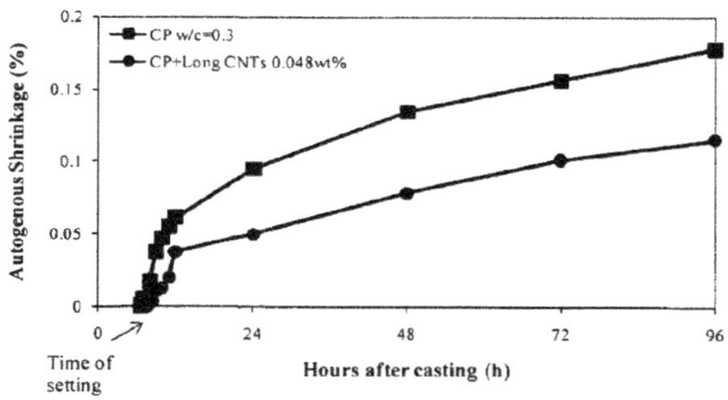

Fig. 19 The autogenous shrinkage results of cement paste (w/c=0.3) and cement paste reinforced with 0.048 wt. % long CNTs [Shah et al. 2009].

Furthermore, the addition of CNTs fines pore size distribution and decreases the porosity (or nanoporosity) of the nanocomposites by filling the gaps (or pores) between the hydration products such as C-S-H and ettringite (Fig. 20). Therefore, the nanocomposites become much more compacted. [Li et al. 2005, Konsta-Gdoutos et al. 2010, Chaipanich et al. 2010].

(4) Strong bonding between CNTs and matrix

CNTs act as nucleating agents for C-S-H, which preferentially form on the surface of CNTs as opposed to the surface of the adjacent unhydrated cement grains

Fig. 20 SEM images of CNTs-fly ash cement paste

(Fig. 21). The nucleation appears to occur along the entire length of the CNTs, rather than at specific locations that might be associated with functional groups. The result is a dense C-S-H formation that appears to be tightly bonded to CNTs, hence producing reinforcing behavior [Makar et al. 2009].

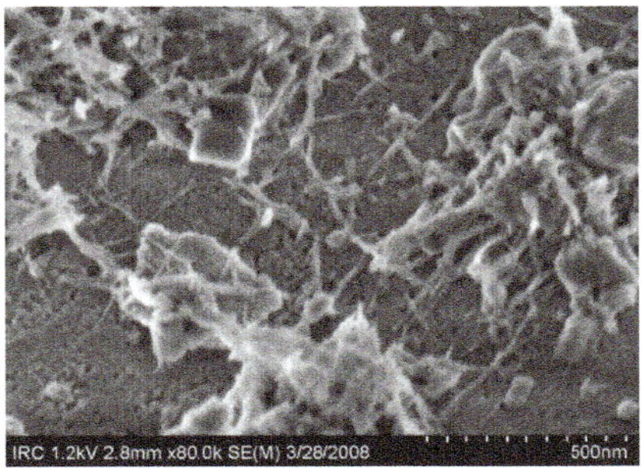

Fig. 21 Growth of C-S-H around SWNTs.

For functionalized CNTs reinforced cement-based materials, the interfacial interactions between the groups of functionalized CNTs' surfaces and hydrations (such as C-S-H and calcium hydroxide) of cement will produce a high bonding strength, and increase the load-transfer efficiency from cement matrix to the reinforcement [Li et al. 2005, Cwirzen et al. 2008]. Moreover, the modified chain groups of surfactants absorbed to CNTs' surface like CNTs' "root" (Fig. 22), and their "rooting" in matrix produces strong interfacial adhesion between the CNTs and the matrix. This thereby enhances nanocomposites' mechanical properties [Vaisman et al. 2006].

For the nanocomposites made of cement/CNTs hybrid, CNTs are partly chemically bonded to matrix through a catalyst particle originating from cement [Cwirzen et al. 2009].

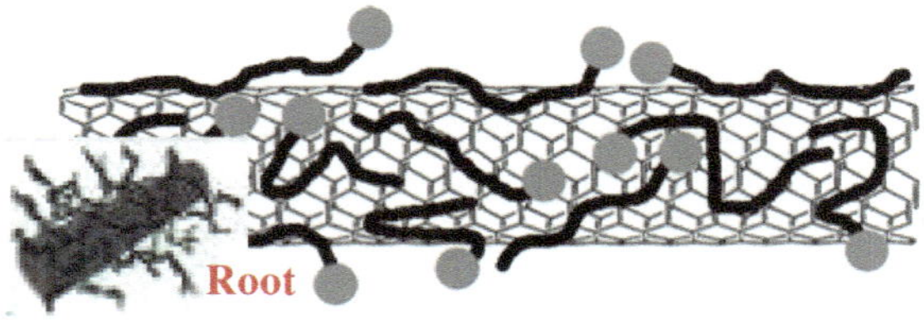

Fig. 22 CNTs absorbed by surfactant.

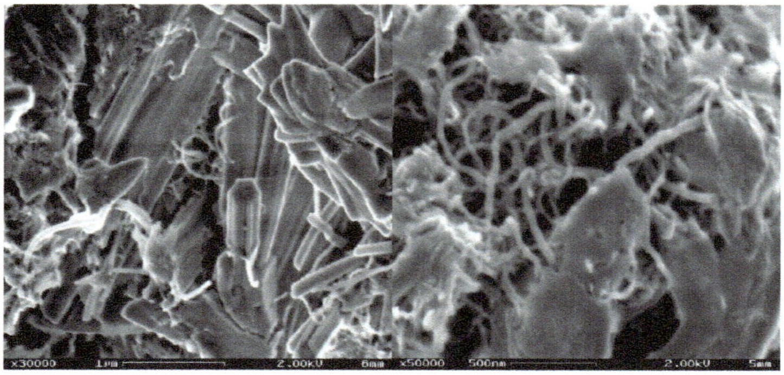

Fig. 23 SEM images of the nanocomposites made of cement/CNTs hybrid

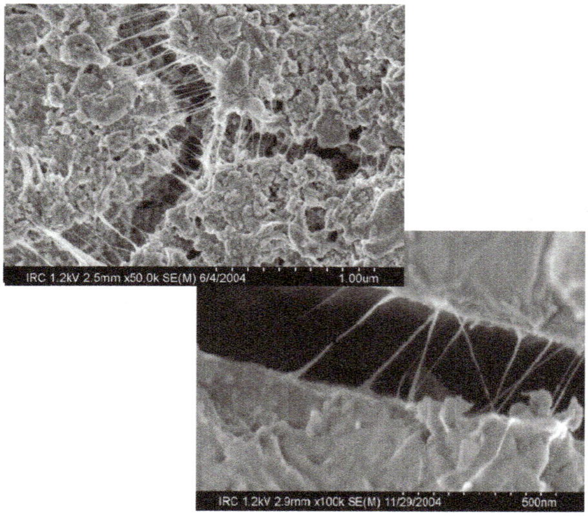

Fig. 24 Crack bridging of CNTs

(5) Crack bridging, pinning effect, fiber pull-out, crack deflection, fiber debonding and fiber breaking

CNTs are anchored well inside the hydration products and cross crack (Fig.23 and Fig.24) [Makar et al. 2005, Nasibulin et al. 2009]. When the cracks in the matrix encounter well-distributed CNTs, the pinning effect and the efficient crack bridging can inhibit the crack growth at the very preliminary stage of crack propagation within composites [Makar et al. 2005].

Additionally, the CNTs form a stitching on fracture surfaces, diverting crack energy into a matrix and inhibiting crack propagation [Veedu 2010]. The bridge coupling effect of CNTs guarantees the load-transfer across voids and cracks [Li et al. 2005]. Furthermore, other reinforcing behaviors such as fiber pull-out (Fig. 25), crack deflection, fiber debonding and fiber breaking also contribute to mechanical enhancement of composites. This enhancement effects result from the energy release caused by these reinforcing behaviors.

Fig. 25 CNTs' pull-out [Raki et al. 2010].

It is worth noting that the complex chemistry and physical structure of CNTs reinforced cement-based materials mean that some fundamental issues need further to be explored at the nanoscale.

4 Electrical and Piezoresistive Properties of CNTs Reinforced Cement-Based Materials

4.1 Electrical and Piezoresistive of CNTs

The unique electrical properties of CNTs are to a large extent derived from their 1-D character and the peculiar electronic structure of graphite. Resistance occurs when an electron collides with some defect in the crystal structure of the material through which it is passing. The defect could be an impurity atom, a defect in the crystal structure, or an atom vibrating about its position in the crystal. Such collisions deflect the electron from its path. But the electrons inside a CNT are not so easily scattered. In a 3-D conductor, electrons have plenty of opportunity to scatter, since they can do so at any angle. Any scattering gives rise to electrical resistance. In a 1-D conductor, however, electrons can travel only forward or backward. Under these circumstances, only backscattering (the change in electron motion from forward to backward) can lead to electrical resistance. But backscattering requires very strong collisions and is thus less likely to happen. So the electrons have fewer possibilities to scatter. This reduced scattering gives CNTs very low resistance. In addition, they can carry the highest current density of any known material, measured as high as 10^9A/cm^2 [Khare and Bose 2005].

Theoretical calculations have shown early on that the electronic properties of the CNTs are very sensitive to their geometric structure. Although graphite is a zero-gap semiconductor, theory has predicted that the CNTs can be metallic, semi-metallic or semi-conducting with different size energy gaps, depending very sensitively on the diameter and helicity of the tubes (Fig. 1) [Dresselhaus et al. 2000].

Field emission is an alternative mechanism to extract electrons. It is a quantum effect where under a sufficiently high external electric field, electrons can tunnel through the energy barrier and escape to the vacuum level. All the field emission sources rely on field enhancement due to sharp tips/protrusions, so they tend to have smaller virtual source sizes because of the primary role of the field enhancement factor. The larger the field enhancement factor, the higher the field concentration, and therefore the lower the effective threshold voltage for emission. CNTs possess the right combination of properties: nanometer-size diameter, structural integrity, high electrical and thermal conductivity, and chemical stability, so they have excellent emission characteristics such as a low threshold field for emission and a high current density. The emission has been observed at fields lower than 1V/m, and high current densities of over 1A/cm^2 have been obtained [Meyyappan 2005, Cheng and Zhou 2003].

When CNTs are subjected to external force such as tension/compression, torsion and squashing, they are deformed. Deformations of CNTs' structures will affect the electron transport condition. As a result, CNTs present extraordinary piezoresistive properties (Fig. 26) [Tombler et al. 2000]. The change in electrical conducitivity of CNTs in response to strain can reach 0.02 S/cm with per 1% change in compressive strain [Pushparaj et al. 2010].

Dispersing conductive fillers into the nonconductive matrix can form conductive materials. The electrical conductivity of these materials is strongly dependent on the concentration of the conductive fillers. At low concentration, the conductivity remains very close to the conductivity of the pure matrix. When a certain concentration is reached, the conductivity of the materials drastically increases by many orders of magnitude. This phenomenon is known as percolation and can be well explained by a percolation theory. The electrical percolation threshold of conductive reinforcements embedded in an insulating matrix is sensitive to the geometrical shape of the conductive fillers. The small size and large aspect ratio are helpful to lower the percolation threshold. Because CNTs have tremendously large aspect ratios, many researchers have found that CNTs reinforced materials exhibit exceptionally low electrical percolation thresholds. For example, the electrical

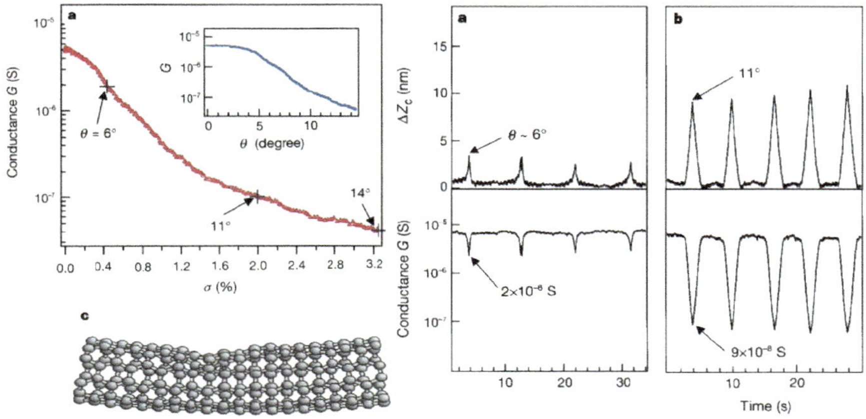

Fig. 26 Conductance change induced by mechanical deformation of metallic CNT using AFM tip (*G* is conductance of CNT, σ is stain of CNT, and ΔZc is vertical displacement).

percolation threshold was reported at 0.0025 wt.% CNTs and conductivity at 2 S/m at 1.0 wt% CNT in CNTs reinforced epoxy materials. CNTs not only can be utilized to tailor conductivity of polymer as reinforcement, but also are excellent conductive fillers to impart piezoresistivity to polymer. Much research efforts have been concentrated on piezoresistivity of CNTs reinforced polymer materials [Khare and Bose 2005].

Therefore, CNTs are expected to be more effective in producing significantly conductive and piezoresistive cement-based materials than traditional conductive fillers (e.g. carbon fibers, steel fibers and carbon black etc).

4.2 Electrical and Piezoresistive Properties of CNTs Reinforced Cement-Based Materials

Like mechanical properties of CNTs reinforced cement-based materials, the electrical and piezoresistive properties of this kind of nanocomposites also depend on many factors. These factors include the type, the concentration and the surface condition of CNTs, the quality of CNTs' dispersion, and the composition of matrix.

Saafi [2009] studied the effect of the concentration of SWNTs on the electrical resistance and piezoresistivity of nanocomposites. He found that there is a sudden decrease in the electrical resistance as the concentration of SWNTs increases from 0 to 0.5 vol. % and a percolation threshold of 0.5 vol. % was assumed for the nanocompostes. The electrical resistance further decreases when the concentration of SWNTs increases from 0.5 to 1 vol. %. However, it appears that the increase of the SWNTs concentration beyond 1 vol. % yields a slight decrease in the electrical resistance (Fig. 27). The electrical resistance of both nanocomposites with 0.5% and 1% SWNTs firstly increases linearly and monotonically and then presents a

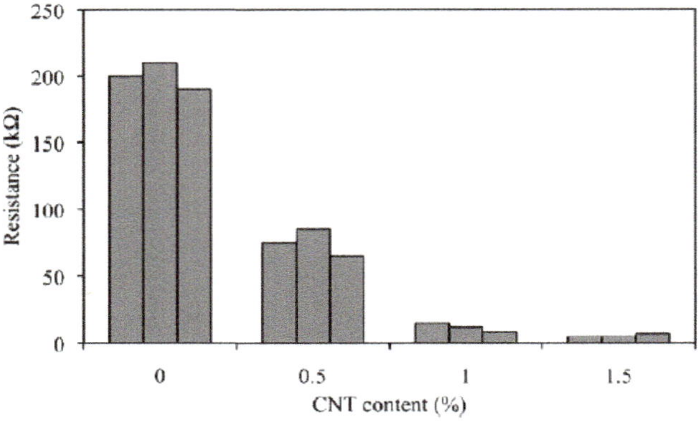

Fig. 27 Effect of SWNTs concentration on electrical resistance of nanocomposites.

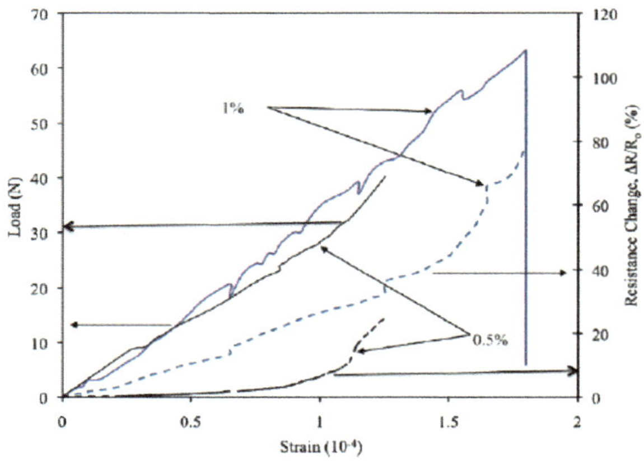

Fig. 28 Typical response of nanocomposites under monotonic direct tensile loading.

non-linear behavior up to failure under monotonic direct tensile loading. However, the nanocomposites with 1% SWNTs exhibit a high sensitivity to the applied stress (Fig. 28). Additionally, the electrical resistance of the nanocomposites with 1% SWNTs firstly increases and then decreases with the increase and decrease of the applied load under direct cyclic tensile loading. The nanocomposites exhibit good piezoresistive repeatability (Fig. 29).

Li et al. [2005] explored the effect of carboxyl functionalization on electrical and piezoresistive properties of nanocomposites. They found that the electrical resistance of SPCNT (nanocomposites with 0.5% carboxyl functionalized CNTs) and PCNT (nanocomposites with 0.5% unfunctionalized CNTs) is 149ohm·cm and 130ohm·cm respectively. The fractional change of resistivity of PCNT is 10%, and that of SPCNT is about 14% (Fig. 30). Both carboxyl functionalized CNTs

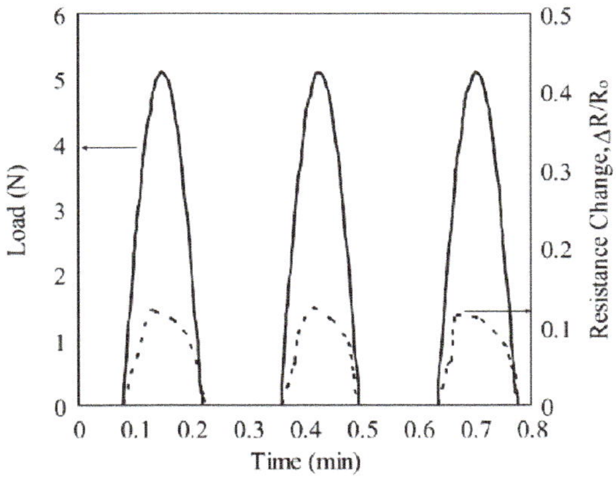

Fig. 29 Typical response of nanocomposites under direct cyclic tensile loading.

and unfunctionalized CNTs can greatly decrease the electrical resistivity and improve piezoresistive properties for the cement paste. The carboxyl functionalized CNTs have a stronger effect on the reinforcement of piezoresistive properties, whereas those unfunctionalized ones have more forceful effect on the reduction of electrical resistivity.

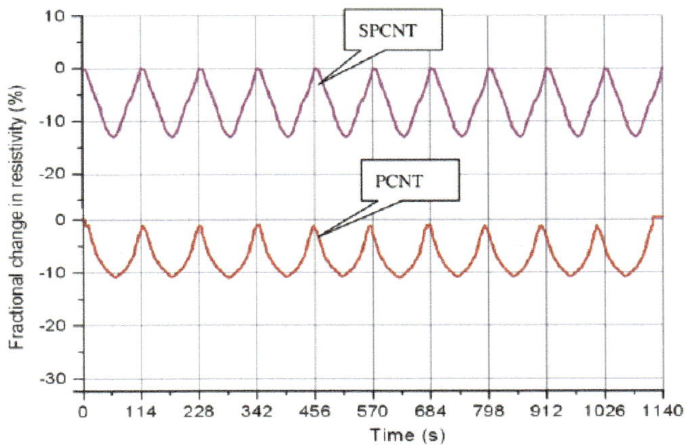

Fig. 30 The fractional change in resistivity vs. time under cyclic compressive loading (0-15KN).

Yu et al. [2009] investigated the piezoresistive responses of the nanocomposites with different dispersion methods and CNTs' concentrations. They found that the electrical resistance of the nanocomposites reinforced with 0.1% non-covalent surface modification MWNTs by using SDS changes synchronously with the compressive stress and the amount of the changes vary with the stress levels (Fig. 31).

The resistance changes of the nanocomposites reinforced with 0.1% carboxyl functionalized MWNTs are also corresponding to the stress load levels (Fig. 32). But the later presents stronger piezoresistive response and higher signal-to-noise ratio than former (Table. 3). The nanocomposites reinforced with 0.1% carboxyl functionalized MWNTs possess similar responses to the stress as from the nanocomposites reinforced with 0.1% carboxyl functionalized MWNTs, but with lower sensitivity (Table. 3).

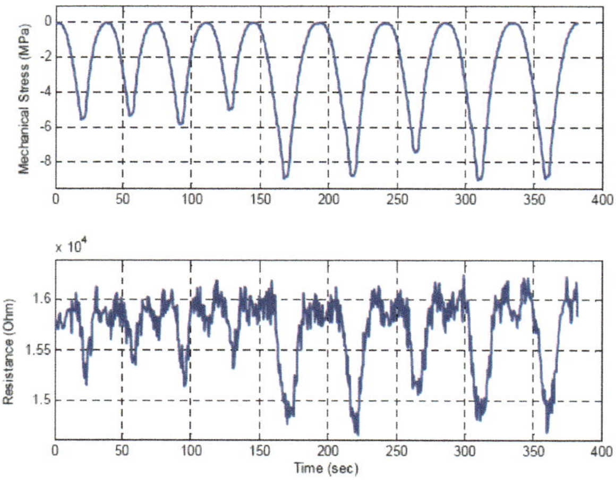

Fig. 31 Piezoresistive response of cement pasete reinforced with non-covalent surface modification CNTs by using SDS, MWNTs are 0.1% of cement by weight.

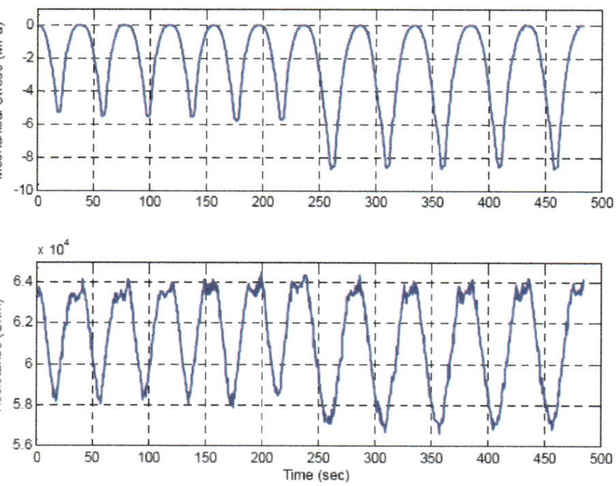

Fig. 32 Piezoresistive response of cement pasete reinforced with carboxyl functionalized CNTs, MWNTs are 0.1% of cement by weight.

Table 3 Comparison of electrical resistance changes of CNTs reinforced cement paste with different CNTs concentrations and under different compressive loads

Types of materials	Resistance change (5.2MPa load)	Resistance change (8.6MPa load)
Plain cement paste	0	0
Nanocomposites reinforced with 0.06% carboxyl functionalized MWNTs	8.8%	10.3%
Nanocomposites reinforced with 0.1% carboxyl functionalized MWNTs	9.4%	11.4%
Nanocomposites reinforced with 0.1% non-covalent surface modification MWNTs	~5.0%	~7.2%

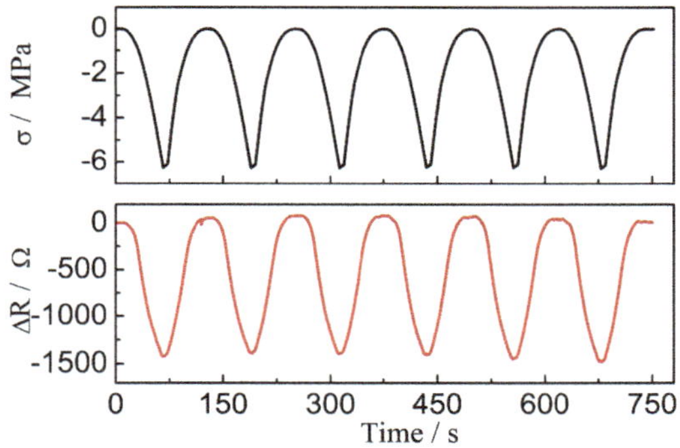

a) Under repeated compressive loading with amplitude

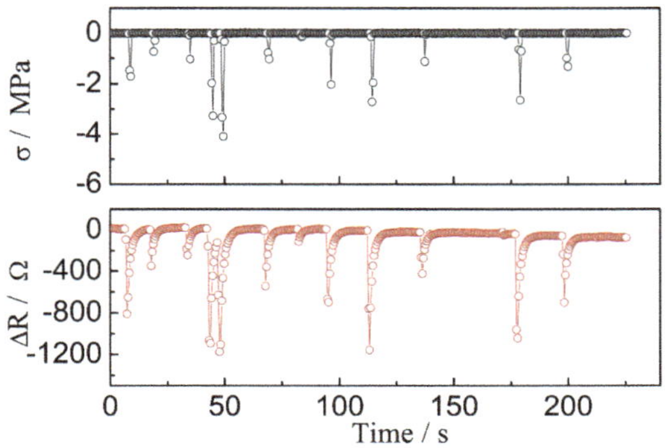

b) Under impulsive loading

Fig. 33 Relationships between compressive stress and electrical resistance of CNTs reinforced cement paste (σ is compressive stress, ΔR is the change in electrical resistance).

Han et al. [2010] studied the variation of the electrical resistance of cement paste and cement mortar reinforced with carboxyl functionalized MWNTs under repeated compressive loading and impulsive loading. They found that the electrical resistances of both cement paste with 0.1% MWNTs and cement mortar with 0.4% MWNTs decrease upon loading and increase upon unloading in every cycle under repeated compressive loading. The response of electrical resistances of these nanocomposites to compressive stress is strongly related under repeated compressive loadings. Additionally, the impulsive loadings also cause strongly correlated changes in the electrical resistance of both nanocomposites. It is worth noting that the addition of fine aggregates (i.e. sand) decreases the piezoresistive response sensitivity of the nanocomposites even though at a higher CNTs' concentration (Fig. 33 and Fig. 34).

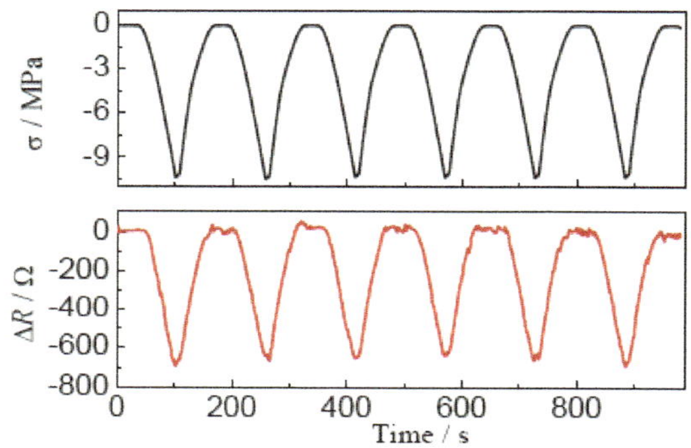

a) Under repeated compressive loading with amplitude

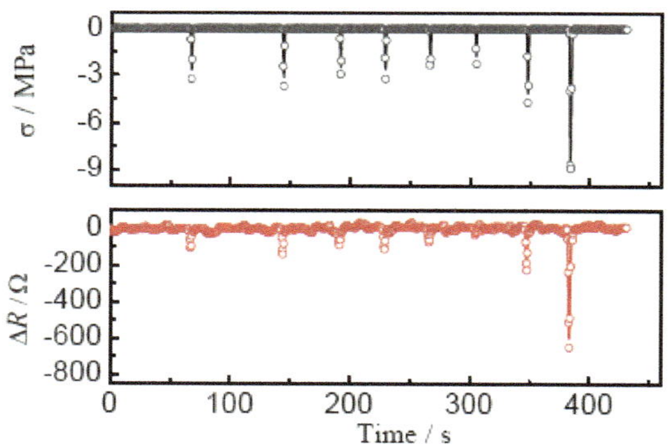

b) Under impulsive loading

Fig. 34 Piezoresistivity of CNTs reinforced cement mortar (σ is compressive stress, ΔR is the change in electrical resistance).

Han et al. [2010] compared the piezoresistive behaviors of MWNTs reinforced cement paste with different water contents under uniaxial compression. They found that the water content has a strong effect on the electrical and piezoresistive properties of the nanocomposites. The electrical conductivity of the nanocomposites increases with water content, while the piezoresistive sensitivities first increase and then decrease with the increase of water content (Fig. 35).

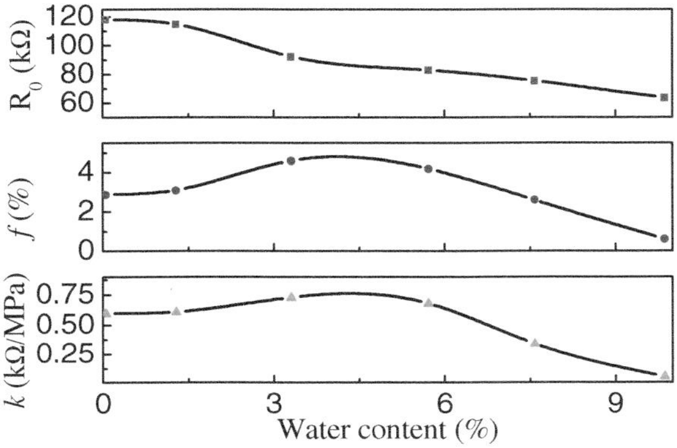

Fig. 35 Comparison of electrical resistances (R_0), maximum change amplitudes of electrical resistance (*f*) and piezoresistive sensitivities (*k*) of MWNTs reinforced cement paste with different water contents.

Luo et al. [2009] compared the electrical conductivity of plain cement paste, and composites with 0.2% MWNTs and using NaDDBS, NaDC, AG, or NaDDBS and TX10 as surfactant. They found that the electrical conductivity of nanocomposites is closely related to the types of surfactants. The nanocomposites fabricated by using NaDDBS as surfactant have the greatest mean value and lowest deviation ($5.314\times10^{-4} \pm 5.107\times10^{-5}$ S/cm) in the electrical conductivity, which is two orders of magnitude with respect to that of plain cement paste. Luo [2009] also studied the effect of concentration of CNTs (from 0.1 wt. % to 2 wt. %), temperature, humidity, type of measurement voltages and compressive stress to electrical resistivity of MWNTs reinforced cement paste. He found that the electrical resistivity of nanocomposites decreases with increase of concentration of CNTs. The percolation threshold is around 1 %. The electrical resistivity increases with measurement time under direct current voltage. The lower water content of nanocomposites is and the higher concentration of CNTs is, the more prominent the increase tend of electrical resistivity with measurement time is. The electrical resistivity approximatively linearly decreases with increase of temperature in the range from -10°C to 80°C. The voltage-current characteristic of nanocomposites is linear above 2.5V of measurement voltage, while is nonlinear below 2.5V. The feature of alternating current impedance module or complex impedance of nanocomposites versus frequency show similar "U" shape. This means that the alternating current impedance module or complex impedance exhibits a three-stage variation trend of fast decrease, keeping stable and fast increase as the measurement frequency increases. In addition, the nanocomposites with 0.5% of MWNTs exhibit sensitive, stable and repeatable piezoresistive response.

Azhari [2008] investigated the effect of different types and concentrations of CNTs and the hybrid of CNTs and CFs on the conductivity of the nanocomposites. She found that the addition of SWNTs at 1% and MWNTs at 1% or 3% slightly decreases the electrical resistivity of the cement paste, while 3% MWNTs and 15% CFs hybrid yields a very low electrical resistivity (Fig. 36). The phase angle of nanocomposites is the most part negative, which confirms the fact that the nanocomposites are capacitive rather than inductive. The nanocomposites with 3% of MWCNT as well as the CNTs/CFs hybrid, exhibit much lower phase angle values especially at frequencies of 100 kHz and below. These nanocomposites seem to show inductive characteristics at some frequencies (where $\theta>0$) (Fig. 37). Results from resistivity tests conducted on samples with 1% MWNTs and 1% SWNTs at two different frequencies of 1 kHz and 100 kHz revealed that the nanocomposites with SWNTs seem to be independent of current frequency as their resistivity values are identical at both frequencies. This, however, did not hold true

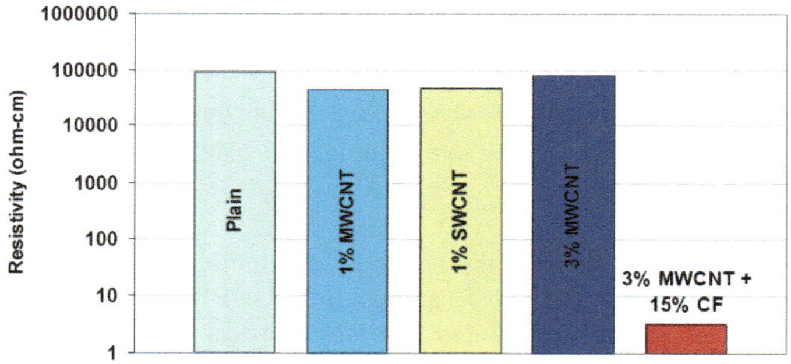

Fig. 36 Resistivity of CNT and hybrid nanocomposites at 28 days.

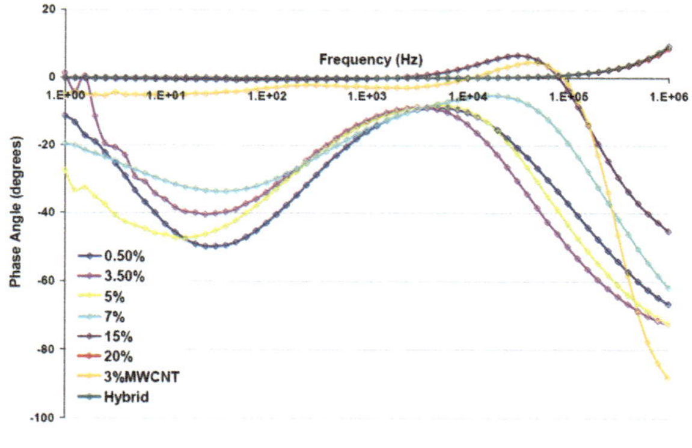

Fig. 37 Phase angle at current frequencies between 1Hz and 1MHz for different cement-based materials.

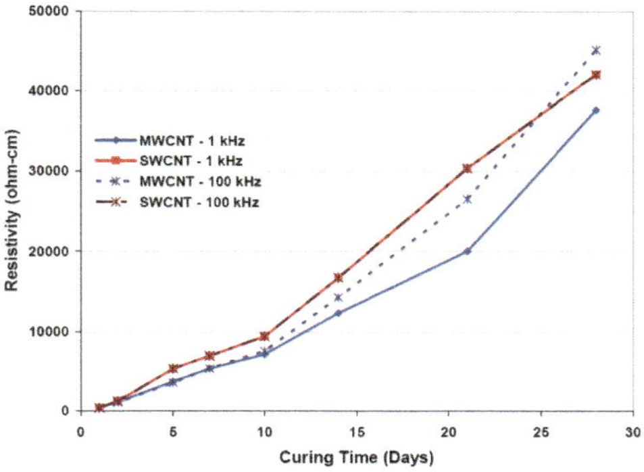

Fig. 38 Effect of current frequency on resistivity values of nanocomposites with MWNTs and SWNTs.

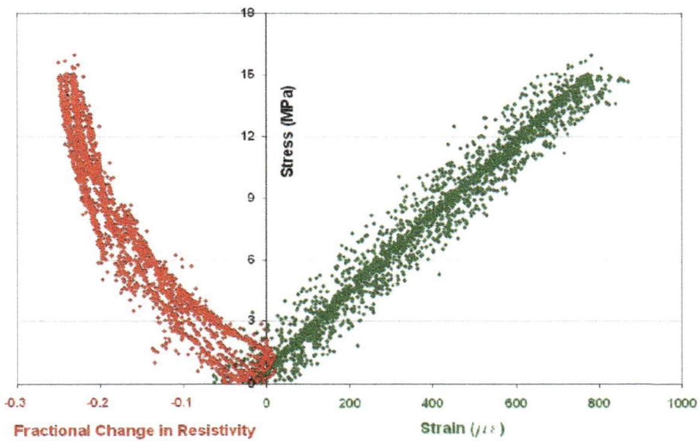

Fig. 39 Stress/strain and stress/fractional change in resistivity relationships of the nanocomposites with 1% MWNTs and 15% CFs hybrid after five cycles of compressive loading.

on the nanocomposites with MWNTs (Fig. 38). Additionally, the relationship between compressive stress and fractional change in resistivity of the nanocomposites with 1% MWNTs and 15% CFs hybrid appears to have a sensible correspondence, i.e. a stable and repeatable piezoresistivity (Fig. 39).

Wansom et al. [2006] combined AC-impedance spectroscopy (AC-IS) with time-domain reflectometry (TDR) to investigate the impedance response of cement paste with different concentrations of MWNTs at initial stage from 15 min to 1 day. They found that the AC-IS Nyquist plots for nanocomposits with 0.75 vol. % and 1 vol. % CNTs show a trend toward decreasing DC resistances at early times (owing to a release of ions to participate in hydration reaction) and a shift back toward higher DC resistances at later times (reflecting the initiation of set between 6 and 8 h). Pre-set, the agreement in DC conductivity of the nanocomposites and the matrix is the result of current traveling through the matrix only

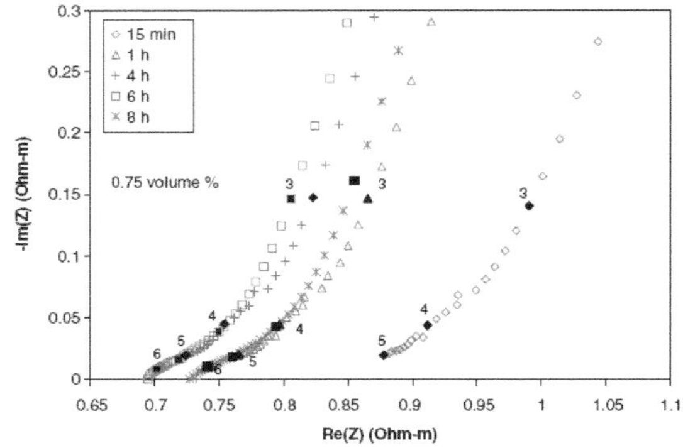

Fig. 40 Experimental Nyquist plots for a nanocomposites with 0.75 vol. % of MWNTs at early times from 15min to 8h.

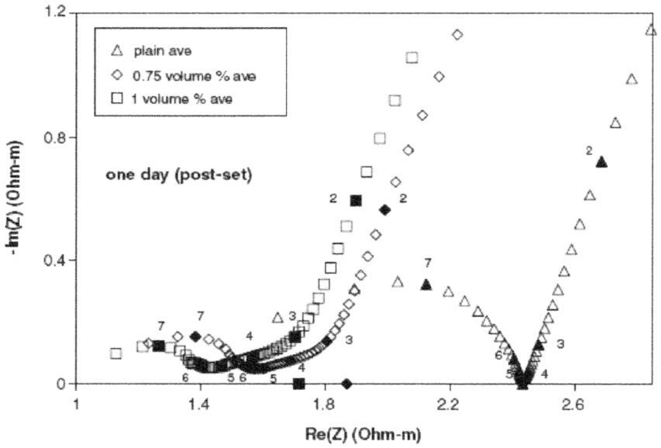

Fig. 41 Nyquist plots for plain cement paste and two nanocomposites with 0.75 vol.% and 1 vol. % of MWNTs based on point-by-point average (for each frequency) over three replicates for each composition.

(i.e., the resistance of the matrix path is much less than that of the percolating fiber path pre-set) (Fig. 40). In Nyquist plots (-imaginary impedance vs. +real impedance) three impedance arcs/features at 1 day of hydration were observed, similar to Nyquist plots for macrofiber and microfiber reinforced cement-based materials (Fig. 41). The intersection of the electrode arc and the intermediate frequency feature (R_{DC} (nanocomposites)) corresponds to the DC resistance of the nanocomposites. The intersection of the two bulk features (Rcusp) corresponds to the AC resistance of the nanocomposite (Fig. 42). Reductions in R_{DC} (nanocomposites) from the matrix resistance are indicative of a nanotube percolating network. Reductions in Rcusp from the matrix resistance are indicative of a discontinuous fiber-fiber path. Both shifts increased with fiber loading. AC-IS measurements can be used to discriminate percolation vs. discontinuous fiber effects in nanocomposites, with the potential for characterizing dispersion issues (e.g., clumping/aggregation) in nanocomposites. Additionally, the dielectric constant vs. frequency plots for the plain paste and the two nanocomposites show that the presence of CNTs clearly enhances the dielectric constant in the 10^3-10^6 Hz range above the baseline dielectric constant of the paste (The baseline dielectric constant of the plain paste is ~300 and significantly larger than the dielectric constant of water (~80).) (Fig. 43).

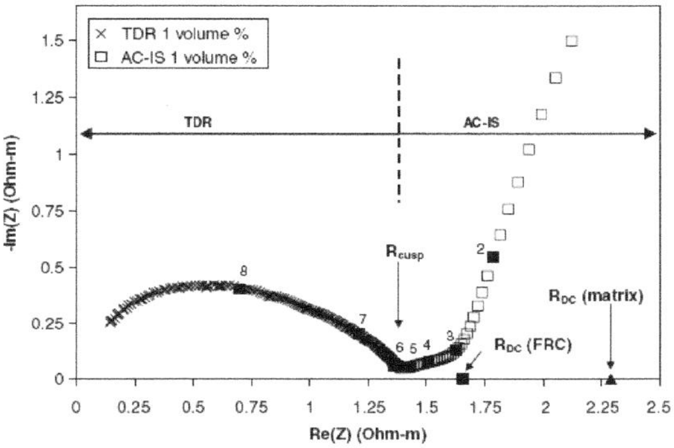

Fig. 42 Nyquist plot for nanocomposites containing 1 vol. % of MWNTs at 1 day with TDR data superimposed on the AC-IS data at high frequencies.

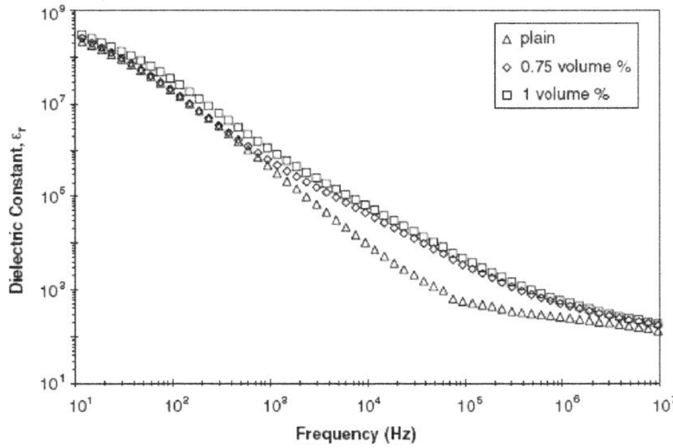

Fig. 43 Dielectric constant vs. frequency plots for plain cement paste and the two nanocomposites with 0.75 vol. % and 1 vol. % of MWNTs.

Nasibulin et al. [2009] and Cwirzen et al. [2009] studied of electrical resistivity of CNTs reinforced cement paste made of cement/CNTs hybrid. They found that the nanocomposites reveal as high as a 40 times increase in the electrical conductivity compared plain cement paste (Table 4).

Table 4 Compressive strength of cement paste prepared by adding cement/CNTs hybrid

Fraction of cement/CNTs hybrid	Synthesis conditions				Electrical resistivity (MΩcm)
	Temp (ºC)	Gas flow rate (cm^3min^{-1})			
		C_2H_2	CO_2	CO	
0	-	-	-	-	9.7
100	550	860	0	177	0.23
100	575	660	660	0	1.3
100	500	500	500	0	1.7
100	525	660	660	0	4.0

4.3 Conductive and Piezoresistive Mechanisms of CNTs Reinforced Cement-Based Materials

The electrical resistance of CNTs reinforced cement-based materials comes from two sources, i.e., the intrinsic resistances of nanotubes and matrix and the contact resistance (i.e. the resistance between connecting CNTs or the resistance of the matrix connecting the crossing nanotubes and through which electrical tunneling occurs). The electric conductivity of individual CNTs is in the order of 10^4~10^7s/m. The electrical resistivity of a plain cement matrix commonly is in a range from 10^6 to 10^9 Ω·cm. The contact resistance at nanotube junctions is rather complicated and depends on concentration and physical properties of nanotubes, tunneling gap at contact points and conductive properties of matrix filling the tunneling gap[Li and Chou 2008]. There are three basic types of conduction mechanisms (i.e. contacting conduction, tunnelling conduction and field emission conduction) contributing to the contact resistance [Lu et al. 2006, Wang et al. 2009, Chen et al. 2008, Li et al. 2005, Han et al. 2010]. Therefore, the conductive and piezoresistive behaviors of the nanocomposites are subjected to four basic aspects as below.

(1) Intrinsic resistance of CNTs
The intrinsic resistance of CNTs reflects the electron transmission behavior in CNTs, which will affect the transmission behavior of electron within the whole nanocomposites. The intrinsic resistance of CNTs is governed by CNTs' types, CNTs' surface condition, load condition, and so on. For examples, the carboxyl functionalization might introduce structural defects resulting in the decrease of conductive properties of CNTs and nanocomposites. When nanocomposites are deformed under external loading, the nanotube length and diameter will alter, resulting in the change of nanotube intrinsic resistance. Therefore, the piezoresistivity of CNTs has a contribution to the piezoresistivity of the nanocomposites[Li et al. 2005].

(2) Intrinsic resistance of cement-matrix
The ionic conduction is a basic type of electrical conduction in cement-matrix. The ions come from hydrated cement. Electrical polarization refers to the phenomenon in which the centers of positive and negative charges do not coincide. As

an electrical field is present during electrical resistance measurement, the movement and concentration of the ions such as K^+, Na^+, Ca^{2+}, OH^- and SO_4^{2-} etc in cement-matrix lead to the electrical polarization [Saad Morsy 1999, Han et al. 2007, Tumidajski 1996]. As a results, the electrical resistivity of nanocomposites increases with measurement time under direct current voltage [Luo, 2009]. The polarization is diminished when conductive fillers are incorporated into the matrix. The more conductive cement-based materials are, the less the tendency to polarize is. Therefore, the increase of CNTs concentration can decrease the electrical resistivity of nanocomposites and increases the stability of electrical resistivity of nanocomposites [Luo, 2009]. In addition, the increase of water content enhances ionic conduction, which can reduce their electrical resistivity [Han and Ou 2008]. Furthermore, the nanocomposites present alternating current characteristics because of electrical properties of cement-matrix [Luo, 2009, Wanson et al. 2006].

(3) Contacting conduction

The forceful effect on reducing electrical resistivity for the CNTs is mainly due to the formation of a well three-dimensional meshwork (Fig. 44). The high concentration and effective dispersion of CNTs are beneficial for forming the contac points of CNTs in the meshwork. The more the contact points between CNTs are, the lower electric resistivity will be [Saafi 2009, Azhari 2008]. In the case of the CNTs/CFs hybrid, the combination of CNTs and CFs provides charge transport over long and short distances and enhances the contacting conduction. This CNTs/CFs hybrid multiscale composite makes the nanocomposites more conductive compared to that only with CNTs [Azhari 2008]. The number of contact points of CNTs increases under compressive load, which can cause an enhancement of conductivity [Li et al. 2005, Yu and Kwon 2009, Han et al. 2009]. However, the tensile load can separate CNTs each other, so the electrical resistivity of the nanocomposites increases under tensile load [Saafi 2009]. When the CNTs concentration exceeds the percolation threshold, the contacting conduction dominates the electrical conductivity of the nanocomposites. As a result, the three-dimensional contacting meshwork of CNTs becomes more stable and the electrical resistivity of the nanocomposites is insensitive to the increase of CNTs' concentration and external load [Saafi 2009]. Additionally, the carboxyl functionalization of CNTs can enhance the bond between CNTs and matrix because of the interfacial interactions between the groups of functionalized CNTs' surfaces and hydrations (such as C-S-H and calcium hydroxide) of cement. This will prevent the contact of CNTs, resulting in an increase in electrical resistivity of the nanocomposites[Li et al. 2005]. For the CNTs reinforced cement mortar, the large-size sands cut off some straight connects of CNTs in matrix. However, the contacting conductive meshwork can be formed at a high concentration of CNTs. As a result, the effective CNTs concentration for CNTs reinforced cement mortar will be much higher than that for CNTs reinforced cement paste [Han et al. 2010].

(4) Tunneling conduction

Unlike contacting conduction, tunneling conduction has no strong enhancement to the conductivity of nanocomposites, while it has a significant contribution to the piezoresistivity of nanocomposites. This means that the tunneling conduction can

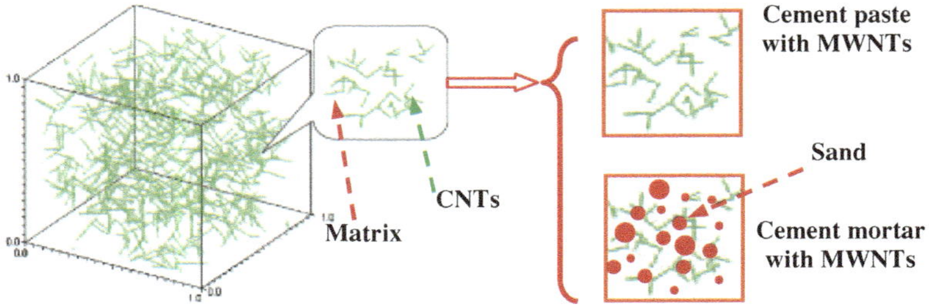

Fig. 44 Morphology of nanotube meshwork in matrix.

occur at a moderate concentration of CNTs. The separation between CNTs is large at a low concentration of CNTs, the electron is hard to generate transition through tunneling condition. However, the contacting conduction becomes dominant in the conductivity of the nanocomposites at a high concentration of CNTs. When the nanocomposites are deformed under compressive or tensile loading, the separation between CNTs will be reduced or increased, i.e. the tunneling conduction is enhanced or weakened. This leads to the change of electrical resistivity of the nanocomposites under external load, expressing strong piezoresitive responses [Li et al. 2005, Yu and Kwon 2009, Han et al. 2009, Saafi 2009, Azhari 2008]. Furthermore, the carboxyl functionalization of CNTs is covered by matrix, so the tunneling junction and the separation between CNTs in nanocomposites are increased. Because the tunneling junction and the separation between CNTs mightily change with the variation of compressive load, the piezoresistivity of the cement paste with carboxyl functionalized CNTs is more intensive than that of the cement paste with unfunctionalized CNTs [Li et al. 2005]. The electrical conductivity of matrix filling the tunneling gaps can be enhanced by the adsorption of water molecules, so the tunneling conduction can easily occur in the nanocomposites with high water content. This can make the conductivity of the nanocomposites easier or harder to change under external load. Therefore, the water content is benefit for the enhancement of conductivity of the nanocomposites, but it has dual effect on the piezoresistivity of the nanocomposites [Han et al. 2005].

(5) Field emission conduction

CNTs have a strong field emission property under electric field. The field emission can induce strong tunneling conduction because of local filed enhancement. This makes the CNTs reinforced cement-based materials more conductive. However, the carboxyl functionalization leads to the opening of the capped ends, and the opened CNTs are far less efficient emitters than closed CNTs due to the change in the work function that aroused from the state of the tip. Field emission of CNTs may decrease due to chemical modification, so the carboxyl functionalization may cause an increase of conductivity of the nanocomposites [Li et al. 2005]. In addition, the field emission effect on the nanotube tips can be enhanced by the adsorption of water molecules [Grujicic et al. 2003, Chen et al. 2004, Qiao et al. 2007], so the field induced tunneling conduction can easily occur in the nanocomposites with high water content. This also can make the conductivity of the nanocomposites easier or harder to change under external load. Therefore, the high water content is also benefit for the enhancement of conductivity of the nanocomposites, but it also has dual effect on the piezoresistivity [Han et al. 2005].

It is worth noting that the four aspects motioned above are not individual, but jointly in charge of the conductivity and piezoresistivity of CNTs reinforced cement-based materials. Just, one or several among four aspects may be dominant for a specific CNTs reinforced cement-based material. In addition, the concentration of CNTs above percolation threshold is beneficial for enhancing the sensitivity of piezoresistive response of nanocomoposites under tensile load, while that below percolation threshold is beneficial for enhancing the sensitivity of piezoresistive response under compressive load.

5 Thermal Conductive and Damping Properties of CNTs Reinforced Cement-Based Materials

Besides mechanical, electrical and piezoresistive properties of CNTs reinforced cement-based materials, researchers also preliminarily explored other two functional properties: thermal conductive property and damping property.

5.1 Thermal Conductive Property of Nanocomposites

Prior to CNT, diamond was the best thermal conductor. CNT have now been shown to have a thermal conductivity at least twice that of diamond [Khare and Bose 2005]. In addition, the negative coefficient of thermal expansion of the CNTs results in a higher thermal stability [Veedu 2010]. Therefore, CNTs is expected to improve the thermal stability of cement-based materials. Veedu (2010) compared the thermal performance of CFs reinforced cement-based materials and CNTs reinforced cement-based materials, and observed that the thermal conductivity of CNTs reinforced cement-based materials is at least 35% and 85% greater than that of CFs reinforced cement-based materials and unreinforced cement-based materials (typically 0.8W/mK), respectively (Table 5).

Table 5 Comparison of thermal conductivity of CNTs reinforced cement-based materials and CFs reinforced cement-based materials

Type of materials	Temperature (°C)	Thermal conductivity (W/mK)
CFs reinforced cement-based materials	32.7	1.1
CNTs reinforced cement-based materials	32.6	1.5

5.2 Damping Property of Nanocomposites

One of the important material characteristic needed for civil engineering structures is the ability to absorb vibrations, typically referred to as the damping capacity. Damping capacity is the dissipation of vibration by converting the mechanical energy introduced by vibration into heat. One of the materials that showed promise in this aspect is fiber reinforced materials. The ability to tailor the interface strength between the fibers and matrix enhances the ability to design for a specific damping capacity. It was observed that a low interfacial shear strength between the fibers and the matrix and a high interfacial area, lead to an increase in damping capacity [Gibson et al. 1980, Luo 2009]. Such damping capacity can be further increased if discontinuous fibers were used rather than continuous fibers to reinforce the matrix [Gibson et al. 1980, Mclean et al. 1975]. The remarkable mechanical properties of CNTs, large specific surface area (i.e. larger interfacial contact area between the nanotubes and the surrounding matrix) of CNTs, and frictions among multiple inter-tubes will be able to provide an increased damping capacity at nanoscale [Slosberg et al. 2003, Zhou et al.2004]. Luo [2009] studied the damping capacity of cement paste with different concentrations of MWNTs (from 0.1 wt. % to 2 wt. %). He found that the critical damping ratio of nanocomposites increase with concentrations of CNTs. The critical damping ratio of nanocomposites with 2% of MWNTs is 1.6 times than that of plain cement paste.

6 Potential Structural Applications of CNTs Reinforced Cement-Based Materials

The goal of developing multifunctional and smart CNTs reinforced cement-based materials is to set up multifunctional and smart structures. Because the study on the piezoresistivity of CNTs reinforced cement-based materials begins latter, less work is focused on the research and application of multifunctional and smart structures fabricated using CNTs reinforced cement-based materials.

Han et al. [2009] developed a self-sensing pavement with piezoresistive MWNTs reinforced cement-based materials (Fig. 45) and investigated the feasibility of using self-sensing pavement for traffic monitoring with vehicular loading experiments. They found that the vehicular loads can lead to remarkable change in electrical resistance of piezoresistive MWNTs reinforced cement-based materials in the self-sensing pavement (Fig. 46), so traffic flow monitoring and even possible identification of different vehicular loadings can be realized by measuring the electrical resistance of the composites. These findings indicate that multifunctional and smart CNTs refinforced cement-based material and structures have great potential for traffic-monitoring applications such as vehicle detection, weighing and speed measurement.

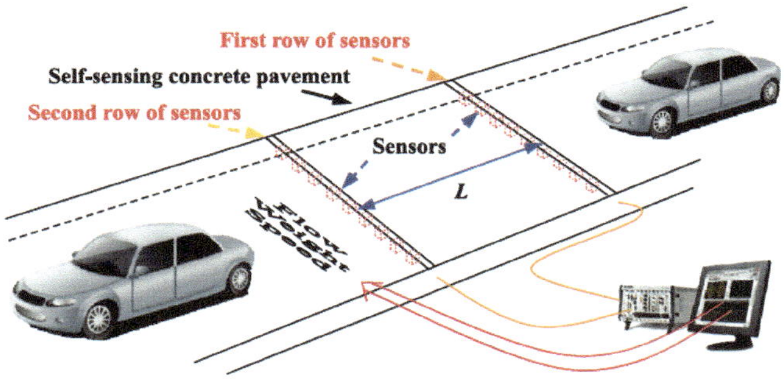

Fig. 45 Illustration of self-sensing concrete pavement for traffic flow detection.

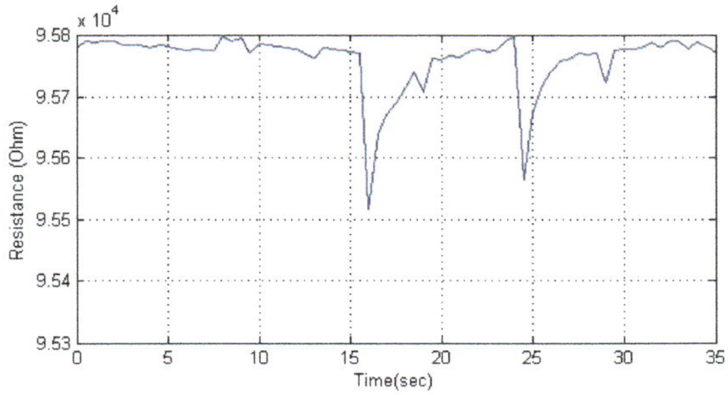

a) Two middle size passenger vehicles pass over the sensor

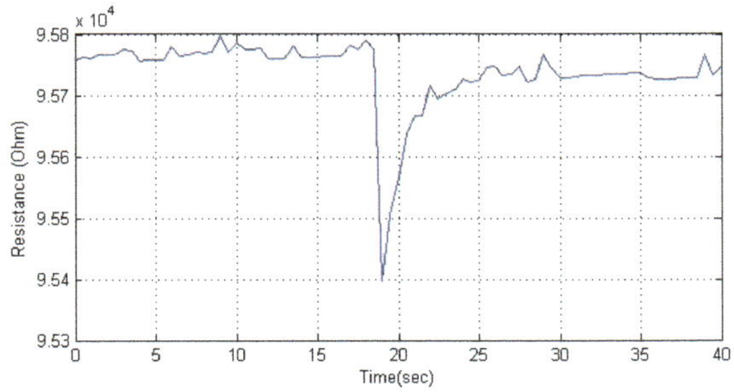

b) A mini-van passes over the sensor

Fig. 46 Variation in electrical resistance of piezoresistive MWNTs reinforced cement-based materials under different vehicular loadings.

Saafi [2009] developed a kind of in situ wireless and embedded sensors for damage detection in concrete structures by using piezoresistive SWNTs reinforced cement-based materials (Fig. 47). These sensors were embedded into concrete beam to set up self-sensing concrete structures subjected to monotonic and cyclic bending. The fabricated self-sensing concrete structures can wirelessly detect crack propagation and damage accumulation during loading by using the change in the electrical resistance of the embedded sensors (Fig. 48). The research findings prove the feasibility of using multifunctional and smart CNTs reinforced cement-based material and structures to monitor the integrity of concrete structures.

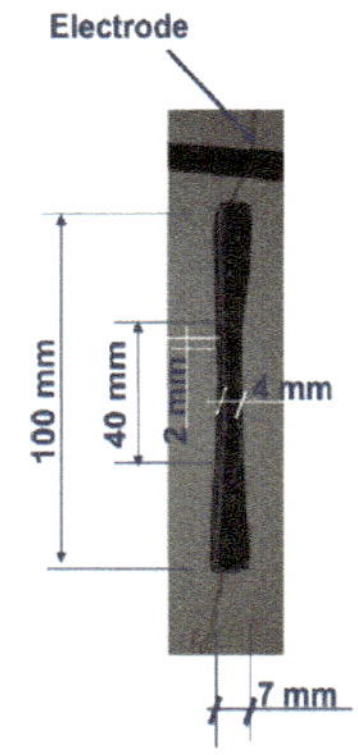

Fig. 47 Embedded SWNTs reinforced cement-based sensors.

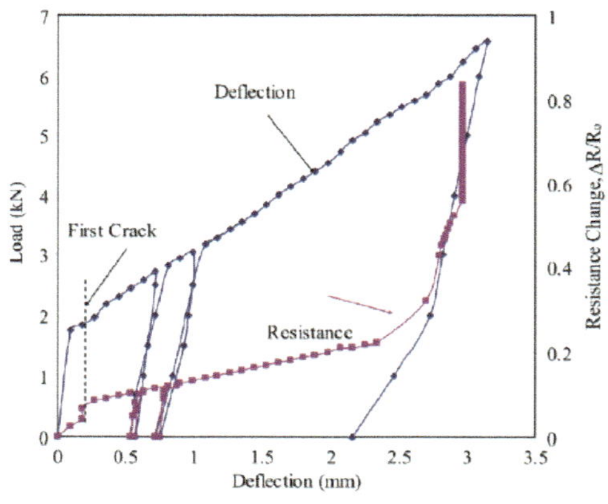

Fig. 48 Typical wireless response of sensors embedded into concrete beams.

Veedu (2010) presented two types of potential application styles of multifunctional and smart CNTs reinforced cement-based materials. A bridge made with these materials can achieve in-situ monitoring for structural health and integrity and for the weight, speed, and location of traffic over the bridge (Fig. 49). Another potential application for the nanocompostes is in the creation of smart highways that will potentially track the location, weight and speed of traffic.

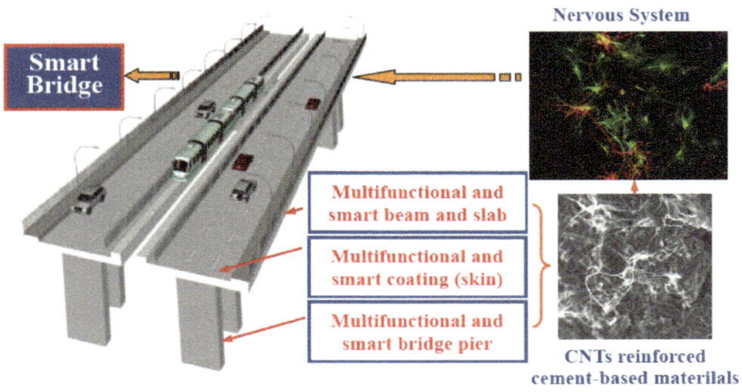

Fig. 49 Smart bridge made with multifunctional and smart CNTs reinforced cement-based materials.

7 Challenges for Development and Deployment of Multifunctional and Smart CNTs Reinforced Cement-Based Materials

Although the CNTs reinforced cement-based materials exhibit unique multifunctional and smart properties, the following challenges are believed to be critical in the future research and development of this kind of materials.

(1) Cost of CNTs
The high production costs of the high quality CNTs will primarily limit the large-scale application of multifunctional and smart CNTs reinforced cement-based materials. However, since 1990 the cost of producing CNTs has fallen 100-fold and can be expected to fall further. Additionally, a very low rate of CNTs can provide a huge contribution to properties of cement-based materials, and the nanocomposites with excellent properties will allow existing structural designs to be produced with reduced material volumes. Furthermore, it is expected that the multifunctional and smart CNTs reinforced cement-based materials will feature high performance price ratio and low life cycle cost. Therefore, the cost of CNTs should no longer be the critical issue.

(2) Adverse impact of aggregate
Concrete and cement mortar are much more useful in real applications than cement paste. But aggregate, especially coarse aggregate, will make it more difficult and complex to enhance the properties of the nanocomposites by using a low concentration of CNTs. For this challenge, the nanotube/fiber hybrid multiscale composite technology may be a potential solution.

(3) Fabrication of high performance cement-based materials reinforced with low concentration of CNTs
Considering the high cost of CNTs, the number of CNTs used in the cement-based materials should be as low as possible to decrease the effective product cost. Therefore, a simple technology should be developed to fabricate high performance

cement-based materials reinforced with low concentration of CNTs. A key problem related with this issue is to find a simple, repeatable, large-scale and low-energy consumption method for distributing CNTs in cement-based materials without altering the manufacturing process of cement-based materials. It is necessary to develop a simple and convenient evaluation method of dispersion quality of CNTs. In addition, it is suggested that the subsequent studies should be aimed at establishing the optimum CNTs' types and CNTs' concentration values as the mix design parameters.

(4) Durability of CNTs reinforced cement-based materials
CNTs have a significant enhancement to microstructure of cement-based materials, so it is expected that CNTs are benefit for long term durability of cement-based materials. Future research should include exploring the durability of CNTs reinforced cement-based materials, such as permeability resistance, frost resistance, erosion resistance, drying shrinkage, abrasion resistance and fire resistance.

(5) Multifunctional and smart properties of CNTs reinforced cement-based materials containing aggregates
It deserves extensive investigation on the effect of aggregates on the multifunctional and smart properties of CNTs reinforced cement-based materials. In this work, the effective nanotube/fiber hybrid multiscale composite method should be explored.

(6) Enhancement mechanisms of CNTs to cement-based materials
Previous research works only qualitatively explain the enhancement mechanisms of CNTs to cement-based materials. Future research should seek a quantitative explanation for describing and forecasting the properties of nanocomposites based on experiment and numerical simulation.

(7) Piezoresistivity of CNTs reinforced cement-based materials under complex external force
Previous research efforts concentrate on the piezoresistivity of CNTs reinforced cement-based materials under uniaxial compressive or tensile loading. In real structures, cement-based materials subject to complex external force, such as bending, shear and torsion. Therefore, the overall piezoresistive characteristics of the nanocomposites should be determined for structural application.

(8) Effect of environment on piezoresistivity of CNTs reinforced cement-based materials
In practice application, the environment condition surrounding cement-based materials and structures is unstable. Temperature and moisture are two main environmental parameters. Their fluctuation will lead to the change in conductivity of the nanocomposites. It is necessary to explore the effects of temperature and moisture on the piezoresistivity and eliminating these effects.

(9) Acquisition and processing technology for sensing signals of smart CNTs reinforced cement-based materials and structures
Unlike microscale fibers, CNTs can form an extensively conductive meshwork in matrix at nanoscale. It is necessary to develop new signal acquisition and

processing method for mining deep information reflecting structural condition. In addition, the passive wireless acquisition technology is deserved to develop.

(10) Other functional and smart properties of CNTs reinforced cement-based materials
CNTs not only have excellent mechanical and electrical properties but also possess superior magnetic and electromagnetic properties. It is expected that CNTs can enhance the magnetic and electromagnetic properties of cement-based materials. In addition, some smart properties (e.g. piezoelectric, ferroelectric, temperature-sensing, humidity-sending properties) can be induced by coupling effects between mechanical, electrical, thermal and magnetic properties of the nanocomposites. There should some profound investigation on other functional and smart properties of CNTs reinforced cement-based materials.

(11) Structural application of CNTs reinforced cement-based materials
Even at very low rates of addition, current prices of CNTs are high enough that the production of significant structures is cost prohibitive. Even with large reductions in CNTs prices, the materials are most likely to be used in niche application. Depending on the application and design requirements, CNTs reinforced cement-based materials can be used in key position of significant engineering structures. In addition, according to the specific application, scenario and environment, the embedded, coating or samdwich structural styles can be used to further decrease the consumption of CNTs reinforced cement-based materials.

8 Concluding Remarks

Recent advances in nanotechnology have opened a door for developing new cement-based materials. CNTs are molecular-scale tubes of graphitic carbon with outstanding properties. They are among the stiffest and strongest fibers known, and have remarkable electronic properties and many other unique characteristics. As a result of these properties coupled with the high aspect ratio, small diameter, lightness, and excellent chemical and thermal stability, CNTs can be used as reinforcement to produce multifunctional and smart cement-based materials. The CNTs reinforced cement-based materials present superior mechanical (e.g. high compressive, tensile and flexural strengths, and fracture toughness), electrically conductive, piezoresistive, thermal conductive and damping properties. A prerequisite for the successful realization of CNTs enhancement effect to cement-based materials is the effective utilization of their high aspect ratio, for which their effective dispersion is essential. In addition, it is expected that CNTs reinforced cement-based materials should exhibit significant durability and other functional or smart properties (e.g. thermal property, magnetic property, and temperature-sensitive property). Therefore, the multifunctional and smart cement-based materials have a wide application foreground in civil infrastructures such as high-rise buildings, highway, bridges, runways for airport, continuous slab-type sleepers for high speed trains, dam and nuclear power plant, and they especially have great potential in the field of structural health monitoring, highway traffic monitoring, electro-magnetic wave shielding, electrostatic elimination, structural vibration control, and deicing or snow-melting. Small CNTs will bring a large revolution in the field of conventional cement-based materials, which should have a beneficial impact on economics, social and environment.

References

Aitcin, P.C.: Cements of yesterday and today Concrete of tomorrow. Cement and Concrete Research 30, 1349–1359 (2000)

Azhari, F.: Cement-based sensors for structural health monitoring. Dissertation for the Master Degree of Applied Science. University of British Columbia, Vancouver, Canada (2008)

Banthia, N.: Fiber reinforced concrete for sustainable and intelligent infra structure. In: First International Conference on Sustainable Built Environment Infrastructures in Developing Countries, Algeria, pp. 337–350 (2009)

Blanchet, G.B., Subramoney, S., Bailey, R.K., Jaycox, G.D., Nuckolls, C.: Sel-assembled three-dimensional conducting network of single-wall carbon nanotubes. Applied Physics Letters 85, 828–830 (2004)

Branner, M., Kavi, A.M., Li, M.G.: Carbon nanotube-fiber reinforced cement and concrete. Patent US 20081034942A1 (2008)

Cao, J., Wang, Q., Dai, H.: Electromechanical properties of metallic, quasimetallic, and semiconducting carbon nanotubes under stretching. Physical Review Letters 90, 157601–157604 (2003)

Chaipanich, A., Nochaiya, T., Wongkeo, W., Torkittikul, P.: Compressive strength and microstructure of carbon nanotubes-fly ash cement composites. Materials Science and Engineering A 527(4-5), 1063–1067 (2010)

Chen, C.W., Lee, M.H., Clark, S.J.: Gas molecule effects on field emission properties of single-walled carbon nanotube. Diamond and Related Materials 13, 1306–1313 (2004)

Chen, K., Xiong, C.X., Li, L.B., Zhou, L., Lei, Y., Dong, L.J.: Conductive mechanism of antistatic poly (ethylene terephthalate)/ZnOw composites. Polymer Composites 30, 226–231 (2008)

Cheng, Y., Zhou, O.: Electron field emission from carbon nanotubes. Comptes Rendus Physique 4, 1021–1033 (2003)

Chung, D.D.L.: Piezoresistive cement-based materials for strain sensing. Journal of Intelligent Material Systems and Structures 13, 599–609 (2002)

Cwirzen, A., Habermehl-Cwirzen, K., Penttala, V.: Surface decoration of carbon nanotubes and mechanical properties of cement/carbon nanotube composites. Advances in Cement Research 20, 65–73 (2008)

Cwirzen, A., Habermehl-Cwirzen, K., Nasibulina, L.I., Shandakov, S.D., Nasibulin, A.G., Kauppinen, E.I., Mudimela, P.R., Penttala, V.: CHH Cement Composite. Nanotechnology in Construction 3 (2009)

De Backer, H., De Corte, W., Van Bogaert, P.: A case study on strain gauge measurements on large post-tensioned concrete beams of a railway support structure. Insight: Non-Destructive Testing and Condition Monitoring 45-12, 822–826 (2003)

De Ibarra, Y.S., Gaitero, J.J., Campillo, I.: Atomic force microscopy and anoindentation of cement pastes with nanotube dispersions. Physica Status Solidi A 203, 1076–1081 (2006)

Dresselhaus, M.S., Dresselhaus, G., Avouris, P.: Carbon nanotubes: synthesis, structure, properties and applications. Springer, Heidelberg (2000)

Dunens, O.M., MacKenzie, K.J., Harris, A.T.: Synthesis of multiwalled carbon nanotubes on fly ash derived catalysts. Environmental Science and Technology 43, 7889–7894 (2009)

Gibson, R.F., Yau, A.: Complex moduli of chopped fiber and continuous fiber composites: comparison of measurements with estimated bounds. Journal of Composite Materials 14, 155–167 (1980)

Girifalco, L.A., Hodak, M., Lee, R.S.: Carbon nanotubes, buckyballs, ropes, and a universal graphitic potential. Physical Review B 62, 13104 (2000)

Grunlan, J.C., Mehrabi, A.R., Bannon, M.V., Bahr, J.L.: Water-based single-walled-nanotube-filled polymer composite with an exceptionally low percolation threshold. Advanced Materials 16, 150–153 (2004)

Grujicic, M., Gao, G., Gersten, B.: Enhancement of field emission in carbon nanotubes through adsorption of polar molecules. Applied Surface Science 206, 167–177 (2003)

Han, B.G., Guan, X.C., Ou, J.P.: Electrode design, measuring method and data acquisition system of carbon fiber cement paste piezoresistive sensors. Sensors and Actuators: A physical 135, 360–369 (2007)

Han, B.G., Ou, J.P.: Embedded piezoresistive cement-based stress/strain sensor. Sensors and Actuators: A physical 138, 294–298 (2007)

Han, B.G., Ou, J.P.: The humidity sensing property of cements with added carbon. New Carbon Materials 23, 382–384 (2008)

Han, B.G., Yu, X., Kwon, E.: A self-sensing carbon nanotube/cement composite for traffic monitoring. Nanotechnology 20, 445–501 (2009)

Han, B.G., Yu, X., Ou, J.P.: Dispersion of carbon nanotubes in cement-based composites and its influence on the piezoresistivities of composites. In: ASME 2009 Conference on Smart Materials, Adaptive Structures and Intelligent Systems (SAMASIS 2009), Oxnard (2009)

Han, B.G., Yu, Y., Han, B.Z., Ou, J.P.: Development of a wireless stress/strain measurement system integrated with pressure-sensitive nickel powder-filled cement-based sensors. Sensors and Actuators: A physical 147, 536–543 (2008)

Han, B.G., Yu, X., Kwon, E., Ou, J.P.: Piezoresistive MWNTs filled cement-based composites. Sensor Letters 8, 344–348 (2010)

Han, B.G., Yu, X., Ou, J.P.: Effect of water content on the piezoresistivity of CNTs/cement composites. Journal of Materials Science (2010), doi:10.1007/s10853-010-4414-7

Hiroshi, I., Yoshiki, O., Hitoshi, K.: Experimental study on structural health monitoring of RC columns using self-diagnosis materials. In: Proceedings of SPIE, vol. 5391, pp. 609–617 (2004)

Hou, T.C., Lynch, J.P.: Conductivity-based strain monitoring and damage characterization of fiber reinforced cementitious structural components. In: Proceedings of SPIE, vol. 5765, pp. 7–10 (2005)

Hu, C.Y., Xu, Y.J., Duo, S.W., Zhang, R.F., Li, M.S.: Non-covalent functionalization of carbon nanotubes with surfactants and polymers. Journal of the Chinese Chemical Society 56, 234–239 (2009)

Iijima, S.: Helical microtubes of graphitic carbon. Nature 354, 56–58 (1991)

Kakade, B.A., Pillai, V.K.: Tuning the wetting properties of multiwalled carbon nanotubes by surface functionalization. Journal of Physical Chemistry 112, 3183–3186 (2008)

Khare, R., Bose, S.: Carbon Nanotube Based Composites- A Review. Journal of Minerals and Materials Characterization and Engineering 4, 31–46 (2005)

Kim, Y.J., Shin, T.S., Choi, H.D., Kwon, J.H., Chung, Y.C., Yoon, H.G.: Electrical conductivity of chemically modified multiwalled carbon nanotube/epoxy composites. Carbon 43, 23–30 (2005)

Konsta-Gdoutos, M.S., Metaxa, Z.S., Shah, P.S.: Highly dispersed carbon nanotube reinforced cement based materials. Cement and Concrete Research (2010), doi:10.1016/j.cemconres2010.02.015

Leng, J.S., Winter, D., Barnes, R.A., Mays, G.C., Fernando, G.F.: Structural health monitor-ing of concrete cylinders using protected fiber optic sensors. Smart Materials and Structures 15(2), 302–308 (2006)

Li, C.Y., Chou, T.W.: Modeling of damage sensing in fiber composites using carbon nanotube networks. Composites Science and Technology 68, 3373–3379 (2008)

Li, G.Y., Wang, P.M., Zhao, X.H.: Mechanical behavior and microstructure of cement composites incorporating surface-treated multi-walled carbon nanotubes. Carbon 43, 1239–1245 (2005)

Li, G.Y., Wang, P.M., Zhao, X.H.: Pressure-sensitive properties and microstructure of carbon nanotube reinforced cement composites. Cement and Concrete Composites 29, 377–382 (2007)

Li, H., Liu, Z.Q., Li, Z.W., Ou, J.P.: Study on damage emergency repair performance of a simple beam embedded with shape memory alloys. Advances in Structural Engineering 7(6), 495–501 (2004)

Lourie, O., Cox, D.E., Wagner, H.D.: Buckling and collapse of embedded carbon nanotubes. Physical Review Letter 81, 16–38 (1998)

Lu, J.R., Chen, X.F., Lu, W., Chen, G.H.: The piezoresistive behaviors of polyethylene/foliated graphite nanocomposites. European Polymer Journal 42, 1015–1021 (2006)

Ludvig, P., Ladeira, L.O., Calixto, J.M., Gaspar, I.C.P., Melo, V.S.: In-situ synthesis of multiwall carbon nanotubes on portland cement clinker. In: 11th International Conference on Advanced Materials, Rio de Janeiro, Brazil (2009)

Luo, J.L.: Fabrication and functional properties of multi-walled carbon nanotube/cement composites. Dissertation for the Doctoral Degree in Engineering, Harbin Institute of Technology, Harbin, China (2009)

Luo, J.L., Duan, Z.D., Li, H.: The influence of surfactants on the processing of multiwalled carbon nanotubes in reinforced cement matrix composites. Physica Status Solidi 206(12), 2783–2790 (2009)

Makar, J.M., Beaudoin, J.J.: Carbon nanotubes and their application in the construction industry. In: Proceedings of the 1st International Symposium on Nanotechnology in Construction, Paisley, p. 331 (2003)

Makar, J.M., Chan, G.W.: Growth of cement hydration products on single walled carbon nanotubes. Journal of the American Ceramic Society 92, 1303–1310 (2009)

Makar, J., Margeson, J., Luh, J.: Carbon nanotube/cement composites-early results and potential applications. In: Third International Conference on Construction Materials: Performance, Innovations and Structural Implications, Vancouver, pp. 1–10 (2005)

Mather, B.: Concrete durability. Cement and Concrete Composites 26, 3–4 (2004)

Meyyappan, M.: Carbon nanotubes science and applications. CRC Press, Boca Raton (2005)

Mclean, D., Read, B.E.: Storage and loss moduli in discontinuous composites. Journal of Material Science 10, 481–492 (1975)

Musso, S., Tulliani, J.M., Ferro, G., Tagliaferro, A.: Influence of carbon nanotubes structure on the mechanical behavior of cement composites. Composites Science and Technology 69, 1985–1990 (2009)

Nasibulin, A.G., Shandakov, S.D., Nasibulina, L.I., Cwirzen, A., Mudimela, P.R., Habermehl-Cwirzen, K., Grishin, D.A., Gavrilov, Y.V., Malm, J.E.M., Tapper, U., Tian, Y., Penttala, V., Karppinen, M.J., Kauppinen, E.I.: A novel cement-based hybrid material. New Journal of Physics 11, 023013 (2009)

Ou, J.P., Han, B.G.: Piezoresistive cement-based strain sensors and self-sensing concrete components. Journal of Intelligent Material Systems and Structures 20, 329–336 (2009)

Odom, T.W., Huang, J.L., Kim, P., Lieber, C.M.: Structure and electronic properties of carbon nanotubes. Journal of Physical Chemistry B 104, 27–94 (2000)

Park, S., Ahmad, S., Yun, C.B., Roh, Y.: Multiple crack detection of concrete structures using impedance-based structural health monitoring techniques. Experimental Mechanic 46(5), 609–618 (2006)

Pushparaj, V.L., Nalamasu, O., Manoocher Birang, M.: Carbon nanotube-based load cells. Patent US2010/0050779 A1 (2010)

Qiao, L., Zheng, W.T., Wen, Q.B., Jiang, Q.: First-principles density-functional investigation of the effect of water on the field emission of carbon nanotubes. Nanotechnology 18, 155–707 (2007)

Raki, L., Beaudoin, J., Alizadeh, R., Makar, J., Sato, T.: Cement and concrete nanoscience and nanotechnology. Materials 3, 918–942 (2010)

Saad Morsy, M.: Effect of temperature on electrical conductivity of blended cement pastes. Cement and Concrete Research 29, 603–606 (1999)

Saafi, M.: Wireless and embedded carbon nanotube networks for damage detection in concrete structures. Nanotechnology 20, 395–502 (2009)

Sanchez, F.: Carbon nanofibre/cement composites: challenges and promises as structural materials. International Journal of Materials and Structural Integrity 3(2/3), 217–226 (2009)

Shah, S.P., Konsta-Gdoutos, M.S., Metexa, Z.S.: Highly-dispersed carbon nanotube-reinforced cement-based materials. Patent US 20090229494A1 (2009)

Slosberg, M., Kari, L.: Testing of nonlinear interaction effects of sinusoidal and noise excitation on rubber isolator stiffness. Polymer Testing 22, 343–351 (2003)

Terrones, M.: Science and technology of the twenty-first century: synthesis, propertypes, and applications of carbon nanotubes. Annual Review of Materials Research 33, 419 (2003)

Thess, A., Lee, R., Nikolaev, P., Dai, H., Petit, P., Robert, J., et al.: Crystalline ropes of metallic carbon nanotubes. Science 273, 483 (1996)

Tombler, T.W., Zhou, C., Alexseyev, L., Kong, J., Dai, H., Liu, L., Jayanthi, C.S., Tang, M., Vaisman, L., Wagner, H.D., Marom, G.: The role of surfactants in dispersion of carbon nanotubes. Advances in Colloid and Interface Science 128-130, 37–46 (2006)

Tumidajski, P.J.: Electrical conductivity of Portland cement mortars. Cement and Concrete Research 26, 529–543 (1996)

Veedu, V.P.: Multifunctional cementitious nanocomposite material and methods of making the same. Patent: US 7666327 B1 (2010)

Wang, L.H., Ding, T.H., Wang, P.: Influence of carbon black concentration on piezoresistivity for carbon-black-filled silicone rubber composite. Carbon 47, 3151–3157 (2009)

Wansom, S., Kidner, N.J., Woo, L.Y., Mason, T.O.: AC-impedance response of multi-walled carbon nanotube/cement composites. Cement and Concrete Composites 28, 509–519 (2006)

Wu, S.Y.: Reversible electromechanical characteristics of carbon nanotubes under local-probe manipulation. Nature 405, 769–772 (2000)

Yakovlev, G., Kerienė, J., Gailius, A., Girnienė, I.: Cement based foam concrete reinforced by carbon nanotubes. Materials Science 12(2), 147–151 (2006)

Yu, X., Kwon, E.: Carbon-nanotube/cement composite with piezoresistive property. Smart Materials and Structures 18, 055010 (2009)

Zhou, X., Shin, E., Wang, K.W., Bakis, C.E.: Interfacial damping characteristics of carbon nanotube-based composites. Composites Science and Technology 64, 2425–2437 (2004)

Applications of Nanotechnology in Road Pavement Engineering

Wynand JvdM Steyn

Abstract. The application of nanotechnology in various applied fields is receiving widespread attention. It is important to ensure that these applications address real questions to allow the technology to improve general well-being of the public, especially when evaluating application in the area of civil engineering. This chapter focuses on the specific applications of nanotechnology in the field of road pavements. The main objectives of pavements are to provide a safe and durable surface on which vehicles can travel, while protecting the underlying layers of material during all environmental conditions. As such, pavements are thus exposed to two main types of loads, namely traffic and the environment. Although good pavements can be constructed using existing materials and techniques, there are a number of areas where the judicious application of nanotechnology techniques should be able to improve the longevity and performance of the service provided by the pavement facility. These include improved and smart materials and characterization of materials. In this chapter the specific current needs that are addressed through these applications are discussed.

There are some concerns in the application of nano-materials regarding the safety of these materials for the environment and humans. These concerns need to be addressed specifically in the road pavement arena, as people are travelling on these facilities and living next to them and may be in direct contact with nano-materials without their knowing. The potential concerns are addressed and evaluated on a macro level.

Nanotechnology is focused on materials in the nano-scale while civil engineering infrastructure (especially road pavements) is focused on the macro scale (100s of kilometers in length). A valid question often posed in this matter is how the changes on a nano-scale can affect the macro scale properties and behavior, and whether scaling of effects and quantities from the nano- to the macro-scale is feasible. This also affects some of the nano-effects that may be affected by the inherent mass of a material, where the same effect will not be visible on a macro-scale due to simple issues such as the effect of gravity. A macro-level analysis of this issue is included, addressing the general concern of scaling effects and indicating where these effects may be an issue in the pavement field.

Wynand JvdM Steyn
Department of Civil Engineering, Faculty of Engineering, Built Environment & IT, University of Pretoria, Pretoria, South Africa
e-mail: Wynand.steyn@up.ac.za

K. Gopalakrishnan et al. (Eds.): Nanotechnology in Civil Infrastructure, pp. 49 – 83.
springerlink.com

1 Introduction

1.1 Chapter Structure

This chapter focuses on providing the required background and information on the status of nanotechnology applications in road pavements. It starts with background on pavements and nanotechnology, and then focuses on materials, characterization and concerns and issues.

The chapter was developed based on published research, with appropriate analysis and synthesis of the available information to provide a detailed understanding of the current status of nanotechnology in road and airfield pavements. The body of knowledge on this topic is constantly developing, and therefore there will always be aspects that may not be covered in as much detail as they are covered in specific research studies. The number of journals and conferences covering the topic is also expanding, and, although attempts were made to incorporate information from most of the relevant journals and conferences, there may be some of those that are not included. This is purely due to the material having been covered already through published information from another source. The reader is urged to evaluate the available body of knowledge consistently to remain updated on the latest developments in the field.

1.2 Pavements

Pavements consist of a combination of layers of engineered materials that generally provide all-weather access to vehicles to travel in a safe en economical way. The layers of materials used are selected and engineered to provide a structure which can withstand the applied vehicular loads under a range of environmental conditions for a defined minimum life. Typically, materials deeper in the pavement structure will be less affected by the application of vehicular loads and environmental variations, while materials forming the surfacing of the pavement are directly exposed to the environment (diurnal and seasonal changes) as well as the applied vehicular loads. It is the intention of a good pavement design to enable vehicles to travel safely and economically.

A typical pavement consists of up to 5 specific layers, as illustrated in Fig. 1. The subgrade, selected and subbase layers typically consist of granular materials, while the base can either be a bound or unbound granular material. Bound granular base layers can be engineered using typically either cement or bitumen. The surfacing of the pavement typically consists of bituminous or cementations materials. These can consist of surfacing seals (a layer of aggregate bound by a layer of bitumen, asphalt (a manufactured mixture of aggregate, fillers and bitumen) or concrete (a manufactured mixture of aggregate, fillers and cement).

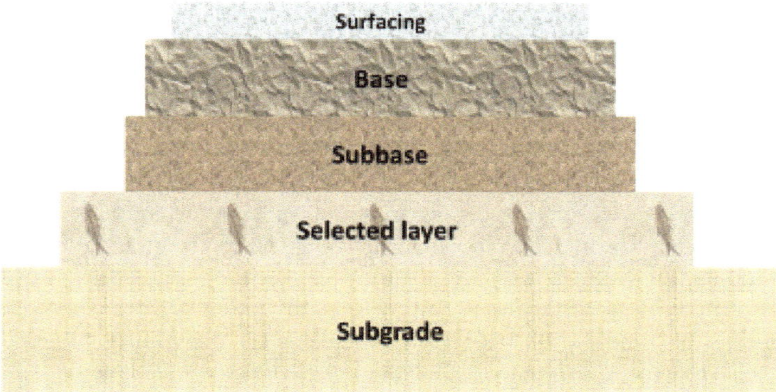

Fig. 1 Typical pavement structure showing various pavement layers.

The behavior of pavement materials is influenced by the environment in which they operate. Typically, bituminous materials are affected by changes in temperature (due to their temperature dependent viscous nature) while granular materials are affected by changes in moisture content. Concrete is affected by both temperature and moisture content. As the various layers in the pavement operate as a unit when loads are applied to them, it is important to view the pavement structure as a system where the behavior of one layer will affect the behavior of the other layers.

Pavements are designed to withstand a specified number of standard loads, applied over a design period. Such design periods typically range between 15 and 30 years. In order to keep the pavement in a serviceable condition during this period, regular maintenance and rehabilitation of the pavement structure is required.

1.3 Nanotechnology

Nanotechnology is the terminology used to cover the design, construction and utilization of functional structures with at least one characteristic dimension measured in nanometers [Kelsall et al 2004]. The nanotechnology field is constantly developing with developments mainly driven by factors such as dedicated initiatives into the field (e.g. the National Nanotechnology Initiative) [Goddard et al 2007], improvements in the characterization equipment and a new understanding into the chemistry and physics of matter on the nano scale. The study of nanotechnology employs the concept that the physical behavior of matter display substantial changes as the size decreases to the nano-meter scale [Wang 2003].

The Federal Highway Administration (FHWA) in the United States of America (USA) supports developments in nanotechnology in highways through various research efforts. Their focus is on developments to face broad challenges and to work towards crash-free, longer lasting and lower impact highways. They focus on the ability of nanotechnology research in various areas to measure, manipulate and model highway matter at the nano-scale. Typical areas of anticipated nanotechnology breakthroughs is seen as in the concrete field (itself being a nano-material) and efforts to develop smart self-healing concretes that are more resistant to environmental and chemical attacks (i.e. from deicing salts). Nanosensors for embedment into the pavement layers are anticipated to not only report on the condition of the material into which it is embedded, but also to communicate with vehicles to ensure safer driving conditions. The abilities of nano-scale materials to bind with pollutants and thereby improve air quality around transportation facilities have already proved to be viable. The improved characterization of both new and recycled materials through nano-characterization procedures can also lead to more optimal use of these materials. Four long-term anticipated effects of the use of nanotechnology in highway research include the development of novel structural materials that can withstand both traffic and environmental loads better, the incorporation of functions such as the ability to generate and transmit energy into pavement structures, improved vehicle-pavement communication regarding road conditions [FHWA 2009].

1.4 Pavements and Nanotechnology

Chemistry is the science concerned with the composition, structure, and properties of matter, while physics is the science that deals with concepts such as force, energy, and mass with the goal of understanding the natural world. Engineering seeks the creative application of scientific principles to design, develop, manufacture and operate structures, machines, manufacturing processes or works while acknowledging issues around economics of operation and safety to life and property. Pavement engineering is the engineering discipline that deals with the design and construction and maintenance of infrastructure such as roads, airfields and bridges.

It is clear that the objectives of science (chemistry and physics) and engineering differ in that the sciences are more concerned with the composition, fundamental properties, structure and interrelationships of matter, while engineering is more concerned with application of these principles in support of humanity (engineering properties). Engineering is thus dependent on the principles developed in the sciences, and focus on the application of these principles in support of humanity. While chemistry and physics generally focus on the smaller scales to enable a more detailed understanding of matter, engineering typically focuses on the scale where the matter works together to perform a certain function (e.g. cement particles that combine with aggregate particles to form a concrete road pavement or building). Scale dimensions thus differ from the nano-scale understanding (sciences) to the macro-scale understanding (engineering).

As an example, Meyyapan et al [2007] presented a summary of the expectations from emerging alternative technologies for Carbon NanoTube (CNT) electronic devices. It was indicated that a clear understanding of the expectations of a novel technology is required in order to evaluate the technology and ensure that the technology can deliver optimally. Once the current needs are well defined, potential applications can be developed. Current major needs of pavement engineering (based on proceedings from recent conferences in the pavement and transport arena) are summarized in Table 1 (based on information from the International Conference on Asphalt Pavements [ICAP – www.asphalt.org], PIARC [www.PIARC.com], Transportation

Research Board annual meeting [TRB – www.trb.org], Conference on Asphalt Pavements in Southern Africa [CAPSA – www.capsa-events.co.za] and South African Transportation Conference [Steyn 2008a]).

Table 1 Summary of broad areas of major research needs in pavement engineering (based on ICAP, PIARC, TRB, CAPSA, SATC conference themes).

Pavement area	Identified Issues	Potential strategies
Road pavements	Selecting adequate pavement types and road techniques	Develop long life / perpetual / sustainable pavements
	Maintaining pavements	Recycling of materials in existing pavements / sustainability
Road bridges and related structures	Increasing durability and safety of structures	Focus on methods to postpone maintenance and prolong life
Earthworks, drainage and subgrade	Promoting optimal use of local materials	Identify methods for treating soils and application of local / in situ materials

Partl et al [2003] anticipates nanotechnology to provide great potential in advancing asphalt pavement technology in the fields of materials design, manufacturing, properties, testing, monitoring and modeling. Specifically, focus areas in asphalt pavement analysis should include the bonds between aggregate, bonds between layers, properties of the mastic, self-repair and rejuvenation of binder, ageing (oxidation) effects and improvements in surface to tire properties. Nanoparticles for pavement materials is required to be non-hazardous low-cost products, due to them being spread over large volumes of material and being in almost direct contact with human activities. The reduction of energy requirements during construction of asphalt pavements through development of improved emulsions and reduction in binder viscosity at ambient temperatures through the introduction of micro-bubbles will not only lead to a potential energy cost saving, but also assist in the lowering of emissions during construction. The typical bitumen binder thickness coating around aggregate is in the order of a few microns. However, most studies on binder properties do not focus on this small dimension. Nanotechnology characterization processes should lead to enhanced understanding and modeling of the interfacial properties between the aggregate and the binder.

Evaluating the various viewpoints, it can be postulated that nanotechnology can potentially play a major role in the improved use of existing and available materials in pavements and the processing of these materials to enable them to fulfill the required specifications of sustainable and perpetual pavement structures. In terms of the need for sustainable pavements (which is a major current need), Maher et al [2006] defined the main criteria for a sustainable pavement as follows:

- Minimizing the use of natural resources;
- Reducing energy consumption;
- Reducing greenhouse gas emissions;

- Limiting pollution (air, water, earth, noise);
- Improving health, safety, and risk prevention, and
- Ensuring a high level of user comfort and safety.

Steyn [2009a] indicated that the research levels for different aspects of nanotechnology in pavements can be defined as follows (using CNT as an example):

- Basic research – discovery of the buckyball and CNT [OECD 2002];
- Oriented basic research – research into the basic properties of CNTs [Goddard et al 2007] and the potential impact on the environment [Tong et al 2007];
- Applied research – evaluation of the manufacturing processes and compatibility and effects of CNTs with cement and aggregate [Makar et al 2005; Yakovlev et al 2006], and
- Experimental development - application of CNTs as fibers in fiber reinforced concrete and as sensors in roads [Shi and Chung 1999].

All four steps are required to ensure success in nanotechnology research and applications. The basic building blocks are required to ensure that the pavement does not fail prematurely through inadequate understanding and knowledge of the CNT (in the above example).

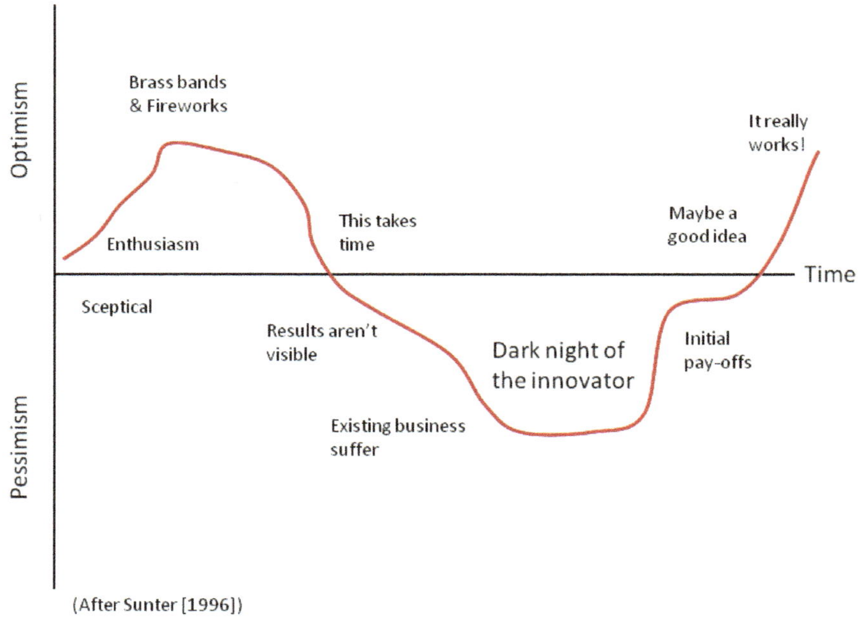

Fig. 2 The path of innovation (adapted from Sunter [1996].

Fig. 2 (adapted from Sunter [1996]) provides an indication of the typical road to innovation. It indicates that new innovations take time and effort to get to fruition. It is significant that optimism regarding potential breakthroughs is very strong during the first part of the innovation, and that this is often dampened by reality. Once the detailed work has been completed, the real potential of an innovation only starts to be realized. Currently, for many nanotechnology innovations the initial period of optimism and excitement still prevails and many major breakthroughs are anticipated based on initial basic and oriented basic research results that may not be realizable at larger scales and cost-effectively. It is important to distinguish between what is possible and what is practical. Ultimately, any innovation will only provide real benefit in the pavement engineering field if there is a real positive benefit/cost ratio.

Based on the introductory discussions, the problem statement for potential applications of nanotechnology in pavement engineering can be summarized in terms of the following three points:

- Identify the current needs that cannot be addressed effectively using current technology;
- Identify the potential nanotechnology solutions that may be applicable in the pavement engineering field, and
- Marry the two concepts to identify nanotechnology solutions with the highest potential benefit/cost ratios and focus on specific developments in those fields.

2 Materials

2.1 Introduction

This section deals with the major pavement materials and the ways in which nanotechnology can be utilized to enhance their performance or to develop novel materials for specific applications. It incorporates the following:

- Improved materials - Better resistance to applied traffic load and environment (resistance to load-related damage as well as environmentally related (moisture and temperature) related damage) – this is inclusive of bituminous, cemented and granular materials and combinations thereof, as well as novel materials – i.e. rework of waste streams
- Smart materials - Materials that can sense inputs (traffic and / or environmental) and either address these actively (i.e. fix a crack when it forms) or store the information to be obtained at a later stage for active input repairs (as a normal sensor / data acquisition system unit)

2.2 Typical Pavement Materials

Typical pavement materials can be summarized as bituminous, cemented and natural soils and gravels. Each of these types of materials has specific properties and applications in a pavement structure. It is important to appreciate these materials' different roles in the pavement structure. These can be summarized as follows:

2.2.1 Bituminous Materials

Bituminous materials consist of a selection of aggregates mixed with specific bitumen to form asphalt (in the US the terminology is slightly different, with bitumen typically termed asphalt and the asphalt mixture termed asphalt concrete). Asphalt is typically used as a surfacing layer, but can also be used as base material. Temperature changes have a major influence on the properties of asphalt materials, as the bitumen is a temperature sensitive viscous material. Water can also affect the behavior of these materials causing stripping of the bitumen from the aggregate [Read and Whiteoak 2003]. Often, a relatively thin surfacing layer consisting of a layer of aggregate and bitumen is constructed as surfacing seals. This is typically done on pavements with lower traffic volumes.

2.2.2 Cemented Materials

Cemented materials can consist of various combinations of natural soil or aggregate and cement. Typically, selected aggregates and cement are combined to form concrete, which is used as a surfacing for pavements. When lower levels of cement (i.e. 3 to 4 per cent) are used together with soils, cemented pavement layers can be provided that are typically used in base and subbase layers of the pavement. The behavior of cemented materials is dependent on temperature (affecting properties such as curling) and moisture changes.

2.2.3 Natural Soils and Gravels

Natural soils and gravels can also be termed unbound materials, and they consist of soil material selected for its specific properties. Typically, water is added to these materials to ensure optimal moisture contents, and then the material is compacted and forms the subgrade, subbase or (in selected cases) base layer of the pavement structure. The behavior of these materials is typically affected by changes in moisture content.

In order to understand the potential application of nanotechnology in pavement structures, it is important to firstly understand the behavior of the initial / standard material, and the way in which the environmental and traffic stresses are affecting this behavior. There exists a host of good references covering tese issues, and therefore it is not covered in more detail in this chapter. The focus in this chapter is on accepting that certain problems exists in the performance of road pavement materials, and the ways in which nanotechnology can potentially assist in solving these problems. The typical expectation of nanotechnology-based innovations in pavement materials is that these innovations will change the properties of the materials such that the negative effects of environmental and traffic loading (i.e. temperature and moisture sensitivity, fatigue cracking and rutting) can be minimized or reversed.

2.3 Improved Materials

2.3.1 Introduction

Nanotechnology partly focuses on the creation of new materials at the molecular level with significantly improved properties on the mechanical and related levels [Chong 2003]. Through judicious application of nanotechnology techniques material behavior problems can be addressed through engineering techniques on the nano-scale. This allows the molecular properties of materials to be better understood, and the molecular properties of materials to be changed, leading to improved behavior of the materials under normal environmental and loading conditions.

The development of improved materials using nanotechnology techniques is one of the areas where probably the most can be achieved to enable beneficial impacts from nanotechnology in pavement engineering. Pavement engineers use a wide range of materials for the construction and maintenance of road pavements. Most of these materials are natural materials that are modified using products such as bitumen, cement and other chemical admixtures. The bulk of the material, typically up to 95 per cent by volume, remains naturally occurring aggregates and soils. Problems often exist in the application of these materials for specific conditions, i.e. incompatibility between certain aggregates and binders (bituminous or cementitious), deterioration of the material during certain environmental conditions (i.e. water susceptibility of granular materials and temperature sensitivity of bituminous materials) and deterioration with use (i.e. fatigue cracking due to overloading). Specific potential nanotechnology solutions for each of these materials are discussed in this section.

2.3.2 Bitumen

Issues addressed in this section on bitumen include refraction of bitumen from crude oil, use of nano-clays, alternatives to bitumen, crack healing, Ultra Violet (UV) protection, and self assembly.

Bitumen is classified as a nano-material. The morphology as well as the interfaces between organic and inorganic materials are intriguing and may be of importance for bitumen and aggregate composition. Kotlyar et al [1998] has shown that the solids associated with bitumen in crude oil can be described as mainly ultrafine (nano-sized) aluminosilicate clays coated with a strongly bound toluene insoluble organic material having asphaltene characteristics. Investigation of the colloidal structure of bitumen indicates that bitumen can also be described as a complex mixture of mostly hydrocarbons whose structure is well described by a colloidal model with solid particles (asphaltenes) with a radius of a few nano-meters dispersed in an oily liquid matrix (maltenes). The critical shear rate of bitumen is an intrinsic property of a given bitumen, directly related to its nanostructure [Lesueur 2009].

Ozin and Arsenault [2005] pointed out the following five prominent principles, which must be considered when working with materials self-assembly of a targeted structure from spontaneous organization of building blocks beyond the sub-nanometer scale:

- Building blocks, scale, shape, surface structure,
- Attractive and repulsive interactions,
- Reversible association-dissociation and adaptable motion of building blocks in assembly, lowest energy structure,
- Building block interactions with solvents, interfaces templates, and
- Building dynamics, mass transport and agitation.

The RILEM Technical Committee for Nanotechnology-based Bituminous Materials (NBM) focuses on the identification of material characterization and modeling issues and challenges at the nano- and microstructural level [Rilem 2009]. Issues around temperature and humidity control during sample preparation and characterization also receives attention to ensure that measurements and observations are not affected by adverse material reactions. The expected outcome of this effort is improved bituminous materials with superior environmental and traffic loading resistance.

Bitumen is a by-product of the production of fuel from crude oil. In typical asphalt pavements the bitumen comprises in the region of 0.5 per cent of the mass and between 5 and 17 per cent of the cost of the road. However, if the bitumen is not available, an alternative binder is required for the bonding of the aggregates in asphalt. Internationally, the reserves of crude oil are viewed as being decreasing with a range of indications of potential reserves. The direct implication for pavement engineering is that the price of bitumen may increase drastically in future (as is already experienced) as the availability decreases. Various international efforts at development of alternative and sustainable binders are on-going [Paige-Green and Steyn 2005]. In this regard the application of manufacturing characterization techniques developed in the nanotechnology field may assist in the development of new sustainable alternatives that can provide the required properties to ensure that cost-effective construction and maintenance of pavements.

Bhasin et al [2009] conducted molecular simulations to evaluate the self healing of cracks in bitumen in asphalt. These simulations focused on determining the energy density of an ensemble of molecules under varying conditions. Analysis of the self-diffusivity of the bituminous binder molecules indicated that a good correlation existed between molecular chain lengths and branching of molecules to the self-diffusivity of molecules at the crack interface and consequently on the self healing properties of the binder. This is an example of the application of molecular analysis in describing a macro-level asphalt problem.

Liao et al [2008] described a novel application where treatment with TiO_2 was evaluated as a treatment against UV and ageing deterioration of bituminous binders. There analyses indicated that nano-sized TiO_2 can affect the ageing of bitumen positively if applied at low rates, leading to lower ageing rates. The effect on the stiffness of the binder need to be evaluated, as the addition of the nano-scale powders stiffens up the binder considerably. Research by Steyn [2009b] evaluated the effect of the addition of TiO_2 and ZnO on potential ageing of bitumen binder. Varying percentages of TiO_2 and ZnO were added to bitumen and applied to the surface of asphalt slabs (similar to a fog spray). These slabs were exposed to direct sunlight for a long enough period to allow potential ageing to take part in the binders. Analysis of the temperatures measured below the applied bitumen layers indicated that the temperature under the untreated section were on average 1.8°C higher than that under the TiO_2 and ZnO treated section, leading to potentially lower ageing rates.

Various researchers evaluated the effect of the addition of nano-clay to bitumen [Polacco et al 2008; Jahromi and Khodaii 2009; Van de Ven et al 2008]. Nano-clay is clay that can be modified to make the clay complexes compatible with organic monomers and polymers. These nano-composites consist of a blend of one or more polymers with layered silicates that have a layer thickness in the order of 1 nm and a very high aspect ratio. Common clays are naturally occurring minerals and subjected to natural variation in their formation. Separation of clay discs from each other results in a nano-clay with a large active surface area (up to 700 to 800 m^2/g). This helps to have an intensive interaction between the nanoclay and the bitumen. The proper selection of modified clay is essential to ensure effective penetration of the polymer into the interlayer spacing of the clay and so resulting in the desired exfoliated or intercalated product. In intercalate structure, the organic component is inserted between the clay layers in a way that the interlayer spacing is expanded but the layers still bear a well-defined spatial relationship to each other. In an exfoliated structure the layers of the clay have been completely separated and the individual layers are distributed throughout the organic matrix.

Various physical properties (such as stiffness and tensile strength, tensile modulus, flexural strength and modulus thermal stability) of the bitumen can be enhanced when it is modified with small amounts of nano-clay, on condition that the clay is dispersed at nano-scopic level. In polymer modified bitumen, the main differences appeared at low frequencies or high temperatures. Generally, the elasticity of the nanoclay modified bitumen is much higher and the dissipation of mechanical energy much lower than in the case of unmodified bitumen [Jahromi and Khodaii 2009].

Van de Ven et al [2008] found that while the stiffness and viscosity of a specific bitumen was not affected by the addition of one specific type of montmorillonite nano-clay, another type of montmorillonite nano-clays did affect stiffness and viscosity. The ageing resistance of the bitumen was improved in both the short and the long term for one of the nano-clays. They concluded that the improvements observed in the experiment are not at a level yet to justify commercial application on a large scale.

Kennepohl [2008] states that the ageing processes of bitumen are generally considered irreversible. While testing sulfur–bitumen binders for pavements in 1977 an encouraging observation was made during fatigue testing of sulfur extended asphalt pavement samples. Flexural stress recovery of a pavement specimen occurred and indicated that rejuvenation of the sulfur–bitumen binder might be possible. The specimen recovered completely overnight in strength after a fatigue test was interrupted due to equipment failure. SEM studies indicated that mono-clinic, needle–like crystals of sulfur deposited in the hairline cracks which were produced by the fatigue test. Since sulfur is uniformly dispersed throughout the asphalt and elemental sulfur has a high vapor pressure, a likely explanation is that sulfur vapors moved around and deposited as fiber-like reinforcement in the open spaces left by the fatigue testing (Fig. 3). A fine dispersion of sulfur globules in liquid asphalt (2 to 3µm diameter) was visible under SEM imaging. This occurrence may very well be termed self assembly (on a limited scale) of asphalt material.

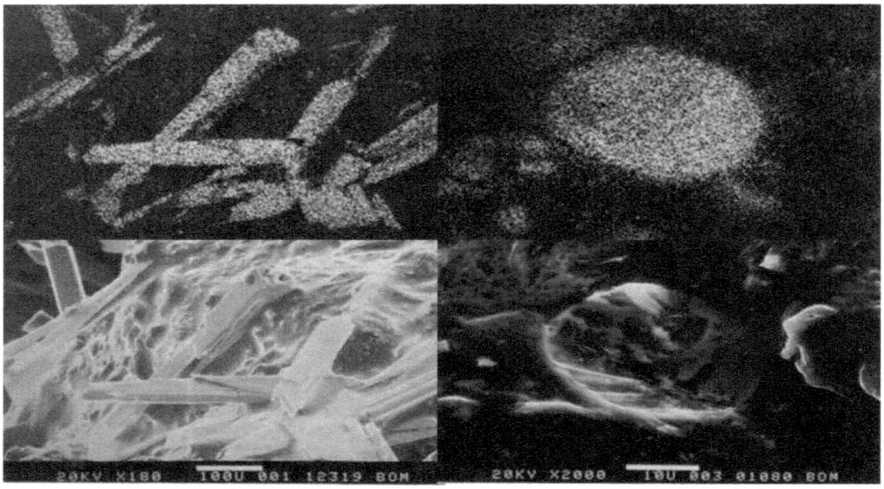

Fig. 3 SEM images showing monocline sulfur crystals (from Kennepohl and Millar [1977]).

2.3.3 Cement and Concrete

Issues addressed in this section on cement and concrete include improvements in strength, environmental resistance, and bonding.

Sobolev and Gutiérrez [2005] described a range of possible effects of nanotechnology developments on concrete in general. The bottoms-up manufacturing process that occurs naturally in nature provides opportunity to possibly obtain cost-effective nano-materials for use in concretes. CNTs are typically viewed as one of the most promising developments impacting on concrete performance. The need for the concrete to remain sustainable and cost- and energy-efficient is vital.

The majority of research using CNTs focuses on reinforcing various types of concrete through incorporation of the CNTs into the cement paste. Fiber Reinforced Concrete (FRC) is produced when fibers (typically steel or polypropylene fibers) are added to a concrete mix to control plastic shrinkage cracking and drying shrinkage cracking. The fibers also lower the permeability of the FRC and produce greater impact, abrasion and shatter resistance in the FRC. The vital typical properties required for the fibers in FRC include diameter, specific gravity, Young's modulus and tensile strength [Brown et al 2002; Shah 2009]. Various researchers have investigated the option of using CNTs as fibers in fiber reinforced concrete.

Makar et al [2005] showed that the addition of CNTs to a concrete increase the hydration rates and that strong bonds develop between the CNT and the cement paste, while Yakovlev et al [2006] measured increases of up to 70 per cent in the compressive strength of CNT reinforced concrete and decreases of up to 12 per cent in the heat conductivity of the concrete.

Robertson [2004] stated that the potential benefits of CNT reinforced materials have been rather disappointing as the moduli measured for CNT reinforced materials are typically much lower than would be expected based on theoretical calculations. Possible reasons for this may include dispersion problems through the host matrix and the small surface area available for bonding between individual CNTs and the matrix.

One of the potential benefits of CNT reinforcement of concrete that was not found in literature is the fact that CNTs will not corrode in corrosive environments as happens in steel FRC. This benefit may specifically be evaluated for application in marine environments.

TiO_2 has been incorporated in concrete to render the concrete a material that can perform certain photocatalytic activities [Cassar 2005; PICADA 2006; CAMDEN 2007; Ballari et al 2009]. The photocatalytic reaction is typically applied for the provision of self-cleaning surfaces and also the removal of NO_x, SO_x, NH_3 and CO pollution from urban areas through a chemical reaction triggered by naturally occurring Ultra Violet (UV) light. Various studies showed the application through the use of concrete blocks or concrete surfacing and indicated localized decreases in NO_x. Hassan [2009] conducted a life cycle impact assessment on TiO_2 coatings for concrete pavements and found that the TiO_2 coating has a positive effect on acidification, eutrophication, criteria air pollutants, and smog formation. The production phase of TiO_2, however, causes an increase in fossil fuel depletion, water intake, ozone depletion, and impacts on human health. The life cycle assessment indicated that these coatings have an overall positive effect on the environment. Evidence do exist that amorphous TiO_2 nano-particles can be hazardous and thus it is important to consider hazard reduction options in applications where TiO_2 is incorporated into concrete pavements. Effective options for hazard reduction include better fixation of nano-particles in nano-composites, changes of nano-particle surface, structure or composition, and design changes leading to the release of relatively large particles [Reijnders 2009].

Thin films of nano-sized material can be deposited on the surface of a host material through various techniques. This includes Self Assembled Monolayer (SAM) and sol-gel methods. The objective of such a treatment is to change the properties of the surface of the host material. Typical reasons requiring such a change include incompatibility between the aggregate and the binder used (e.g. cement or bitumen) and the need to improve the bonds between the aggregate and binder. In some cases the need may also exist to prevent bonding between two materials, and in such cases changes in the surface morphology / chemistry of the two materials may be required.

The adhesion between aggregate and bitumen receives ongoing research attention in pavement engineering. A relatively good understanding exist regarding the fundamental forces acting on these materials, and it is known that different levels of adhesion exist between different types of aggregate and bitumen [Hefer et al 2005]. Stripping of bitumen from aggregate occurs under certain environmental (mostly higher moisture contents) and loading conditions. The potential of applying SAMs onto aggregates to prevent stripping between the two materials and thereby improving the performance of the asphalt should be evaluated.

Sanfilippo and Munoz [2009] evaluated the effect of such surfacing layers on concrete processing with enhanced mechanical properties and prevention of deleterious reactions between aggregate and binder some of the main outcomes. Some siliceous aggregates react with the alkali hydroxides in cement, causing expansion and cracking over a period of many years [Addis 2001]. This reaction has two forms - alkali-silica reaction (ASR) and alkali-carbonate reaction (ACR). In ASR, aggregates containing certain forms of silica (amorphous micro-crystalline) react with alkali hydroxide in cement to form a gel that swells as it adsorbs water from the surrounding cement paste or the environment. These gels can swell and induce expansive pressures large enough to damage concrete. Pre-coating of these aggregates can prevent the type of reaction (similar to the work described by Sanfilippo and Munoz [2009] and Sanfillipo et al [2009], rendering currently unsuitable aggregate economically suitable for concrete applications.

It is possible to measure the nanoscale mechanical properties of the interfacial transition zone (ITZ) in concrete [Sanfilippo et al 2009; Shah 2009]. Although it is widely accepted that the properties of the ITZ have to be taken into account in modeling the overall mechanical properties of concrete, current models assume the modulus without much theory or experimental data to support these assumptions. This is due to the practical constraint of doing direct experimental measurement on the narrow region (around 14 μm) around a course aggregate. A Triboindenter has been successfully used to overcome this difficulty.

The development of Slip Form Self Compacting Concrete (SF-SCC) has been made possible by combining the concepts of particle packing, admixture technology and rheology in order to control the flow behavior of concrete. Specifically, through addition of different materials such as nano-clays and fly ash to the composition, successful SF-SCC mixes have been developed that maintains a balance between flowability during compaction and stability after compaction. The particle microstructure of the nano-clays used in the SF-SCC was evaluated using SEM [Shah 2009].

A combination of nano-mechanical testing with imaging at the nanoscale has been used to determine local mechanical properties of different phases present in cement paste and concrete microstructure. The indenter tip of a Triboindenter was used to capture AFM like images of a representative area on a sample. Indents were made with a specified maximum load and loading rate. Observed mechanical properties of the C-S-H gel in different areas varied (between 10 GPa and 35 GPa) while the modulus of the residual cement particles were between 100 GPa and 130 GPa. In many cases, paste around unhydrated particles hade a higher modulus [Shah 2009].

Livingston [2005] applied Quasi-Elastic Neutron Scattering (QENS) characterize the effects of temperature, chemical composition and cement/water ratio of concrete and developed a mathematical model of hydration reaction kinetics. It was illustrated that the critical features that affect the macroscopic properties of the concrete can be characterized through an improved understanding of the calcium-silicate-hydrate (C-S-H) gel which consists of colloidal particles in the 1 to 100 nm range.

2.3.4 Soils and Gravel

Although it was not officially termed as such, many geotechnical materials can be defined as nano-materials and their behavior has for many years been studied on a nano level. This includes the study of mineral crystal and structure, water- and soil-water interactions and specifically the behavior of clay minerals. The study of water on soil particle surfaces occur in the nano-scale with the layer of water of hydration around soil particles typically being around 0.5 nm thick. Clay minerals are also characterized on the nano-scale with typical plate thicknesses of between 0.7 nm and 1 nm [Taha et al 2005]. Through nanotechnology the structure of clays can be manipulated and specific properties obtained. The improved understanding of materials and their bonds to water molecules improve the understanding of the behavior of these materials when applied in pavement layers, and therefore the expected changes that the materials may undergo when moisture contents change in a pavement structure.

One of the changes that a material undergoes when used at the nano-scale is that the surface area to volume ratio typically increase drastically. One of the materials often encountered in pavement engineering is naturally occurring clays. These clays pose very specific problems to the pavement engineering field as their response to changes in moisture content and the platelet structure of the material cause most clays to have low friction angles and some to be expansive. Various traditional treatments of the problem include the stabilization of the clay using materials such as lime, treatment with sulphonated petroleum products that affects the ion surface properties (exchangeable ions) of the material or the removal of the material from the pavement structural layers. However, research into the behavior of clay minerals in the nanotechnology field [Larsen-Basse and Chong 2005] is providing new insights into the fundamental properties of clays. Application of this new knowledge may lead to alternative methods for the stabilization of the clays found in road reserves, and may thus lead to less expensive methods for using these materials in the pavement structure. This is also an area where the characterization of materials using high resolution equipment such as the SEM and AFM has led to new insights into the sub-microscopic properties of the materials, identifying potentially better methods for the treatment of these materials.

Nugent et al [2009] evaluated the interactions between exo-polymers and kaolinite clay to determine effects on the behavior of the kaolinite. The nano-scale interactions between the kaolinite and the cations and biopolymers used in the study were evaluated and it was found, inter alia, that biopolymer-induced aggregation of clay particles decreased the liquid limit of the kaolinite, formation of a clay-polymer interconnected network through cation bridging and hydrogen bonds increased the kaolinite liquid limit and reduction in the thickness of the electric double layer on the kaolinite clay surfaces decreased the liquid limit. It was concluded that the macroscopic response of the exo-polymer and kaolinite interactions competing with each other at the nanoscale was reflected in the observed variations in liquid limit.

Analysis of surface properties of various types of aggregate used in pavement construction can assist in an improved understanding of the behavior of the aggregate. In Fig. 4 an indication of the surface properties of two typical South African pavement materials are shown, indicating that, although both originates from the same size fraction, their geological origins cause their surface characteristics (and therefore potential behavior) to differ.

Fig. 4 Indication of SEM images of Andesite and Granite surface characteristics.

2.4 *Smart Materials*

2.4.1 Introduction

The development of smart materials that are enhanced to provide specific properties is one of the potential applications of nanotechnology in pavements. This includes materials that can harvest energy, resist ageing, reinstate cracks and sense parameters from their environment. This section evaluates some of these developments

2.4.2 Energy

Energy can potentially be harvested from matter through the judicious application of nano-generators. Although the efficiency and durability of harvesting materials such as piezoelectric nano-wires is steadily improving, the voltage and power produced by a single nano-wire are insufficient for real devices. Integration of large numbers of nano-wire energy harvesters into a single power source is necessary, requiring alignment and synchronization (of charging and discharging processes) of the nano-wires. Sheng Xu et al [2010] demonstrated the possible vertical and lateral integration of ZnO nano-wires into arrays that are capable of producing sufficient power to operate real devices. Lateral integration of 700 rows of ZnO nanowires produced a peak voltage of 1.26 V at a strain of 0.19 per cent (potentially sufficient to recharge an AA battery). Vertical integration of three layers of ZnO nano-wire arrays produced a peak power density of 2.7 mW cm^{-3}. This nanogenerator was used to power a nano-wire pH sensor and a nano-wire UV sensor, demonstrating a self-powered system composed entirely of nano-wires. Strain rates of less than 2 per cent per second was shown to generate up to 1.2 V and peak power densities of 2.7 $mWcm^{-3}$.

2.4.3 High-Strength Materials

Amorphous metallic alloys (metallic glass), are lucrative engineering materials owing to their superior mechanical properties such as high strength and large elastic strain. Their main drawback is their propensity for highly catastrophic failure through rapid shear banding, significantly undercutting their structural applications. By reducing dimensions to 100 nm, Dongchan and Greer [2010] demonstrated how Zr-based metallic glass nano-pillars attained ceramic-like strengths (2.25 GPa) and metal-like ductility (25 per cent) simultaneously. A phenomenological model for size dependence and brittle-to-homogeneous deformation is provided. This may have applications in pavement structures such as bridge components in future.

2.4.4 Self-Healing Materials

Buehler and Ackbarow [2007] discussed evaluation of fracture mechanics to improve the ability to perform structural engineering on macro scale and introduced the concept of how dimensional scales interact with each other. Kumar and Curtin [2007] evaluated crack and microstructure interaction with a discussion on the understanding of initiation of cracks in metals. The concepts discussed in these two references can similarly be adapted to improve the nano-scale understanding of crack development in the bituminous binder / aggregate interface as well as cracking of concrete on pavements.

Balazs [2007]and Kessler et al [2003] investigated the modeling of self-healing materials, starting with biological examples which may potentially be expanded to infrastructure where cracks that develop in the pavement layers may self-heal based on the introduction of microcapsules in the cement matrix. The scale requirements and the potential benefit to determine whether the development of such cracks is life-threatening or whether this may be more of a nice-to-have than essentiality is still required. Self-healing of fly-ash in the pavement materials has been observed before [Barstis and Crawley 2000] and can be seen as a nano-scale effect that is utilized already.

The CRack Arrest and Self-Healing in COMPosite Structures (CRASHCOMPS) [Crashcomps 2010] project aims to develop crack arrest/redirection concepts in polymer composite structures under monotonic compressive loading with the aim of inhibiting the growth of rapid, unstable cracks, such that global failure of the infrastructure is hindered. Further, incorporation of self-healing capability whereby the arrested crack will autonomously heal is evaluated. Together, these techniques should provide a unique ability to arrest and heal critical cracks in infrastructure. The research focuses on carbon fiber polymer composites. Although these materials are typically used in the aerospace and transport industries (aircraft wings, helicopter rotor blades and ship hulls) potential applications exist in pavement related structures such as bridge decks.

2.4.5 Sensors

One of the well-publicized possibilities of nano-materials is the development of sensors that act as part of the substrate that is being observed, thereby allowing very fine measurements on a small scale and obviating the need to add external sensors to a system [Goddard et al 2007]. While external sensors tend to interfere with the mechanics of the system being monitored, the incorporation of sensors as part of the matrix of the system provides the potential that the matrix actually provides feedback and issues around bonding between sensors and the matrix are canceled. The application of CNTs in traffic monitoring [Shi and Chung 1999] is an example of such application. In this type of applications the transfer of data from the sensor to a data acquisition device and the analysis of data still require further work, especially if the sensors are being distributed inside a pavement of several kilometers length. Although Micro-ElectroMechanical Sensors (MEMS) devices can often have features that are measured in nanometers, MEMS and nanotechnology are fundamentally different. MEMS devices (which are typically millimeter scale) operate on the principle that the properties of matter as a continuum is observed, while nanotechnology deals with matter's quantum mechanical behavior. The potential applications of MEMS devices in pavements can thus provide valuable information on a micro-scale, bit not yet on a nano-scale.

2.4.6 Nanophosphor

A major road safety need in rural Africa is the illumination of road pavements to improve visibility and road safety. The potential use of nanophosphors combined with road surfacing materials or paints for this purpose was evaluated at CSIR [Steyn 2008b]. Nano-phosphors are nano-scale crystalline structures with a size dependent bandgap that can be altered to change the color of light [Kelsall et al 2004]. If the road can act as the source of the light (be made luminescent) it can play a role in improving road safety as the source of the light will not be dependent on external power and the use of a motorized vehicle any more. Current research outputs have shown that nano-phosphors can be added to traditional pavement materials such as concrete, bitumen and road paint to enable these materials to become luminescent after exposure to light, while research is ongoing to further investigate issues such as the increase in luminescence duration, the type of bonds that form between the nano-phosphors and the substrate materials and the up-scaling of manufacturing techniques to enable practical amounts of nano-phosphor to be manufactured. Fig. 5 indicates a SEM image of a nano-phosphor concrete sample while Fig. 6 shows SEM images of the nano-phosphor embedded in road paint, bitumen and concrete.

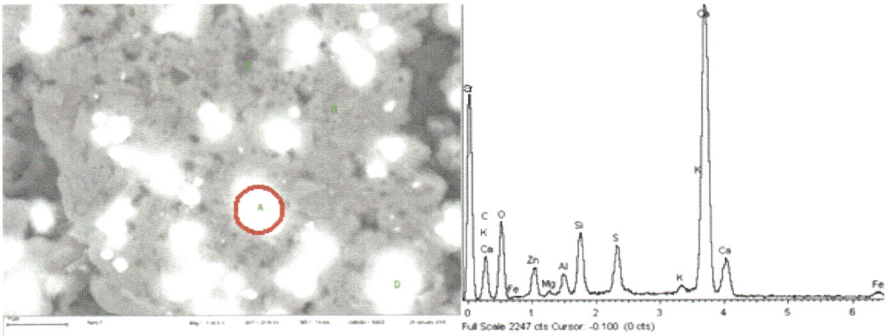

Fig. 5 SEM image of 10 per cent nano-phosphor concrete sample with EDX analysis output showing constituent elements of the nano-phosphor.

a) Neat green nano-phosphor

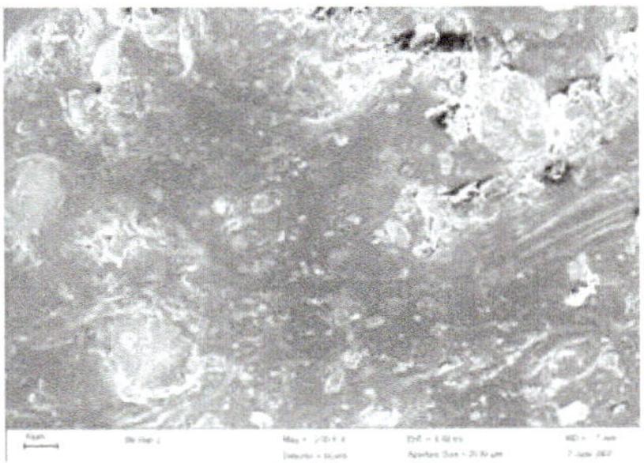

b) Bitumen with 10% nano-phosphor

c) Red road paint with 10% nano-phosphor

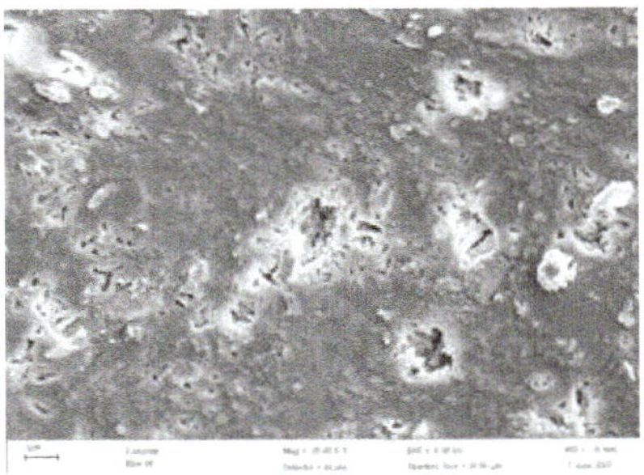

d) Concrete with 10% nano-phosphor

Fig. 6 Four typical SEM images a) neat green nano-phosphor, b) 10 per cent nano-phosphor in bitumen, c) 10 per cent nano-phosphor in red road paint concrete and d) 10 per cent nano-phosphor in concrete.

3 Characterization of Materials

3.1 Introduction

Accurate characterization of materials is important in order to evaluate their expected behavior and properties, specifically if these materials are nano-sized or if their macro-scale properties are similar and an understanding of their nano-scaled properties is required to apply them adequately in pavement engineering. Often, traditional engineering tests and properties do not sufficiently distinguish between different materials.

Typical nanotechnology instrumentation used in pavement engineering includes imaging techniques such as the Scanning Electron Microscope (SEM) and Atomic Force Microscope (AFM) and analysis techniques such as spectroscopy. It is important to consider the type of information required as well as the required resolution before a specific technique is selected.

Kelsall et al [2004] indicated that characterization techniques can be divided (based on the required information and resolution) into morphology (nanostructural architecture), crystal structure (detailed atomic arrangement contained within the microstructure), chemistry (elements present) and electronic structure (nature of bonding between atoms). Microscopy techniques available for characterization include the various types of electron microscopy. These include Scanning Electron Microscopy (SEM) (used for examination and analysis of surface and subsurface nano-structured systems) and Transmission Electron Microscopy (TEM) and Scanning Transmission Electron Microscopy (STEM) (analysis of bulk structure of thin samples). Scanning probe techniques such as Atomic Force Microscopy (AFM) are used to study the surface characteristics of a sample through magnitudes of atomic forces. While SEM can be used for resolutions of down to 1 μm, AFM can provide resolution of down to 0.1 nm. Detailed descriptions of these various techniques falls outside of the scope of this chapter, and can be found in Kelsall et al [2004] and other technical texts.

It appears from literature (and conference themes) that a majority of current nano-characterization work is conducted on cement and concrete samples, with soils and gravels and bitumen receiving less attention. This may be due to the nature of the samples and the possibilities of observing structural changes inside the mortar due to hydration. Further, the various parameters that can be changed to affect the properties of cement mortars (i.e. water/cement ratio, cement content, cement type, etc) leads to a wide range of effects that can be studied.

The developments around characterization techniques that have occurred during the past decade in nanotechnology have opened the door to observations of pavement engineering materials on a scale not available before. As for other fields of study, this brought about a major change in the way that materials are being observed and also the understanding of the behavior of these materials.

3.2 *Observational Techniques*

Observational techniques are used to observe phenomena in engineering and to support the explanation of such phenomena. For pavement engineering this ranges from the macro scale where satellite images are used to observe macro scale phenomena (mainly in terms of geographical, geological and developmental issues) down to the nano-scale where matter is studied in its smallest observable form. Steyn [2007] has shown the hierarchy of possible observations (Table 2) including the smaller scales obtained from the Scanning Electron Microscopy (SEM) and Atomic Force Microscopy (AFM) techniques. These are starting to yield useful results in the understanding of the behavior of engineering materials [Mgangira 2007a].

Table 2 Hierarchy of possible pavement engineering observational techniques.

<table>
<tr><th>General wave-length bands</th><th>X-ray wavelength 10 to 10⁻² nm</th><th>Visible wavelength 400 to 700 nm</th><th>Infrared wave-length 103 to 106 nm</th></tr>
<tr><td rowspan="6">Indication of currently used technology and resolution of features</td><td rowspan="3">NCI</td><td>Satellite images [Mm resolution]</td><td>Infrared satellite images [Mm resolution]</td></tr>
<tr><td>Aerial photographs / LIDAR [km resolution]</td><td>Infrared airborne images [km resolution]</td></tr>
<tr><td>Standard photos [m resolution]</td><td rowspan="4">NCI</td></tr>
<tr><td>Computerized Tomography (CT) scans [mm resolution]</td><td>Microscope images [mm resolution]</td></tr>
<tr><td>SEM images [µm resolution]</td><td>NCR</td></tr>
<tr><td>NCR</td><td>AFM images [nm resolution]</td></tr>
</table>

NCI – No Current Images used typically in pavement engineering.

3.3 *Characterization Methods*

3.3.1 Introduction

In order to have better control over material properties, it is necessary to understand their nano-structure and to relate it with the macro-properties. The development of the Scanning Electron Microscope (SEM) and Atomic Force Microscope (AFM) allowed the ability to see and manipulate atoms, molecules and nano-sized objects, making it possible to explore the structure of materials at a scale of 40 to

50 nm. Most publications on characterization of pavement-related make use of either the SEM or AFM. In the following sections the basic principles of use of the two techniques are briefly discussed, followed by references to existing work where the applications of these characterization methods have lead to an improved understanding of the properties and behavior of pavement materials.

3.3.2 Scanning Electron Microscopy

The SEM is typically used for imaging surface and subsurface microstructure of samples. An accelerating voltage of between 1 and 30 kV is typically used. It images the sample by scanning it with a high-energy beam of electrons in a raster scan pattern. The electrons interact with the atoms that make up the sample producing signals that contain information about the sample's surface topography. A schematic diagram of the SEM is shown in Fig. 7 [Kelsall et al 2004].

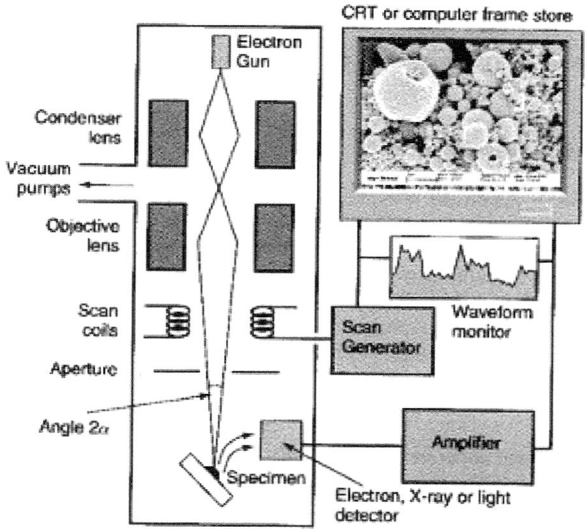

Fig. 7 Schematic diagram of the layout of a Scanning Electron Microscope (SEM) [Kelsall et al 2004].

3.3.3 Atomic Force Microscopy

The AFM consists of a cantilever with a sharp probe that is used to scan the sample surface. When the probe is brought into proximity of a sample surface, forces between the probe and the sample lead to a deflection of the cantilever according to Hooke's law. Forces that are measured in AFM include mechanical contact force, van der Waals forces, capillary forces, chemical bonding, electrostatic forces, magnetic forces etc. A schematic diagram of the AFM is shown in Fig. 8 [Kelsall et al 2004].

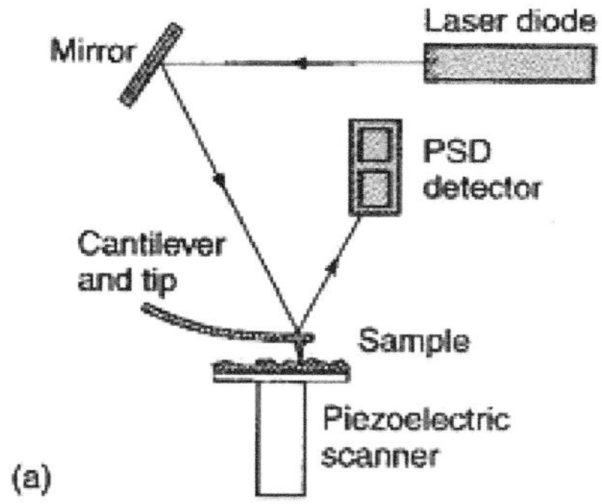

Fig. 8 Schematic diagram of the layout of an Atomic Force Microscope (AFM) [Kelsall et al 2004].

3.4 Specific Examples of Applications

3.4.1 Introduction

In this section some specific examples of applications using the various characterization techniques are discussed to illustrate the potential applications and benefits in pavement engineering.

3.4.2 Soil Characterization

The chemical and mineralogical composition of soils has long been characterized on the nano level using techniques such as Electron Probe Micro-Analysis (EMA), Scanning Electron Microscopy (SEM) and Energy Dispersive X-ray Analyzers (EDXRA) [Bullock 1981; Esawaran and Shoba 1981; Bisdom 1981; FitzPatrick 1984]. The application of these techniques has led to improved models and understanding of the macro-behavior of soils, based on an interpretation of their nano- and micro-structures. This knowledge is applied in pavement engineering through improved decisions regarding stabilization of specific materials and potential for compactability of materials.

3.4.3 Scanning Electron Microscopy (SEM) Applications

Mgangira [2007b] used the SEM to evaluate the micro-texture of two similar sands that behaved differently when being stabilized using various non-standard soil stabilizers. None of the typical engineering tests provided insight into the reason why the stabilization did not work similarly for the two types of sands evaluated. The SEM investigation showed the difference in terms of fines that were not identified in the standard grading analysis of the sands, which allowed the stabilizers to bond better with the sand grains. In Fig. 9 SEM images of the two types of sands are shown, with the fines visible on the right image.

Fig. 9 SEM images of two sands showing traces of small fractions on the right [Mgangira 2007b].

Mgangira [2010] used a SEM to evaluate the interaction between two different soil samples (black clay and reddish brown residual chert) and enzyme-based liquid stabilizer solutions. The results showed that the addition of the enzyme-based products modified the microstructure of the material matrix and that the type of microstructure changes that take place after treatment will be dependent on the product which in turn is a function of the compositional make up of the product itself. Whether this leads to significant improvement in engineering properties is another question. The observed connectivity of the microstructure and the presence of a surface network binding the particles, following treatment with product A differs from that seen on samples treated by product B and is considered to be the underlying explanation in the difference in performance of the two products. This is considered to be the result of the compositional characteristics of the products. The SEM images of the untreated and treated chert is shown in Fig. 10 to indicate the differences observed in the structure, and the presence of a tentacle-like network of structures on the surface of the sample.

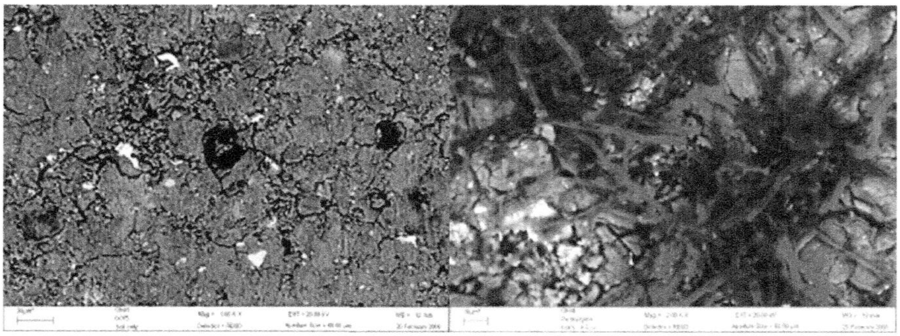

Fig. 10 Microstructure image of untreated chert sample (magnification 1000x) (Left) and chert sample treated with enzyme-based liquid stabilizer (magnification 2000x (Right).

Muniandy [2002] used SEM to evaluate the differences between Malaysian cellulose oil palm fiber and traditional commercially available cellulose fibers used in Stone Mastic Asphalt (SMA) mixes. It was found that the Malaysian fiber sizes appeared to be less than 500 μm compared to the traditional fibers which had thicknesses of around 50 μm. Use of the SEM analysis assisted in the process of defining the differences between the two types of fiber.

Stulirova and Pospisil [2008] discussed procedures to enable SEM to be conducted on bitumen samples. The oil fraction (maltenes) in the samples needs to be eliminated prior to characterization as the resolution of the images is too low if the samples are non-conductive. This process leaves the asphaltene fraction to be studied. It is important to understand that only changes in the asphaltene structure can thus be studied using the SEM technique.

Peethamparan et al [2008] evaluated the physicochemical behavior of cement kiln dust treated kaolinite clay using, inter alia, SEM and EDX techniques. Analysis of the samples using SEM indicated that the most important morphological modification of the kaolinite due to the cement kiln dust treatment was the change from clay flake stacks observed in untreated kaolinite clay to a disaggregated and randomly oriented form. This detail assisted in the development of a model for the stabilization mechanism of cement kiln dust treated kaolinite clay.

Steyn and Jones [2006] evaluated the performance of an in-service pavement with a cemented base layer that was recycled and stabilized with bitumen emulsion using Accelerated Pavement Testing (APT). Upon analysis of the material properties (that behaved exceptionally well after 25 years of trafficking) SEM was used to evaluate the microstructure of the material. By comparing carbonated and un-carbonated material from the stabilized base layer using a SEM, it was possible to conclude that the matrix of the base appeared to have residual cement from the original recycled cement stabilized layer that was able to hydrate and form cementitious bonds (although no cement was added during the recycling). Although it is difficult to detect bitumen using the SEM, small quantities were detected in the base materials by the presence of high carbon concentrations identified using the EDX facility. An indication of the SEM images showing the cementitious bonds and the bitumen is shown in Fig. 11.

Fig. 11 SEM images showing location of bitumen in matrix and new cement reaction crystals.

3.4.4 Atomic Force Microscope (AFM) Applications

Pauli et al [2001] used AFM in characterizing bituminous binders and their respective properties. They correlated the surface morphology with the constituents in the bitumen and concluded that the AFM data may be used to improve the understanding of the precipitation of the asphaltenes in the bitumen and that this information may lead to an improved understanding of the interaction between the surface of the aggregate and the bitumen in asphalt.

The AFM was used to investigate the structural morphology of crumb rubber bitumen at the interfacial regions, especially during aging [Huang et al 2006]. Masson et al [2006] used phase-detection AFM to evaluate bitumen morphology and found that bitumen can be classified into three groups, based on the different domains or phases visible.

Steyn [2009a] evaluated the correlation between the aging of bituminous binders on the road (a property that severely affects the deterioration of the pavement surface) and the elastic stiffness of the binders as measured using the AFM. Initial data focused on the surface morphology and a clear difference could be observed between the surface morphology of a bituminous binder that was aged at different temperatures. In Fig. 12 the overall difference in the surface morphology for

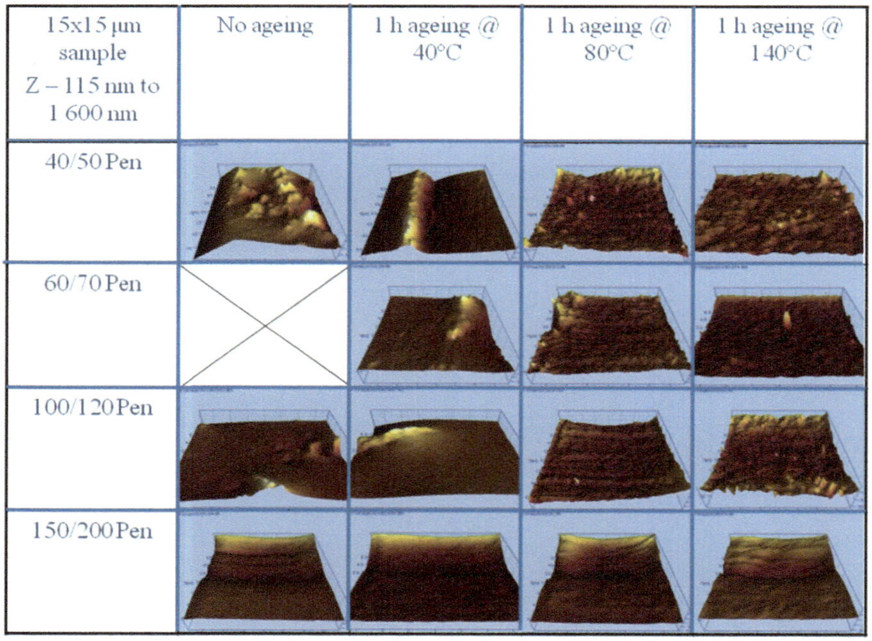

Fig. 12 The surface morphologies of four bituminous binders as aged at three different temperatures and observed using the AFM [Steyn, 2009b].

four binder types aged at three temperatures can be observed. Research in this area is continuing, with the focus on improving the understanding and potential correlation between surface morphology and performance of bitumen.

Masson et al [2007] described a novel application of the AFM where they evaluated the structure of various bitumens at low temperatures using cryo-microscopy and related the results to material stiffness. AFM were conducted at imaging temperatures of 22°C, -10°C, -27°C, -55°C and -72°C. These temperatures were selected based on glass transition temperatures of the bitumens. At these low temperatures the bitumens did not pollute the tips with sticky residues. The effect of temperature on volume was specifically noteworthy. The AFM observations of new domains that became visible upon cooling of the bitumen samples were consistent with segregation. The microscopy work at -10°C to -30°C showed that the bitumen contracted, although not all bitumen phases contracted equally. Further, topographic features of around 85 nm in height were visible at the low temperatures and not at room temperature.

4 Concerns and Issues

4.1 Introduction

Although a large number of potential ideas and applications for nanotechnology and pavement engineering exist, it is also important to remain realistic and identify and accept the current limitations and challenges inherent in this field. In this section a short summary of those challenges and limitations potentially affecting these applications are provided. These concerns and challenges include environmental and health and safety concerns, as well as issues around scaling of nano-effects and costs. It is important to realize that some issues are not new, as it was indicated earlier that bitumen and cement (which are both traditional pavement materials) can both be classified as nano-sized materials, and therefore the same concerns will exist with these traditional pavement materials.

4.2 Environment

The effect of various nano-materials on the natural environment is a hot topic in nanotechnology and environmental research. Uncertainty regarding the potential effects of materials that exist on the nano-scale with properties that are different than when using the material on a micro or macro scale has led to various investigations [NNI 2003]. Some work in this regard shows that the potential effects may be minimal [Tong et al 2007]. As pavements are constructed in the natural environment, all materials (including nano-materials) used in the construction and maintenance of pavements need to be compatible to the natural environment and their effects on the natural environment should be minimal. Typical potential problems in this regard include the leaching of materials into groundwater, release of materials into airways through the generation of dust on unpaved roads and the exposure to potentially harmful materials during construction and maintenance operations.

However, the reasons for using nano-materials often include specifically the different performance of the material on the nano-scale, and also the different effect that it has on the environment when used on the nano-scale. In cases such as improved methods for purification of water (which is a relevant example, although not related to pavement engineering), the different effect on potentially deadly microbes in the water is specifically the reason why the nano-materials are being used [Hassan 2005; Savage and Diallo 2005]. Also, the application of nano-materials in the fight against deadly diseases such as tuberculosis (TB), malaria and HIV [Maclurcan 2005] should also not be misjudged, as the focus of the application is specifically to have a distinct effect on natural organisms. A clear distinction is thus required between controlled and sought-after effects on the environment and uncontrolled and unexpected effects on the environment.

4.3 Health and Safety

The health risk of nano-particles is a function of its hazard to human health and its exposure potential. Users of nano-particles should endeavor to mitigate the potential risks of nano-particles during the design stage rather than downstream during manufacturing or customer use (i.e. when the material is already embedded in the pavement). Five general principles that can be used as framework to address mitigation of health risks are [Morose 2010]:

- Size, surface and structure (change the size, surface, or structure of the nano-particle so that the desired product functionality is preserved, but the hazard potential of the nano-particle is decreased);
- Alternative materials (identify an alternative material that can be used to replace the hazardous nano-particle);
- Functionalization (intentionally bond molecules to nano-particles to change the properties in a manner that the desired product properties are preserved, but the hazard potential of the nano-particle is reduced or eliminated);
- Encapsulation (enclose a potentially hazardous nano-particle within a material that is less hazardous), and
- Reduction (evaluate the option of using smaller quantities of the hazardous nano-particle in the product while maintaining product functionality).

4.4 Costs

The costs of most nanotechnology equipment and materials are currently relatively high. This is partly due to the novelty of the technology, but also due to the complexity of the equipment. However, in the case of the nano-materials, costs have been shown to decrease over time and the expectations are that, as manufacturing technologies improve, the costs of the materials will decrease. Whether such decreases will render the materials as run-of-the-mill pavement engineering materials will have to be seen. Current opinion is that in special cases, the materials will enable unique solutions to complicated problems that cause them to be cost effective, which will lead to large scale application of these specific technologies. In other cases the traditional methods for treating the problem may still remain the most cost effective. As indicated earlier, the job of the engineer is to solve real-world problems and provide a facility to the general public at a reasonable cost.

4.5 Scale Effects

Chong [2003] highlighted one of the major challenges in nanotechnology application in civil engineering (and therefore pavement engineering) to be the several orders of magnitude in time and space scales from the nano level to micro and meso levels. The physical scales ranging from nano- to systems-level are illustrated in Table 3 (after Chong [2003]).

Table 3 Physical scales from nano-to systems-level for pavement engineering (after Chong [2003]).

MATERIALS		STRUCTURES	INFRASTRUCTURE	
Nano (10^{-9})	Micro (10^{-6})	Meso (10^{-3})	Macro (10^{0})	Systems (10^{3})
Molecular scale	Microns	Millimeters	Meters	Kilometers
Nano-mechanics	micromechanics	composites	Pavement layers	Roads and bridges

The unique environment of the pavement engineer who works with large volumes of material should always be appreciated when evaluating potential applications of nanotechnology. The effects on manufacturing capacity and performance of the nano-materials when combined with bulk aggregates and binders should be evaluated to ensure that the beneficial (nano-scale) properties are still applicable and cost- and energy-efficient at these scales.

The scale chasm between typical nanotechnology and pavement engineering has been illustrated by various researchers [Buehler and Ackbarow 2007; Larsen-Basse and Chong 2005]. It can also be illustrated when the potential application of carbon nanotubes (CNTs) as fibers in fiber reinforced concrete is evaluated. Typically, the volume of steel or polypropylene fibers used for such an application may be in the region of 2 per cent. If it is assumed that the application rate of the CNTs in the concrete will be similar to that of the steel fibers, this translates to approximately 270 kg of steel fibers (for a typical steel fiber reinforced pavement of 1 km - 7.4 m wide and 100 mm thick concrete) being replaced by 52 kg of CNTs. A typical pavement rehabilitation project (10 km to 100 km length) may thus require in the order of 520 kg to 5 200 kg of CNTs. The economical production of such volumes of CNTs may not be realistic at this stage. The dimensional jump from nano-scale to macro-scale thus influences the potential usage of nano-materials.

Van Breugel and Guang [2005] demonstrated that virtually all engineering works are multi-scale projects, with typical structural engineering projects covering up to six orders of magnitude. If the various models on the different scales are connected correctly, the macro-level performance of an engineering structure (i.e. pavement) should be predictable based on the fundamental laws governing the material on a nano-level.

Shah [2009] stated that, although cementitious construction materials are mainly used in large quantities, fundamental properties such as strength, ductility, early age rheology, creep and shrinkage, fracture behavior and durability of these materials depend to a great extent on structural elements and phenomena which are effective at micro- and nano-scale. Characterization of cementitious materials at the nanoscale thus provides improved understanding of the material at the fundamental level and leads to the development of new materials based on nano-modification.

5 Concluding Remarks

The objective of this chapter was to provide to the pavement engineer who may not be familiar with the potential applications of nanotechnology in the traditional pavement engineering field an introduction into this area. Information was provided on the various nanotechnology definitions and the specific materials typically used in pavement engineering and potential applications of nanotechnology in improving these materials discussed. Various characterization options was shown, and the current concerns and issues around the environment and scaling effects highlighted. It is the duty of the prudent pavement engineer to keep abreast of developments that can assist in providing more cost-effective and sustainable pavement facilities to the general public.

Based on the information and discussions in this chapter, the following conclusions are drawn regarding the use of nanotechnology in road pavements:

- The application of nanotechnology developments in the field of pavement engineering can potentially lead to advances in solving general engineering problems;
- Most of these applications, however, first need to be scaled to the dimensional applications that are typical for the pavement engineering environment;
- The technical- and cost-effectiveness of available technologies should both be evaluated as part of the evaluation of nanotechnology solutions in engineering;
- Fundamental research into the properties of engineering materials to improve the understanding regarding their performance is an important output of nanotechnology characterization of pavement materials, and
- The development of novel materials and the improvement of existing materials in response to scarcity of natural materials become a possibility through application of nanotechnology techniques on traditional pavement materials.

Based on the information and discussions in this chapter, the following recommendations are made regarding the application of nanotechnology in road pavement engineering:

- The gap between scientists (specifically physicists and chemists) and engineers should be narrowed to enable the potential of nanotechnology innovations to be applied in appropriate areas of engineering to support the general public good;
- Scientists should continue with their focus on basic and oriented basic research, providing the required building blocks for engineers to construct a better environment;
- Engineers should actively and regularly evaluate the available technologies (specifically in the field of nanotechnology, as it evolves rapidly) to identify potential solutions to problems that they are required to address in society, and
- Challenges such as the potential impact of nano-materials on the natural environment around pavements should receive the required attention to ensure that solutions to pavement engineering questions do not cause new hazards to the public utilizing the natural and built environment.

References

Addis, B.J.: Fulton's Concrete Technology. Portland Cement Institute, Midrand, South Africa (2001)

Balazs, A.C.: Modeling self-healing materials. Materialstoday 10(9), 18–23 (2007)

Ballari, M.M., Hunger, M., Hüsken, G., Brouwers, H.H.: Heterogeneous photocatalysis applied to concrete pavement for air remediation. Nanotechnology in Construction 3 (2009)

Barstis, W.F., Crawley, A.B.: The use of fly-ash in highway construction. Report 84-DP59-MS-05, Mississippi Department of Transportation, Jackson, MS (2000)

Bhasin, A., Bommavaran, R., Little, D.N., Greenfield, M.: Intrinsic healing in asphalt binders – Measurement and impact of molecular morphology. In: Proc. 6th Int. Symp. on Maintenance and Rehabilitation of Pavements and Technological Control, Politecnico di Torino, Torino, Italy (2009)

Bisdom, E.B.A.: Microchemical analysis in thin sections of soils. In: Proceedings of the International Working Meeting on Soil Micromorphology, vol. 1, pp. 53–69. AB Academic Publishers, Netherlands (1981)

Brown, R., Shukla, A., Natarajan, K.R.: Fiber reinforcement of concrete structures. URITC Project 536101, University of Rhode Island Transportation Center, Kingston, RI, USA (2002)

Buehler, M.J., Ackbarow, T.: Fracture mechanics of protein materials. Materialstoday 10(9), 46–58 (2007)

Bullock, P.: The changing face of soil micromorphology. In: Proceedings of the International Working Meeting on Soil Micromorphology, vol. 1, pp. 1–18. AB Academic Publishers, Netherlands (1981)

CAMDEN, Pioneering paving to tackle air pollution (2007), `http://www.camden.gov.uk` (accessed March 2007)

CAPSA Home page, `http://www.capsa-events.co.za` (accessed April 2010)

Cassar, L.: Nanotechnology and photocatalysis in cementitious materials. In: Proc. 2nd Int. Symp. on Nanotechnology in Construction, Bilbao, Spain, pp. 277–283 (2005)
Chong, K.P.: Nanotechnology in civil engineering. In: Proc. 1st Int. Symp. on Nanotechnology in Construction, Paisley, Scotland, pp. 13–21 (2003)
Crashcomps (2010), `http://www3.imperial.ac.uk/crashcomps` (accessed on May 1, 2010)
Dongchan, J., Greer, J.R.: Transition from a strong-yet-brittle to a stronger-and-ductile state by size reduction of metallic glasses. Nature Materials 9, 215–219 (2010)
Esawaran, H., Shoba, S.A.: Scanning electron microscopy in soil research. In: Proc. Int. Working Meeting on Soil Micromorphology, vol. 1, pp. 19–52. AB Academic Publishers, Netherlands (1981)
FHWA, Nano-scale approaches for highway research. Exploratory Advanced Research Program, Publication No. FHWA-HRT-10-033 HRTM-04/01-10(1M)E (2009)
FitzPatrick, E.A.: Micromorphology of soils. Chapman and Hall, USA (1984)
Goddard III, W.A., Brenner, D.W., Lyshevski, S.E., Iafrate, G.J.: Handbook of nanoscience, engineering, and technology, 2nd edn. CRC Press, Boca Raton (2007)
Hassan, M.H.A.: Small things and big changes in the developing world. Science 309(5731), 65–66 (2005), doi:10.1126/science.1111138
Hassan, M.M.: Evaluation of the Environmental Impacts of Titanium Dioxide Photocatalyst Coatings for Pavements Using Life-Cycle Assessment. In: Proc. 88th Transportation Research Board meeting, Washington D.C. (2009)
Hefer, A., Little, D.N., Lytton, R.L.: A synthesis of theories and mechanisms of bitumen-aggregate adhesion including recent advances in quantifying the effects of water. J. Ass. of Asphalt Paving Technologists (2005)
Holzer, L., Münch, B., Gasser, P.: New approaches in microstructure analysis of cementitious materials: From micro to nano, from 2D to 3D and from qualitative to quantitative characterization. In: Proc. 2nd Int. Symp. on Nanotechnology in Construction, Bilbao, Spain, pp. 133–141 (2005)
Huang, S.-C., Pauli, A.T., Beemer, A., Robertson, R.E.: Influence of crumb rubber on the fatigue performance of asphalt pavement. In: Proc. 10th Int. Conf. on Asphalt Pavements, ICAP, Quebec City, Canada (2006)
ICAP Home page, `http://www.asphalt.org` (accessed September 2009)
Jahromi, S.G., Khodaii, A.: Effects of nanoclay on rheological properties of bitumen binder. Construction and Building Materials 23, 2894–2904 (2009)
Kelsall, R.W., Hamley, I.W., Geoghegan, M.: Nano-scale science and technology. John Wiley and Sons Ltd., Chichester (2004)
Kennepohl GJA, Miller LJ (1977) Sulfur-Asphalt Binder Technology for Asphalt Pavements. Advances In Chemistry Series 165: New Uses of Sulfur, p135, ACS, New Orleans, La. USA.
Kennepohl, G.J.A.: Asphalt pavements and the environment. In: ISAP Int. Symp. on Asphalt Pavements and Environment, Zürich, Switzerland (2008)
Kessler, M.R., Sottos, N.R., White, S.R.: Self-healing structural composite materials. Composites: Part A 34, 743–753 (2003)
Kuennen, T.: Road Science, Better Roads (July 2004), `http://www.betterroads.com`
Kotlyar, L.S., Sparks, B.D., Woods, J.R., Raymond, S., Le Page, Y., Shelfantook, W.: Distribution and types of solids associated with bitumen. In: Petroleum Science and Technology, vol. 16(1-2), pp. 1–19. Taylor & Francis, Colchester (1998)
Kumar, S., Curtin, W.A.: Crack interaction with microstructure. Materialstoday 10(9), 34–44 (2007)

Larsen-Basse, J., Chong, K.P.: Nano-materials in construction and rehabilitation: Contributions and perspectives of the US National Science Foundation. In: Proc. 2nd Int. Symp. on Nanotechnology in Construction, Bilbao, Spain, pp. 17–19 (2005)
Lesueur, D.: The colloidal structure of bitumen: Consequences on the rheology and on the mechanisms of bitumen modification. Advances in Colloid and Interface Science 145, 42–82 (2009)
Liao, G., Huang, X., Ba Sang, D.: Antiultraviolet Aging and Antithermal-Oxygen Aging Tests of Asphalts Adapting to Environment of Tibetan Plateau of China. In: Proc. 87th Transportation Research Board meeting, Washington D.C (2008)
Livingston, R.A.: Neutron scattering methods for concrete nanostructure characterization. In: 2nd Int. Symp. on Nanotechnology in Construction, Bilbao, Spain, pp. 115–124 (2005)
Maclurcan, D.C.: Nanotechnology and developing countries. Part 1: What possibilities. J. Nanotechnology Online (2005), doi:10.2240/azojono0104
Maher, M., Uzarowski, L., Moore, G., Aurilio, V.: Sustainable pavements – Making the case for longer design lives for flexible pavements. In: Proc. 51st Annual Conf. of the Canadian Technical Asphalt Association, Charlottetown, Prince Edward Island, Canada, pp. 44–56 (2006)
Makar, J., Margeson, J., Luh, J.: Carbon nanotube/cement composites – early results and potential applications. In: Proc. 3rd Int. Conf. on Construction Materials: Performance, Innovations and Structural Implications, pp. 1–10. University of British Columbia, Vancouver, B.C., Canada (2005)
Masson, J.-F., Leblond, V., Margeson, J.: Bitumen morphologies by phase-detection atomic force microscopy. J. Microscopy 221(1), 17–29 (2006)
Masson, J.-F., Leblond, V., Margeson, J., Bundalo-Perc, S.: Low-temperature bitumen stiffness and viscous paraffinic nano-and micro-domains by cryogenic AFM and PDM. Jnl. of Microscopy 227(3), 191–202 (2007)
Meyyapan, M., Iyuke, S.E., Falcon, R.: Carbon engineering and nanotechnology course. Short course presented at the University of the Witwatersrand, Johannesburg, South Africa (2007)
Mgangira, M.: Use of scanning electron microscope in road materials characterization. CSIR BE report CSIR/BE/IE/IR/2007/024/B, Pretoria, South Africa (2007a)
Mgangira, M.: Particle bonding characterization of a sand mixture treated with non-traditional additives. In: Proc. 86th Transportation Research Board (TRB) meeting, TRB, Washington D.C (2007b)
Mgangira, M.: Identification of microscale characteristics of treated subgrade materials and how they relate to macroscopic properties. In: Proc. 29th Annual Transportation Conference, SATC, Pretoria, South Africa (2010)
Mokhatab, S., Fresky, M.A., Islam, M.R.: Applications of nanotechnology in oil and gas E&P. Jnl. of Petroleum Technology (JPT) Online 58(4) (2006)
Morose, G.: The 5 principles of "Design for Safer Nanotechnology". Jnl. Of Cleaner Production 18, 285–289 (2010)
Muniandy, R.: Laboratory evaluation of Malaysian cellulose oil palm fiber for use in Stone Mastic Asphalt mixes. Int. Jnl. of Pavements (IJP) 75(3), 13–21 (2002)
NNI, Nanotechnology and the environment. Report of a National Nanotechnology Initiative Workshop, Arlington, VA, May 8-9 (2003)
Nugent, R.A., Zhang, G., Ganbrell, R.P.: The Effect of Exopolymers on the Liquid Limit of Clays and Its Engineering Implications. J. Transportation Research Board, 210134–210143 (2009)
OECD, Frascati Manual 2002. Organisation for Economic Co-operation and Development (OECD), Paris, France (2002); ISBN 92-64-19903-9

Ozin, G.A., Arsenault, A.C.: Nanochemistry – A Chemical Approach to Nanomaterials. RSC Publishing, The Royal Society of Chemistry, UK (2005)
Paige-Green, P., Steyn, W.J.M.: An investigation into the need for bitumen alternatives for road infrastructure. CSIR BE Report TR-2005/09. CSIR Built Environment, Pretoria, South Africa (2005)
Partl, M.N., Gubler, R., Hugener, M.: Nano-science and –technology for asphalt pavements. In: Proc. 1st Int. Symp. on Nanotechnology in Construction, Paisley, Scotland, pp. 343–355 (2003)
Pauli, A.T., Branthaver, J.F., Robertson, R.E., Grimes, W.: Atomic Force Microscopy investigation of SHRP asphalts. In: Proc. Symp. on Heavy Oils and Residue Compatibility and Stability, 221st National Meeting, American Chemical Society: Division of Petroleum Chemistry, San Diego, California, USA (2001)
Peethamparan, S., Olek, J., Diamond, S.: Physicochemical Behavior of Cement Kiln Dust (CKD)-Treated Kaolinite Clay. J. Transportation Research Board 2059, 80–88 (2008)
PIARC Home Page, `http://www.piarc.org/en/technical-committees` (accessed September 2007)
PICADA, Official presentation, Innovative facade coatings with de-soiling and de-polluting properties. EC GRD1-2001-40449, GTM Construction, Nanterre Cedex, France (2006)
Polacco, G., Kříž, P., Filippi, S., Stastna, J., Biondi, D., Zanzotto, L.: Rheological properties of asphalt/SBS/clay blends. European Polymer Journal 44, 3512–3521 (2008)
Read, J., Whiteoak, D.: The Shell bitumen handbook, 5th edn. Thomas Telford Publishing, London, UK (2003)
Reijnders, L.: The release of TiO2 and SiO2 nanoparticles from nanocomposites. Polymer Degradation and Stability 94, 873–876 (2009)
RILEM, Technical Committee NBM (Nanotechnology-based Bituminous Materials) General Information Web Page (2009)
`http://www.rilem.net/tcDetails.php?tc=NBM` (accessed April 2009)
Robertson, J.: Realistic applications of CNTs. Materialstoday 7(10), 46–52 (2004)
Sanfilippo, J.M., Munoz, J.: Using nanotechnology to play with aggregate mineralogy: Application in concrete processing. Presentation at the Nanotechnology-based Concrete Materials Task Force Meeting (AFN15T), Transportation Research Board Meeting, Washington D.C (2009)
Sanfilippo, J.M., Anderson, M.A., Muňoz, J.F., Cramer, S.M., Tejedor, M.I.: Nanotechnology to Manipulate the Aggregate-Cement Paste Bond: Impacts on Concrete Performance. In: Proc. of the 2009 Mid-Continent Transportation Research Symposium, Ames, Iowa, USA (2009)
Savage, N., Diallo, M.S.: Nano-materials and water purification: Opportunities and Challenges. J. Nanoparticle Research 7(4-5) (2005)
Shah, S.P.: Next Horizon in High Performance Concrete: Self-Consolidating Concrete and Nanotechnology. Nanotecnología en el Hormigón Y Hormigones Autocompactantes. Jornada Técnica JT-01, 131–151 (2009)
Xu, S., Qin, Y., Xu, C., Wei, Y., Yang, R., Wang, Z.L.: Self-powered nanowire devices. Nature Nanotechnology 5, 366–373 (2010)
Shi, Z.-Q., Chung, D.D.L.: Carbon fiber-reinforced concrete for traffic monitoring and weighing in motion. Cem. Concr. Res. 29, 435–439 (1999)
Sobolev, K., Gutiérrez, M.F.: How nanotechnology can change the concrete world. American Ceramic Society Bulletin 84(10) (2005)
Steyn, W.J.M.: Applications of observational techniques in engineering. In: Proc. 26th Annual Transportation Conference, SATC, Pretoria, South Africa (2007)
Steyn, W.J.M.: Research and application of nanotechnology in transportation. In: Proc. 25th Annual Transportation Conference, SATC, Pretoria (2008a)

Steyn, W.J.M.: Development of autoluminescent surfacings for concrete pavements. J. Transportation Research Board 2070, 22–31 (2008b)
Steyn, W.J.M.: Potential applications of nanotechnology in pavement engineering. ASCE Jnl. of Transportation Engineering 135(10), 764–772 (2009a)
Steyn, W.J.M.: Nanotechnology in Pavement Engineering 2008/9. PG Report: CSIR/BE/IE/IR/2009/0098/B, CSIR Built Environment, Pretoria, South Africa (2009b)
Steyn, W.J.M., Jones, D.J.: Technical Memorandum: HVS testing of N12-19 East, Section 2. Contract Report CSIR/BE/IE/ER/2007/0001/B, CSIR Built Environment, Pretoria, South Africa (2006)
Stulirova, J., Pospisil, K.: Observation of bitumen microstructure changes using Scanning Electron Microscopy. Road Materials and Pavement Design 9(4), 745–754 (2008)
Sunter, C.: The high road: Where are we now. Tafelberg Publishers, Cape Town (1996)
Taha, M.R., Ismail, E., Chik, Z.: Some nano aspects and concepts in geotechnology. In: 2nd Int. Symp. on Nanotechnology in Construction, Bilbao, Spain, pp. 373–381 (2005)
Tong, Z., Bischoff, M., Nies, L.: Impact of Fullerene (C60) on a soil microbial community. B. Environ. Sci. Technol. 41, 2985–2991 (2007)
TRB Home page, `http://www.trb.org` (accessed November 2009)
Van Breugel, K., Guang, Y.: Multi-scale modeling in engineering practice and materials research – a vehicle for progress. In: Applications of Nanotechnology in Concrete Design. Thomas Telford Publishing, London (2009)
Van de Ven, M.F.C., Molenaar, A.A.A., Besamusca, J., Noordergraaf, J.: Nanotechnology for binders of asphalt mixtures. In: Proc. Road for Life, Copenhagen, Denmark, pp. 842–853 (2008)
Wang, K.: Size Effect in the Cell Response to Substances in Solid Phase. China Particuology 1(1), 3 (2003)
Yakovlev, G., Keriené, J., Gailius, A., Girniene, I.: Cement based foam concrete reinforced by carbon nanotubes. Materials Science (Medžiagotyra) 12(2) (2006)

Application of Nanoscience Modeling to Understand the Atomic Structure of C-S-H

R. Panneer Selvam, Kevin D. Hall, Vikramraja Janakiram Subramani, and Shanique J. Murray

Abstract. Concrete is used extensively for constructing buildings and highways. In order to enhance macroscopic mechanical properties (tensile strength) it is necessary to understand the structure and behavior of C-S-H gel at the atomic level. Current status in nanoscience modeling of C-S-H is reviewed. The computational tools available and their limitations are also surveyed. Application of nano science in improving the cement paste by studying the interaction of salt, water etc using the crystal structure of Tobermorite and Jennite are discussed first. Method to find the amorphous structure of C-S-H from the crystal structure of Tobermorite is discussed next. An amorphous atomic structure of C-S-H is proposed from the study. Finally studies conducted using the proposed amorphous structure of C-S-H is discussed.

1 Introduction

Concrete is used extensively in constructions for buildings as well as in highways. The durability and long life-cycle of concrete is affected by minute cracks due to shrinkage and load-related stress. This is due to the relative lack of tensile strength compared to the compressive strength of concrete. Cement pastes which exhibit

---

R. Panneer Selvam
BELL 4190 University of Arkansas, Fayetteville, AR 72701, USA
e-mail: rps@uark.edu

Kevin D. Hall
BELL 4190 University of Arkansas, Fayetteville, AR 72701, USA
e-mail: kdhall@uark.edu

Vikramraja Janakiram Subramani
TUHFP, North Campus, University of Tulsa, OK 74104, USA
e-mail: vikramrajajs@utulsa.edu

Shanique J. Murray
BELL 4190 University of Arkansas, Fayetteville, AR 72701, USA
e-mail: sjmurra@gmail.com

K. Gopalakrishnan et al. (Eds.): Nanotechnology in Civil Infrastructure, pp. 85–102.
springerlink.com

low tensile strength tend to exhibit greater shrinkage cracking and reduced durability. If the tensile strength in cement paste can be increased, then the shrinkage cracking potential can be minimized. It is believed that the strength and cohesion of cement paste is controlled by the formation of Calcium Silicate Hydrate (C-S-H) gel.

In order to enhance macroscopic mechanical properties (tensile strength) it is necessary to understand the structure and behavior of C-S-H gel at the atomic level. Even though cement has been used in construction for more than 100 years, there is a lack of understanding of the atomic structure of cement. For crystalline structures like metals the atomic structure is determined from experimentation using X-ray diffraction (XRD) techniques. For amorphous materials like glass there is no clear experimental technique available. In addition, cement has multiple elements (Ca, Si, O, H) contributing to the formation of C-S-H gel, further complicating its structure.

Taylor [1997] reported from his research that the X-ray diffraction patterns of cement pastes reveal no evidence of long range order in C-S-H structures. Fig. 1 shows the shape of diffraction peaks for a perfect crystal, amorphous substance, and liquid. In general, the XRD pattern is a plot of diffracted intensity (I) of X-ray vs angle of incidence (θ). The peak intensities provide information about the lattice parameters. The diffraction peaks or the peak intensities for perfect crystals are narrow and sharp. For imperfect crystals or amorphous materials, the peaks are broadened. In case of liquids, it will be a continuous and gradually varying function. When the XRD plots of cement paste reported in Taylor [1997] are compared with these plots, it is apparent that they are very similar to amorphous materials.

These challenges have prevented researchers from making significant gains in understanding the structure of C-S-H. Recent developments in nanoscience have allowed several researchers to focus on combining experimental and theoretical techniques to better define the atomic structure of cement paste.

Current status in this regard will be reviewed. The computational tools available and their limitations will also be surveyed. Current work to date on the atomic structure of C-S-H will be reported.

2 Nanoscience Modeling Tools

General review of nanoscience modeling tools applied to material science can be found in Lech [2001], Ghoniem and Cho [2002], Li and Liu [2002], Starrost and Carter [2002] and Rafi-Tabar[2004]. Cramer [2002], Kaxiars [2003] and Martin [2004] deal with ab-initio modeling whereas Allen and Tildesley [1987] and Rapaport [2004] discuss the implementation of molecular dynamics (MD), molecular statics (MS) and Monte Carlo (MC) simulation methodology for liquids.

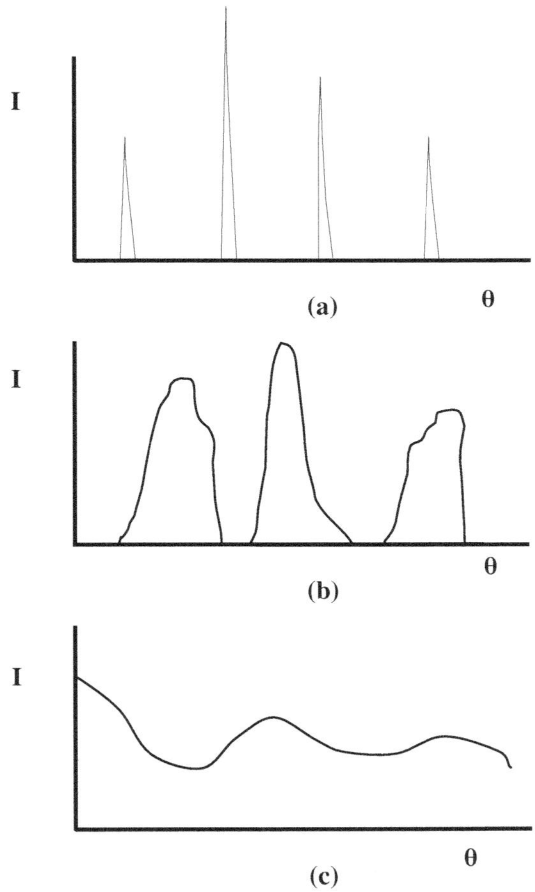

Fig. 1 Diffraction patterns of: a) perfect crystal b) imperfect crystal c) liquid

2.1 Ab-Initio Modeling

When the electronic and other interactions between atoms are considered without any empirical parameters from experiments, then this method is called ab-initio or first principle modeling. At this time only atoms of the order of 1000 is considered for computation due to large cpu consumption. But the method provides ways to understand the interaction forces between atoms from first principle. This approach considers the interaction between electrons of different atoms through Schrodinger equations and the interactions between nucleuses of the atoms using Newtonian dynamics. In addition to fundamental study the method is also useful to determine the forces between atoms or to determine the empirical interatomic potentials between atoms. The interatomic potentials will be utilized in molecular dynamics (MD) and molecular statistics (MS) modeling. Manzano et al. [2007b]

used ab-initio calculations to find the possible atomic structure for C-S-H gel. They identified the probable bond of Tobermorite and Jennite crystals that can be broken to obtain a stable C-S-H basic block. They also determined partial charges of the ions in C-S-H. Gmira et al. [2004] used this approach and found that Hamid's structure [1991] is more stable than Merlino's structure [1999].

2.2 Molecular Dynamics

The atomic interaction study using ab-initio procedures are difficult to apply to a large number of atoms due to intensive cpu time. For a better understanding of physical and chemical interaction between atoms and to obtain basic properties in the nano level MD is used. MD is preferred if one wants to calculate the time-dependant quantities such as transport coefficients. Atoms from 1000 to a billion have been considered so far. In this approach the interaction between atoms are approximated and related through empirical interatomic potentials. This approximation provides means to study several nano phenomena using millions of atoms. The interaction between the atoms is governed by Newtonian dynamics. From the time evolution of the atoms and their interactions at specific temperature the mechanical and thermodynamical properties and stress-strain relationship can be derived. Janakiram Subramani [2008] used MD to identify the weak bond in the Tobermerite structure that can be broken to have C-S-H basic block. Murray et al. [2010] used MD to compute the stress strain relation of possible C-S-H structure.

2.3 Molecular Statics

In molecular static (MS) modeling, energy minimizations are performed on the atomic structure as described in Leach [2001]. This minimization will yield an equilibrium structure based on the closest local minima of the initial atomic structure. However, the optimized structure, thus obtained may characterize one possible equilibrium or stable structure of the many that could exist. The bulk properties of the material like mechanical properties, vibrational properties, electrical properties depend on the curvature of the energy surface i.e., second derivative of the energy with respect to atomic positions and lattice strains. For further details one can refer to Janakiram Subramani [2008]. The bulk properties can be computed from the final atomic structure arrived from energy minimization. Several researchers used GULP code [Gale and Rohl 2003] to compute mechanical properties of C-S-H related crystal structure [Gmira et al. 2004] or C-S-H amorphous structure [Manzanao et al. 2007a, Janakiram Subramani et al. 2009]. This approach is also used by Manzano et al. [2007a] and Janakiram Subramani et al. [2009] to find the possible amorphous atomic structure of C-S-H.

2.4 Monte Carlo

Monte Carlo (MC) methods are used to model the physical events through probabilities [Leach 2001]. This method does not have time evolution and hence it is much faster than MD study. This method is applied to get the equilibrium properties of surfaces and interfaces. Using MC one can evolve a system from non-equilibrium to approach equilibrium. The method is also applied to study film growth type problems and is called MC for non-equilibrium system. Similar to MD, this method can also be used to calculate thermodynamic properties. The method can skip high barriers with several local minima faster than MD.

3 Modeling Study to Find the Atomic Structure of C-S-H

Taylor (1986) made the first attempt to identify some of the characteristics of C-S-H by comparing it with existing crystalline minerals. It was believed that the local atomic structures in the C-S-H gel are similar to those in some of the naturally occurring crystalline minerals such as Tobermorite, Jennite, Clinotobermorite, Foshagite etc. Kirkpatrick et.al [1997] and Yu et al.[1999] also confirmed through their spectroscopic studies that the C-S-H contains Tobermorite and Jennite like structures. Two widely used crystalline minerals that compare with the atomic structure of C-S-H are 1) Tobermorite [Hamid 1981, Merlino 1999] and 2) Jennite [Merlino 1999]. With this understanding several researchers have conducted nano modeling using Tobermorite or Jennite crystal structures as a start to understand the atomic structure or properties of C-S-H.

3.1 C-S-H Understanding Using Crystal Structures

Faucon et al. [1996] used MD simulations to understand the NMR data concerning C-S-H. Using the Tobermorite structure they observed that in the absence of calcium atoms, no break in the chain is observed. The presence of Ca causes partial rupture of Si chains. Faucon et al. [1997] also showed that Al atom can be substituted in the place of Si atom. Such substitution is possible where the bridging Si is energetically less stable. Faucon et al. [1999] used the similar substitutions using Na. Interaction of C-S-H with water, alkali and salt are reported in Faucon et al. [1997, 1998 & 1999]. Similar studies were also conducted by Kirkpatrick and their research group using crystal structures like Tobermorite, Jennite etc. in [2007] and the references there in. Coveney and Humphries [1996] used similar crystal structures to study the phosphonate retarders on hydrating cements. Later Bell and Coveney [1998] explained why borate retarders are efficient at high temperatures using MD and also provided novel approaches to increase their performance further. Sanchez and Zhang [2008] investigated the interaction of graphite with C-S-H using Tobermorite 9 A crystal structure. They demonstrated that the polarity of the functional group can be used as an indicator of affinity to C-S-H.

Gmira et al. [2004] used the Tobermorite structure to calculate the mechanical properties like bulk modulus using GULP and compared it with experimental measurements. The calculations are done using MS. They also used ab-initio calculations to understand the nature of cohesive forces in C-S-H.

Janakiram Subramani [2008, 2009] used MS to compute the mechanical properties like Young's modulus, Poisson's ratio etc. The computed bulk property of the crystal structure is 4 to 5 times greater than the cement paste.

3.1.1 Deterioration of Cement Due to Magnesium (Mg) Salts

Janakiram Subramani [2008] investigated the attack of Magnesium (Mg) salt on concrete structures using Hamid's [1981] Tobermorite 11 A° crystal structure. Upon prolonged exposure to Mg salts, the calcium ions from the cement are leached out, which in turn triggers the deterioration of the concrete structures. The deterioration is initiated by the substitution of the less stable calcium ions in the cementitious materials by Mg ions. Knowledge on the potential energies of Calcium atoms at its site (also called as site potentials) in the Tobermorite structure could shed more light on its exchange/substitution capacities. The site potentials of Calcium atoms in the Tobermorite structure for different Ca/Si ratios (0.66, 0.83 and 1.00) were determined using GULP. Figure 2 shows the stability of calcium atoms in the Tobermorite structure. The potential energy data that are encircled in the graph below are those that correspond to the interlayer Calcium atoms. The number of calcium atoms at the interlayers of the Tobermorite structure varies with Ca/Si ratio. The number of calcium atom in Ca/Si ratio of 0.66, 0.83 and 1.0 is none, one and two respectively. The potential energies of interlayer Calcium atoms (corresponding to site 5 & 6) are much higher than the potential energies of the lattice Calcium atoms (site 1, 2, 3 and 4). This suggests that the lattice Calcium atoms are far more stable than the interlayer calcium atoms. Hence, the interlayer Calcium atoms are most vulnerable for cationic-exchange or substitution.

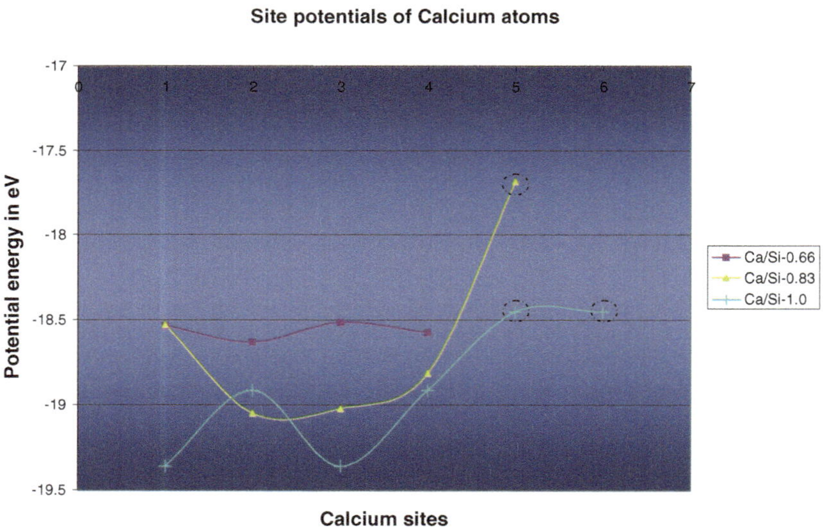

Fig. 2 Site potential of Calcium (Ca) atoms

Magnesium (Mg) ions from the water substitute the energetically less stable interlayer Calcium (Ca) atoms in the C-S-H structure. In this case, only the substitution of interlayer Ca^{2+}ions by Mg^{2+} is considered. Figure 3 shows the Magnesium (Mg) substituted C-S-H.

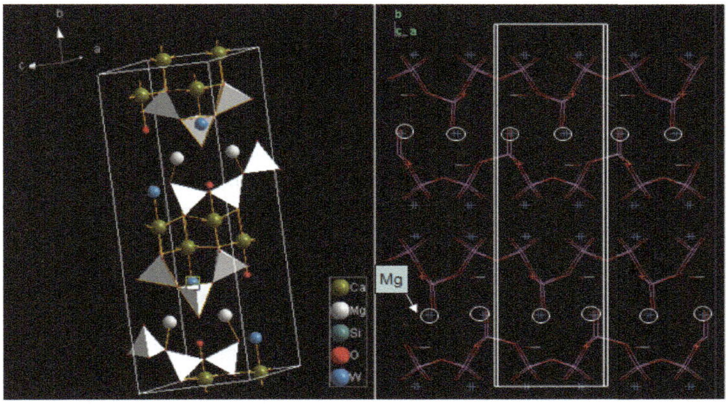

Fig. 3 Magnesium (Mg) substituted C-S-H-Tobermorite crystal structure

Noticeable decrease in Elastic modulus value was observed for Mg substituted C-S-H structures as shown in Table 1. Leaching of Calcium (Ca) leads to the reduction in the strength, stiffness and durability of concrete. This results in the concrete eventually losing its cohesive nature. When all Calcium (Ca) atoms in the C-S-H structures are substituted and occupied by Magnesium (Mg) ions, it results in the formation of Magnesium (Mg) Silicate Hydrates. This results in loss of cementitious characteristics and eventually leads to deterioration of rigid cement paste into a non-cohesive mass.

Table 1 Comparison of bulk properties of Tobermorite with/without Magnesium (Mg) substitution

Species	**Dopant**	**Bulk Modulus K (GPa)**	**Shear Modulus G (GPa)**	**Elastic Modulus E (GPa)**	**Poissons ratio**
Tobermorite 11 A	Without Mg	79	37	96	0.3
Tobermorite 11 A	With Mg	68	27	71	0.32

3.1.2 Physical and Chemical Compatibility of Flyash in Cement

Composite cement contains one or more inorganic materials that takes part in the hydration reactions and contribute significantly to the hydration product. Fly ash is one of the most important sources of such inorganic materials that can readily be mixed with cement. Fly ash is known to be rich in Al_2O_3 and SiO_2. Janakiram Subramani[2008] attempts to explain the mechanism of different possible cationic substitutions that can take place during the hydration of composite cement. Hamid's (1981) Tobermorite 11 A° is considered as the C-S-H model for this study. To perform this study, it is critical to know which of the six silicon (Si) atoms in the C-S-H model need to be substituted. Calculation of the site potentials of all the silicon (Si) atoms in the C-S-H structure will provide a good idea about its exchange capacities.

The potential energy of silicon (Si) atoms at site 3 and 5 for Ca/Si-0.83 were found to be much lesser than those at other sites (1, 2, 3 and 4). Thus, the bridging silicon (Si) atoms have the least cohesive energy and are less stable compared to non bridging silicon (Si) atoms. Hence Si_3 and Si_5 are the preferential sites for substitution. The Aluminum ions available from the fly ash also participate in the hydration mechanism by substituting the silicon (Si) atoms in the C-S-H structures. As Al^{3+}ions tend to form tetrahedral coordination similar to silicon (Si) atoms, it substitutes the energetically less stable bridging silicon (Si) atoms. But this substitution leaves a negative charge induced into the C-S-H structure. The charge compensation can be done in two ways: a) Bridging silicon (Si) atoms can be substituted by two Al^{3+} ions & one Ca^{2+} ion (in interlayer) b) Bridging silicon (Si) atoms can be substituted by two Al^{3+} ions & two protons attached to the oxygen atoms (all oxygen atoms connected to bridging silicon (Si) atoms are hydroxylated). Fig 4 shows the structure of C-S-H with Aluminum substituted only at the bridging silicon (Si) sites.

The aluminum substitution reaction in the silicate chains of C-S-H can also be viewed with another perspective. During the initial stages of hydration of the cement-based materials, the formation of dimeric C-S-H structures is favored. Alumina available from the flyash gets hydrated to Aluminum hydroxide ($Al(OH)_3$). Since Aluminum is also tetrahedrally coordinated, $Al(OH)_3$ acts a bridging molecule between dimeric silicate chains. Thus, the addition of flyash favors the transformation of Dimeric silicate chains to Drierketten type silicate chains. Only that the silicon (Si) in the bridging tetrahedra is substituted by Aluminum in this case. Also, the silica available from the flyash interacts with the Calcium (Ca) hydroxide in the cement paste and thus, forms additional C-S-H.

For Case (a), the two negative charges induced by the displacement of two Si^{4+} ions by two Al^{3+} ions were compensated by addition of one Ca^{2+} ion to the labile layer. Since the addition of a Calcium (Ca) atom in the interlayer was adequately sufficient to compensate the charge deficit induced near the bridging silicon (Si) atoms, this substitution did not provoke rupture in the tetrahedral silica chains. This shows the chemical compatibility of inorganic material additions to cement from the 'atomic structure' point of view. On the other hand, substitution of aluminum ions in non-bridging silicon (Si) sites induced rupture in the tetrahedral chains. Though the charge deficit can be compensated theoretically by adding Calcium (Ca) ions or adding protons to the oxygen atoms for bridging silicon (Si) atoms, the charge compensation was poor because the protons to compensate for the charge deficit are at a relatively greater distance from aluminum. This poor charge compensation near the non bridging silicon (Si) sites substituted by aluminum, induce a rupture in the chains.

However, the calculated properties of Aluminum substituted C-S-H (on bridging silicon (Si) sites) did not differ much from the un-substituted C-S-H (Table 2). The results justify the physical compatibility of inorganic material additions (such as fly ash rich in alumina and silica) to cement. The following two cases were also studied but are not reported here. 1) Substitution of Aluminum in place of Bridging silicon (Si) atoms with two protons attached to the oxygen atoms (hydroxylated) for charge compensation and 2) Substitution of Aluminum in place of non bridging silicon (Si) atoms.

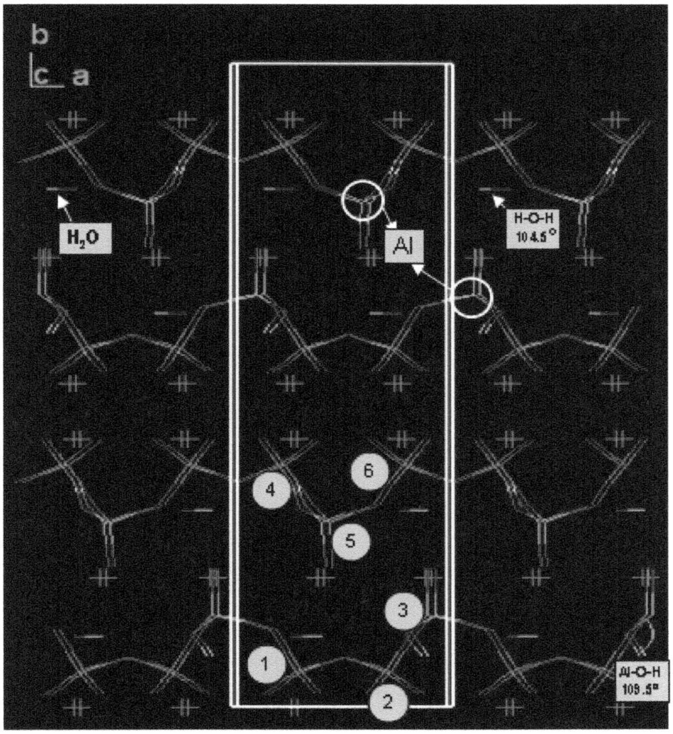

Fig. 4 Aluminium substituted C-S-H

Table 2 Comparison of bulk properties of C-S-H with/without Aluminum substitution

C-S-H model-Ca/si-0.83	No substitution	Si3 & Si5 site substitution
Bulk Modulus K (GPa)	62	69
Shear Modulus G (GPa)	34	36
Elastic Modulus E (GPa)	86	92
Poissons ratio	0.27	0.28

The Tobermorite or Jennite crystal structures are useful to understand the C-S-H structure as well as to study the reaction and properties in the nano level when other elements are substituted. Since C-S-H is amorphous the MD study using crystal structures which are similar to C-S-H provided a starting point for further study.

3.2 Modeling to Find the Amorphous Structure of C-S-H

The computed bulk properties of Tobermorite crystal structure is 4 to 5 times the bulk properties of cement paste as reported by Janakiram Subramani et al. [2009]. Plassard et al. [2004] observed a bridging silicon (Si) atom missing in the silicate chains thereby forming dimers or pentamers. If the missing Si atom occurs after every two Si positions in the chain it is called a dimer whereas if the missing Si atom occurs after every five Si positions in the chain it is called a pentamer. To verify this observation by MD, the site potential of Si atoms were computed and compared. The Si atom at site 3 and 5 of the Ca/Si ratio of 0.66 are energetically less stable than the corresponding one in Ca/Si ratio of 0.83. From this observation the two bridging Si atoms are removed and one Ca atom is added at the interlayer. The details of the procedure is reported in Janakiram Subramani et al. [2009]. The bulk properties of this C-S-H structure were determined using energy minimization methods and found to be K=42.5 GPa, G=26 GPa E = 64 GPa. These values are still higher than experimental and simulation results of K=18 GPa, G=9.7 GPa and E = 20-30 GPa reported by Manzano et al. [2007]. From this one can conclude that the length of silicate chains influence the mechanical properties.

The bulk cement paste has porosity where as the suggested C-S-H structure is non porous. To consider the porosity different possible combination were proposed by Janakiram Subramani et al. [2009] at the atomic level. To consider those models one needs large computational domain with many atoms and hence a porosity of 31% is considered and the mechanical properties were corrected according to Mori-Tanaka [1973] equation. The corrected bulk properties were K=21 GPa, G=13 GPa E = 34 GPa. These values are in reasonable comparison with experimental values of K=18 GPa, G=9.7 GPa and E = 20-30 GPa. Hence, it can be deduced from this study that the length of the silicate chains along with the porosity have a significant effect on the mechanical properties of C-S-H. The dimer or pentamer can become the basic block of C-S-H structure.

3.3 C-S-H Understanding Using Amorphous Structure

3.3.1 Stress-Strain Relationship Using MD

Selvam et al. [2009] and Murray et al. [2010] calculated the stress strain relationship of C-S-H using the basic block of C-S-H structure proposed by Janakiram Subramani et al. [2009]. Since the compressive strength of the cement paste is 10 times greater than the tensile strength in the macro level; an attempt is made to study the strength at the nano level. They used MD to apply different strain rate in time. The strain is applied only in one direction and the corresponding stress is calculated. They used LAMMPS code to perform MD calculations. Before MD modeling conjugate gradient procedure is used to minimize the potentials. They considered both C-S-H crystal and dimer structure for comparison. The size of the unit cell considered for computation is 1.17 nm x 0.739 nm x 2.27 nm and the total number of atoms in the crystal and dimer structure was 144 and 124. The computed elastic modulus, tensile and compressive strength of crystal and dimer structure were (96 GPa, 70 GPa), (12 GPa, 3.5 GPa) and (15 GPA,15 GPa) respectively. The macro level tensile strength of 2 GPa is 57% of the dimer value. This difference can be due to not considering porosity and much larger computational domain. The tensile strength of crystal structure is almost the same as compressive strength where as the dimer structure tensile strength is 23% of compressive strength.

3.3.2 Mechanical Properties of C-S-H Including Porosity

Selvam et al. [2009] and Murray [2009] considered four unit cells of the dimer structure proposed by Janakiram Subramani et al. [2009] to validate the proposed C-S-H structure including porosity. For this work, three simulation cells were created. The names of the simulation cell are Stack 1, Stack 2 and Stacks 3. Each simulation cell contained four Tobermorite dimer unit cell and the number of atoms in each simulation cell is 576 atoms. The dimension of each simulation cell is shown in Table 3. The configuration of atoms in "Stack 1" is created by translating the atoms in a Tobermorite unit cell to 4 regions (corners) of the simulation cell as shown in Fig. 5. The Tobermorite unit cells are spaced one angstrom apart to incorporate "breaks" in the silicate chains.

Table 3 Dimensions of Simulation Cells

Simulation Cell	Number of Atoms	a (Å)	b (Å)	C (Å)
Stack 1	576	10	20	50
Stack 2	576	10	25	44
Stack 3	576	13	18	65

In Fig 5 Tobermorite unit cell 1 is located at the origin. In stack 1, Tobermorite unit cell are 5.22 angstroms apart in the b direction and 6 angstrom apart in the c direction where b and c are in the plane of the paper. In Stack 2, Tobermorite unit cell are 10.22 angstroms apart in the b direction and 0 angstrom apart in the c direction as shown in Fig. 6. Stack 3 was constructed differently in that the Tobermorite unit cells are not placed at the corners. Instead, the first and forth Tobermorite unit cells are centered in the simulation box. The distance between first and forth Tobermorite unit cells is 21 angstroms in the c direction. The second and third unit cells are placed in the space between the first and forth Tobermorite and spaced 3.22 angstrom in the b direction. Fig.7 illustrates a sketch of Stack 3. The

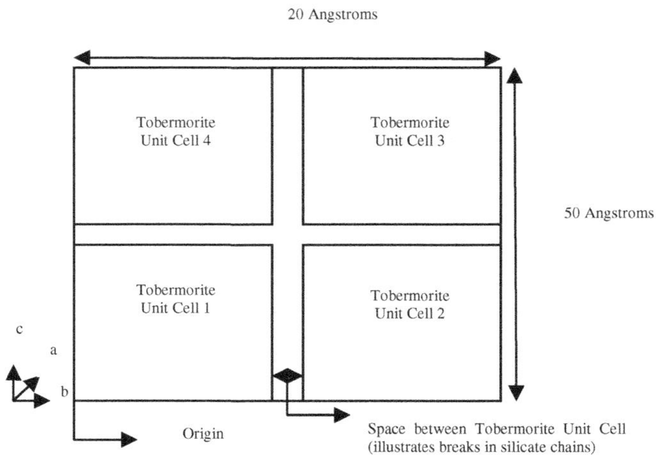

Fig. 5 Sketch of packing C-S-H structure using 4 unit cells Stack 1

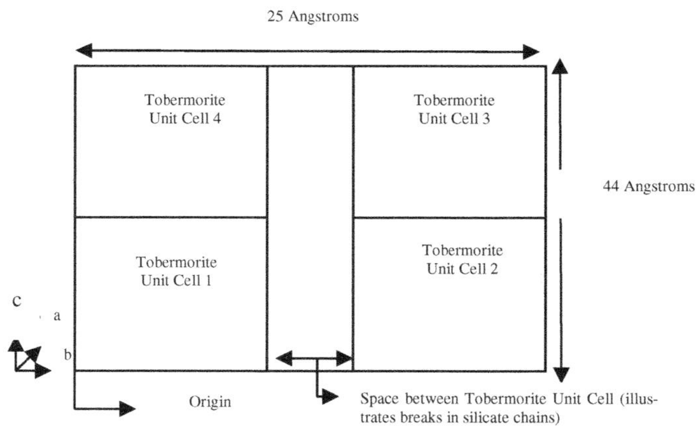

Fig. 6 Sketch of packing C-S-H structure using 4 unit cells Stack 2

atomic coordinates of atoms are in Cartesian coordinates, therefore $\gamma = 90$ degrees. Charges of atoms are the same as the previous study. Fig. 8 shows the atomic view of the three stacks.

Molecular Statics via GULP code was used to calculate the mechanical properties of C-S-H structure based on Tobermorite dimer structure packed together. The Tobermorite structures are spaced a few angstroms apart. This spacing was done to incorporate bond breakage in the silicate chains. The calculated mechanical properties are shown in Table 4. It can be seen that as the density of C-S-H unit

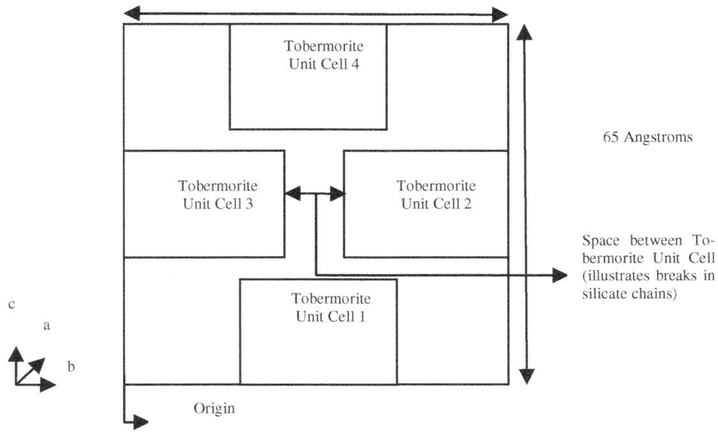

Fig. 7 Sketch of packing C-S-H structure using 4 unit cells Stack 3

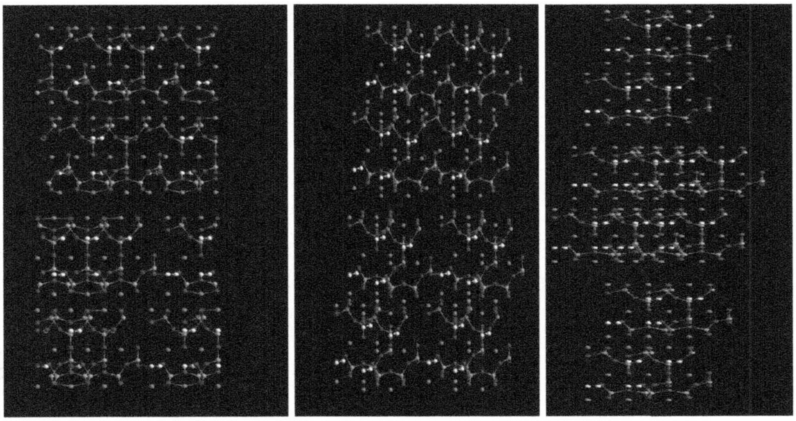

Fig. 8 Atomic view of packing C-S-H dimer structure using 4 unit cells (a) Stack 1 (b) Stack 2 and (c) Stack 3

approached the density of C-S-H found in experiments; the computed elastic modulus are in reasonable agreement with nano indentation experiments. At the nano meter length scale, C-S-H gel exists in two forms; Low Density C-S-H and High Density C-S-H. The dry density values for low density and high density C-S-H gel are 1.44 g/cm^3 and 1.75 g/cm^3 respectively. Nanoindentation experiments from Constantinides and Ulm [2007] show that the range of elastic modulus corresponding to low density C-S-H is 13-26 GPa and the range of elastic modulus corresponding to high density C-S-H is 26 -39 GPa. The computed elastic modulus for Stack 3 is 30 GPa and it is more closer to high density C-S-H. This packing method may be a possible way to account for porosity in C-S-H.

Table 4 Mechanical properties of C-S-H structures packed together

C-S-H model	Density g/cm^3	Bulk Modulus GPa	Shear Modulus GPa	Poissons ratio	Elastic Modulus GPa
Stack 1	2.24	46	23	0.29	59
Stack 2	1.93	39	23	0.25	58
Stack 3	1.55	21	12	0.30	30
Exp. HD-C-S-H	2.13	n/a	n/a	n/a	26-39
Exp. LD-C-S-H	1.44	n/a	n/a	n/a	13-26

4 Conclusions and Future Research

Current status of the understanding of atomic structure of C-S-H is reviewed. Nano modeling tools available at this time is also reviewed. In most of the initial works Tobermorite or Jennite like crystal structures were used and the chemical interaction of materials like salt, flyash, Magnesium (Mg), water etc were investigated. Recently, Janikiram Subramani et al. [2009], Selvam et al. [2009] and Monzano et al. [2007] have proposed that the breaking of the silicon bridging bond in the Tobermorite structure results in the basic C-S-H structure in the form of dimer or pentamer. Using this basic C-S-H block, Murray et al. [2010] computed the stress strain relationship at the nano level. They compared the mechanical properties of amorphous structure to crystal structure and found that the amorphous structure mechanical properties agreed better with the macro cement properties. Further work is reported by including porosity in the computational cell by arranging four C-S-H basic blocks in three different ways and the mechanical properties are computed. Stack 3 mechanical properties are in reasonable agreement with macro cement properties. These studies clearly illustrate that the basic C-S-H structure is amorphous. The amorphous C-S-H structure including porosity may be a starting point for further study.

Fundamental understanding of how C-S-H is formed by chemical reaction at different stages of hydration and how it can be modified to be denser is yet to be achieved. Pellenq et al. [2004, 2008] propose that the cohesion of cement paste is mostly due to chemical reaction. Nano modification due to addition of nano clay

[Tregger et al. 2010] supports this theory. In addition Metaxa et al. [2010] showed that by the addition of 0.048% of carbon nano fibers (CNF) more than 25% increase in tensile strength can be obtained. The interaction of CNF with C-S-H gel in this case is mostly mechanical. Hence cohesion in cement paste is due to mechanical and chemical interaction at the nano level. Further research is underway to include both mechanical and chemical interaction in the modeling by using discrete element method (DEM). Chandler et al. [2010] considered mechanical interaction in their DEM modeling using basic C-S-H structure as spherical balls. We are proposing C-S-H as fibers or sticks so that the mechanical and chemical interaction can be considered similar to Anandarajah [1994] and Yao and Anandarajah [2003]. The chemical interaction properties will be taken from MD modeling. This model will help to better understand the behavior of cement paste in the nano and micro level and will provide more insight in modifying the properties in the nano level.

Acknowledgments

The authors acknowledge the continued support provided by Mack Blackwell Transportation Center at the University of Arkansas and Arkansas Highway Transportation Department for performing research in the area of nanoscience application for concrete. The authors also acknowledge the input provided by Mr. Wayne Hodo graduate student from U.S. Army Corps of Engineers, Vicksburg in improving the paper.

References

Allen, M.P., Tildesley, D.J.: Computer Simulation of Liquids. Clarendon Press, Oxford (1987)

Anandarajah, A.: Discrete-element method for simulating behavior of cohesion soil. Journal of Geotechnical Engineering 120, 1593–1613 (1994)

Ayuela, A., Dolado, J.S., Campillo, I., de Miguel, Y.R., Erkizia, E., Sánchez-Portal, E., et al.: Silicate chain formation in the nanostructure of cement-based materials. J. Chem. Phys. 127, 164710-1–164710-8 (2007)

Bell, I.S., Coveney, P.V.: Molecular Modelling of the mechanism of action of borate retarders on hydrating cements at high temperatures. Molecular Simulation 20, 331–356 (1998)

Bensted, J., Barnes, P.: Structure and performance of cements. Spon Press, New York (2002)

Chandler, M.Q., Peters, J.F., Pelessone: Modeling nanoindentation of calcium silicate hydrate. Transportation Research Record 2142, 67–74 (2010)

Constantinides, G., Ulm, F.J.: The nanogranular behavior of CSH. Journal of Mechanics and Physics of Solids 55, 64–90 (2007)

Coveney, P.V., Humphries, W.: Molecular Modelling of the mechanism of action of phosphonate retarders on hydrating cements. J. Chem. Soc. Faraday Trans. 92, 831–841 (1996)

Cramer, C.J.: Essentials of computational chemistry. John Wiley & Sons, Chichester (2002)

Dolado, J.S., Griebel, M., Hamaekers, J.: A Molecular Dynamic Study of Cementitious Calcium Silicate Hydrate (C-S-H) Gel. Journal of the American Ceramic Society 90, 3938–3942 (2007)

Faucon, P., Delaye, J.M., Virlet, J.: Molecular Dynamics Simulation of the Structure of Calcium Hydrates. Journal of Solid State Chemistry 127, 92–97 (1996)

Faucon, P., Le Bescop, P., Adenot, F., Bonville, P., Jacquinot, J.F., Pineau, F., Felix, B.: Leaching of cement: Study of the surface layer. Cement and Concrete Research 26(11), 1707–1715 (1996)

Faucon, P., Delaye, J.M., Virlet, J., Jacquinot, J.F., Adenot, F.: Study of the structural properties of the C-S-H(I) By molecular dynamics simulation. Cement and Concrete Research 27(10), 1581–1590 (1997)

Faucon, P., Jacquinot, J.F.: Molecular Dynamics simulation of Al3+ and Na+ substitutions in the Tobermorite structure. Philosophical Magazine B 75(5), 769–783 (1997)

Faucon, P., Richet, C., Lefebvre, C., Adenot, F., Jacquinot, J.F., Petit, J.C.: Retention capacity of the Calcium silicates hydrates (C-S-H) for Caesium. Key Engineering Materials 132-136, 2204–2207 (1997)

Faucon, P., Charpentier, T., Henocq, J.C., Petit, J.C., Virlet, J., Adrenot, F.: Interaction of Alkalis (Cs+) with Calcium Silicate Hydrates. In: 21st International Symposium on the Scientific Basis for Nuclear Waste Management, MRS 1997, Davos Suisse, vol. 506, pp. 551–559 (1997)

Faucon, P., Gerard, B., Jacquinot, J.F., Marchand, J.: Water attack of a cement paste: towards an improved accelerated test? Advances in Cement Research 10(2), 67–73 (1998)

Faucon, P., et al.: Interaction between Salts (NaCl, CsCl) and Calcium Silicate Hydrates (C-S-H). J. Phys. Chem. B 103, 5212–5219 (1999)

Faucon, P., Petit, J.C.: Silicon (Si) Substitution for Aluminium in Calcium Silicate Hydrates. J. Am. Ceram. Soc. 82(5), 1307–1312 (1999)

Gale, J.D., Rohl, A.L.: The General Utility Lattice Program. Mol. Simul. 29, 291 (2003)

Ghoniem, N.M., Cho, K.: The emerging role of multiscale modeling in nano and micromechanics of materials. Journal of Computer Methods in Engineering and Science 3, 147–173 (2002)

Gmira, A., Zabat, M., Pellenq, R.J.-M., Van Damme, H.: Microscopic physical basis of the prochemical behavior of cement-based materials. Materials and structures/Concrete Science and Engineering 37, 3–14 (2004)

Gmira, A., Minet, J., Francischini, A., Lequeux, N., Pellenq, J.M., VanDamme, H.: Molecular engineering of the cohesion in neat and hybrid cement hydrates. In: Proceedings of ACI Session on, Nanotechnology of Concrete: Recent Developments and Future Perspectives, Denver, USA, November 7, vol. 254, pp. 29–40 (2006); also available as: American Concrete Institute SP254

Hamid, S.A.: The crystal structure of the 11A° Natural Tobermorite Ca2.25[Si3O7.5 (OH) 1.5].H2O. Zeitschrift fur Kristallographie 154, 189–198 (1981)

Janakiram Subramani, V., Murray, S., Selvam, R.P., Hall, K.: Atomic structure of calcium silicate hydrate (C-S-H) using molecular mechanics. In: Proceedings TRB 88th Annual Meeting, January 11-15, paper no: 09-0200 (2009)

Janakiram Subramani, V.: Potential Applications of Nanotechnology for improved performance of Cement based materials. MSCE Thesis, BELL 4190, Department of Civil Engineering, University of Arkansas (2008)

Kalinichev, A., Kirkpatrick, R.J., Cygan, R.T.: Molecular modeling of the structure and dynamics of the interlayer and surface species of mixed-metal layered hydroxides; Chloride and water in hydrocalumite (Friedel's salt). American Mineralogist 85, 1046–1052 (2000)

Kalinichev, A., Kirkpatrick, R.J.: Molecular Dynamics Modeling of Chloride Binding to the surfaces of Calcium Hydroxide, Hydrated Calcium Aluminate and Calcium Silicate Phases. Chem. Mater. 14, 3539–3549 (2002)

Kalinichev, A., Wang, J., Kirkpatrick, R.J.: Molecular Dynamics Modeling of the Structure, Dynamics and Energetics of Mineral – Water Interfaces: Application to Cement Materials. Cement and Concrete Research 37, 337–347 (2007)

Kaxiras, E.: Atomic and electronic structure of solids. Cambridge University Press, New York (2003)

Kirkpatrick, R.J., Yarger, J.L., McMillan, P.F., Ping Yu, P., Xiandong Cong, X.: Raman spectroscopy of C-S-H, tobermorite and jennite. Advanced Cement Based Materials 5, 93–99 (1997)

Leach, A.R.: Molecular modeling: Principles and applications, 2nd edn. Pearson Education, Harlow (2001)

Li, S., Liu, W.K.: Meshfree and particle methods and their applications. Applied Mechanics Review 55, 1–34 (2002)

Manzano, H., Dolado, J.S., Guerrero, A., Ayuela, A.: Mechanical properties of crystalline calcium-silicate-hydrates:comparison with cementitious C-S-H Gels. Phys. Stat. Sol. (a) 204 (6), 1775–1780 (2007a)

Manzano, H., Ayuela, A., Dolado, J.S.: On the formation of Cementitious C-S-H nano particles. J. Computer-Aided Mater. Des. 14, 45–51 (2007b)

Manzano, H., Dolado, J.S., Griebel, M., Hamaekers, J.: A molecular dynamics study of the aluminosilicate chains structure in Al-rich calcium silicate hydrated (C-S-H) Gels. Phys. Stat. Sol. (a) 205(6), 1324–1329 (2008)

Manzano, H., Dolado, J.S., Ayuela, A.: Elastic properties of the main species present in Portland cement pastes. Acta Materialia 57, 1666–1674 (2009)

Martin, R.M.: Electronic structure: Basic theory and practical methods. Cambridge University Press, New York (2004)

Merlino, S., Bonaccorsi, E., Armbruster, T.: Tobermorites: Their real structure and order-disorder (OD) character. American Minerologist 84, 1613–1621 (1999)

Metaxa, Z.S., Konsta-Gdoutos, M.S., Shah, S.P.: Carbon nanofiber-reinforced cement-based materials. Transportation Research Record 2142, 114–118 (2010)

Mori, T., Tanaka, K.: Average stress in matrix and average elastic energy of materials with misfitting inclusions. Acta Metallurgica 21, 1605–1609 (1973)

Murray, S.J.: Determination of the strength and stiffness of C-S-H using Molecular Dynamics. MSCE Thesis, BELL 4190, Department of Civil Engineering, University of Arkansas (2009)

Murray, S.J., Jankiram Subramani, V., Selvam, R.P., Hall, K.D.: Molecular dynamics to understand the mechanical behavior of cement paste. Transportation Research Record 2142, 75–82 (2010)

Pellenq, R.J.-M., Van Damme, H.: Why does concrete set?: The nature of cohesion forces in hardened cement-based materials. MRS Bulletin 29, 319–323 (2004)

Pellenq, R.J.M., Lequeux, N., VanDamme, H.: Engineering the bonding scheme in C–S–H: The iono-covalent framework. Cement and Concrete Research 38, 159–174 (2008)

Pellenq, R.J.-M., et al.: A realistic molecular model of cement hydrates. Proceedings of the National Academy of Sciences of the USA 106(38), 16102–16107 (2009)

Plassard, C., Lesniewska, E., Pochard, I., Nonat, A.: Investigation of the surface structure and elastic properties of calcium silicate hydrates at the nanoscale. Ultramicroscopy 100, 331–338 (2004)

Rafi-Tabar, H.: Computational modelling of thermo-mechanical and transport properties of carbon nanotubes. Physics Reports 390, 235–452 (2004)

Rapaport, D.C.: The art of molecular dynamics simulation, 2nd edn. Cambridge University Press, New York (2004)

Sanchez, F., Zhang, L.: Molecular dynamics modeling of the interface between surface functionalized graphitic structures and calcium-silicate-hydrate: Interaction energies, structure, and dynamics. Journal of Colloid and Interface Science 323, 349–358 (2008)

Selvam, R.P., Murray, S.J., Jankiram Subramani, V., Hall, K.D.: Potential application of nanotechnology on cement based materials, Report: Mack Blackwell Transportation Center, University of Arkansas, MBTC DOT 2095/3004 (2009), http://www.uark.edu/rd_engr/MBTC/MBTC_2095-3004.pdf

Stassost, F., Carter, E.A.: Modeling the full monty: baring the nature of surfaces across time and space. Surface Science 500, 323–346 (2002)

Taylor, H.F.W.: Proposed Structure for Calcium Silicate Hydrate Gel. J. Am. Ceram. Soc. 69, 464–467 (1986)

Taylor, H.F.W.: Cement Chemistry, 2nd edn. Thomas Telford Ltd., London (1997)

Tregger, N., Pakula, M., Shah, S.P.: Influence of micro- and nanoclays on fresh state of concrete. Transportation Research Record 2141, 68–74 (2010)

Yao, M., Anadarajah, A.: Three-dimensional discrete element method of analysis of clays. Journal Engineering Mechanics 129, 585–596 (2003)

Yu, P., Kirkpatrick, R.J., Poe, B., Mcmillan, P.F., Cong, X.: Structure of calcium silicate hydrate (C-S-H): Near, mid, and far-infrared spectroscopy. Journal of the American Ceramic Society 82, 742–748 (1999)

The Effect of SWCNT and Other Nanomaterials on Cement Hydration and Reinforcement

Jon Makar

Abstract. Additions of nanomaterials to cement pastes and concretes may have a significant effect on the performance of ordinary Portland cement (OPC), OPC blends and concrete. Due to the scale of these particles, they may not only modify the bulk behavior of the matrix in a manner similar to micro- and macroscopic additions, but also influence the formation and structure of the C-S-H and other products formed during hydration. Nanoscale additives also have the potential to add new capabilities to concrete. There is therefore growing interest in producing composite materials that include nanomaterials.

Single walled carbon nanotubes (SWCNT) are of particular interest due to their desirable properties as reinforcing materials. They have recently been shown to nucleate the formation of C-S-H during the hydration of ordinary Portland cement (OPC). A series of complementary studies are presented here. The effects of nano-titania, nano-calcium carbonate and nano-alumina dispersed by sonication with OPC on hydration are discussed and compared to the effect of SWCNT dispersed by the same method. The impact of dispersing the SWCNT in blends of admixture and mix water on the hydration of both tricalcium silicate and OPC are investigated. Fracture surfaces of OPC samples blended with all four nano-materials and hydrated for 7 days are analyzed. Classical reinforcing behaviour was identified in the SWCNT/OPC composite alone. A new, nanoscale reinforcing mechanism based on SWCNT bundle pull out is described. The experimental results are used to draw conclusions about the nucleation mechanism in SWCNT composites and on the impact of SWCNT dispersion method on the performance of those composites.

1 Introduction

The addition of nanomaterials, such as nanotubes, nanoparticles and ultrahigh surface area particles to cement and concrete is one of the most promising areas of research for the application of nanotechnology to construction materials. In

Jon Makar
Institute for Research in Construction, National Research Council Canada
1200 Montreal Road, Ottawa, Ontario, Canada
e-mail: jon.makar@nrc-cnrc.gc.ca

K. Gopalakrishnan et al. (Eds.): Nanotechnology in Civil Infrastructure, pp. 103–130.
springerlink.com

addition to being a bulk material formed by chemical reactions [Taylor 1997], calcium silicate hydrate (C-S-H), the main component in hydrated ordinary Portland cement (OPC) and other cements is known to have a fine structure on the order of a few nanometers [Taylor 1997]. As a result, the addition of nanomaterials to hydrating OPC and other cements has the potential to affect both the physical structure of the C-S-H and the hydration reactions themselves.

While studies of nanomaterials mixed with hydrating OPC may provide fundamental insight into the behaviour of cement and concrete, interest in these blends has been driven by the possibility of producing novel and/or superior properties in concrete that may be directly applicable to the construction industry. As a result, considerable attention has been paid both to adding carbon nanotubes and to adding a variety of nanoparticulates to OPC.

Carbon nanotubes (CNTs) are particularly attractive for use in cementitious systems because they appear to be close to ideal reinforcing materials. Ultrahigh aspect ratios [Zheng et al. 2004], extremely high yield strengths [Yu et al. 2000] and moduli of elasticity [Salvetat et al. 1999], and elastic behaviour [Walters et al. 1999] all point to the potential of CNTs in reinforcing applications. In addition, the nanometric diameters of CNT means that if they are well dispersed in a matrix, cracks will encounter one or more reinforcements soon after formation, inhibiting growth at the earliest stage possible. Most work to date has been done on multi-walled carbon nanotubes (MWCNT), which are less expensive and more readily available than single walled carbon nanotubes (SWCNT). There are a number of reports describing both mechanical [Campillo et al. 2004, Ibarra et al. 2006, Li, et al. 2005, Xiang et al. 2005, Cwirzen et al. 2008, Konsta-Gdoutos et al. 2010] and electrical [Li et al. 2007, Wansom et al. 2006] properties of these composites. A recent summary of carbon nanotube/cement composite research [Makar 2009] and an overview of the topic [Raki et al. 2010] can both be found elsewhere.

Less work has been done on OPC/SWCNT composites. SWCNT, however, have higher aspect ratios than MWCNT and individual tube diameters that are close to estimates of the structural spacing of C-S-H layers in hydrated OPC [Taylor 1997b]. They are therefore potentially more desirable as reinforcing materials than are MWCNT. The work presented here is part of a longer term study on the behaviour of cementitous materials when hydrated in the presence of SWCNT.

Early studies [Makar et al. 2005] used Vicker's microhardness testing to show that the mechanical performance of SWCNT/OPC could be as high as 600% of OPC alone. The improvement in performance was seen to be highest during the first few days of hydration, which suggested that the effect may have been as much due to the SWCNT accelerating the hydration process as due to reinforcing behaviour. A subsequent detailed study [Makar and Chan 2009] that used SWCNT distributed on OPC grains by sonication showed that SWCNT do indeed act to nucleate cement hydration reactions. The work presented here expands on that study. The results of an alternative approach to dispersing SWCNT in cementitious systems using sonication in admixture and mix water are presented. The results of the previous hydration study are also compared to the hydration behaviour of OPC sonicated with 3 different nano-materials (nanoparticulate calcium carbonate and ultra-high surface area alumina and titania). Microscopic evidence for classical reinforcing behaviour is described and the implications of reinforcing by bundles of SWCNT, as opposed to individual nanotubes, are discussed. The results are used to provide further insight into the nucleation mechanisms behind the observed effects of SWCNT on OPC hydration. Finally, the implications of the two different methods of dispersing SWCNT on the ability of the SWCNT to reinforce hydrated cement matrices are discussed.

1.1 Effects of SWCNT on the Hydration of Sonicated OPC

An extensive study of the hydration of OPC sonicated with SWCNT in isopropanol has been published elsewhere [Makar and Chan 2009]. In summary, sonication of OPC alone reduced the maximum heat flow and retarded its timing compared to the as-delivered OPC control. Secondary C_3A and C_4AF reactions [Pratt and Ghose 1983, Makar and Chan 2008] were suppressed or spread out in time compared to the control. The SWCNT bundles accelerated the hydration reaction compared to both the control as-delivered OPC and OPC sonicated alone. They also produced a higher maximum heat flow. The heat generated by the secondary C_3A and C_4AF reactions was similar to that in the as-delivered OPC, not the OPC sonicated alone.

This behaviour was attributed to the SWCNT appearing to act as nucleating sites for the C_3S hydration reactions, with the C-S-H forming directly on the SWCNT. As a result, the C-S-H formed a network on the surface of the OPC grains before growing away from the surface. Evidence of fiber pullout was observed in the sample material by 24 hours of hydration. No evidence was observed for C_3A hydration products forming on the SWCNT bundles, but the ettringite formed in the initial hydration processes did have a reduced aspect ratio compared to the control samples, suggesting that the presence of the SWCNT inhibited their growth.

Two different mechanisms were proposed to explain the nucleating effect of the SWCNT. First, it is possible that electrostatic behaviour drove the nucleation process. Both the attractive force of image charges due to the local presence of ions and trapping of water molecules by polarization effects may have occurred. Secondly, it is also possible that metallic ions (i.e. Ca^{2+}) may have been adsorbed on the SWCNT. In either case, the trapping of some of the constitutents necessary for C-S-H formation against the SWCNT resulted in the observed nucleation effects.

2 Experiment

2.1 Materials

Samples were prepared from GU type (US type 10) ordinary Portland cement (OPC) produced by Lafarge Canada or from monoclinic tricalcium silicate ($3CaO \cdot SiO_2$ or C_3S) supplied by CTL Inc (Skokie, Ill). Details of the OPC composition have been given elsewhere [Makar and Chan 2008]. Single walled carbon nanotubes from two sources were used. Work on evidence for classical reinforcing behaviour was primarily conducted using SWCNT produced by Carbon Nanotechnologies (USA, now Unidym), with a diameter of ~1.0 nm and typical lengths of 0.1 to 1 μm. These SWCNT were produced using the HiPCo method and acid purified [Chiang et al. 2001]. Samples used to investigate reinforcing behaviour had at 2% by mass SWCNT content added to the OPC. Work on hydration behaviour and confirmation work on classical reinforcing behaviour was conducted using SWCNT produced by the National Research Council Canada using laser ablation [Kingston et al. 2004] and purified using a solvent based method [Kingston, et al. In preparation]. Alumina and titania nanomaterials were supplied by Nanoscale Materials Inc (Manhattan, KS). The alumina was amorphous and had a surface area of 550 m^2/g, while the titania had a surface area of greater than 400 m^2/g and a crystallite size of less than 20 nm. Reagent grade $CaCO_3$ nanoparticles were supplied by READE (Riverside, RI). They had typical diameters on the order of 50-100 nm and a surface area of 20.5 m^2/g. The nanomaterial content was 2% of the mass of the OPC in each blend investigated.

2.2 Dispersion of SWCNT for Use in Cementitious Composites

Dispersion of nanomaterials is one of the key issues facing the development of all nanocomposites. Nanomaterials need to be evenly dispersed in the matrix material in order to produce repeatable and consistent effects. Well dispersed nanomaterials may also be expected to produce desirable properties with a lower mass content. Successful dispersion may, however, also be difficult to obtain as the very high surface areas of nanomaterials can cause them to adhere together due to electrostatic effects and van der Waals forces. The latter effect is particularly important in SWCNT due to their highly ordered structure. The van der Waals forces between SWCNT have been estimated to be as high as 11 GPa [Vodenitcharova et al. 2007].

There are additional complications in the dispersion of SWCNT for use in a cementitious matrix. SWCNT are hydrophobic and will not readily disperse in mix water. At the same time, the organic solvents where SWCNT are known to disperse [Ausman et al. 2000] may be undesirable in cementitious systems.

Despite these difficults, at least two methods have been used to disperse SWCNT in cementitious systems. The first, dispersion by sonication, has been discussed in detail elsewhere [Makar and Chan 2009]. The second method, dispersion in cement admixtures, will be addressed below. The results presented later in this chapter are, however, drawn from samples prepared with both methods. The first method is more useful for investigating interactions between the cementitious material and the SWCNT, while the second is likely more practicable for producing larger samples and production materials.

Cement admixtures are particularly desirable as dispersing agents since their behaviour in cementitious systems is already well understood. In addition, once the SWCNT are dispersed in the admixture, the mixture can, in principle, be added to the mix water and cement as would be the case for the admixture alone.

The literature reports that polycarboxylate superplasticizers have been used successfully to disperse multiwalled carbon nanotubes (MWCNT) [Campillo et al. 2004, Ibarra et al. 2006, Li, et al. 2005, Xiang et al. 2005] in cementitious systems. Tests were therefore conducted on dispersing SWCNT using a variety of different cement admixtures (Tab. 1). In all cases sonication was employed to break up bundles of the purified SWCNT and the maximum allowable concentration of the admixture was used.

Table 1 Admixtures investigated as means of dispersing SWCNT

Admixture	Manufacturer	Composition	Application
Air Extra	Euclid Chemicals, Cleveland, OH	Sulfonated fatty acids	Air Entrainment
Eucon 37	Euclid Chemicals, Cleveland, OH	Naphthalene sulfonate	Superplasticizer
Eucoplast 911	Euclid Chemicals, Cleveland, OH	Melamine	Superplasticizer
Glennium 3400	BASF, Cleveland, OH	Polycarboxylate	High-Range Water-Reducing Admixture
Disal	Handy Chemicals, Candiac, QC	Poly-naphthalene sulfonate sodium salt	Superplasticizer

Despite the success of the polycarboxylate superplasticizer in dispersing MWCNT, it was not found to be effective in preventing the SWCNT from clumping together again in bundles once sonication had ceased. The air entraining agent and the two Euclid supplied superplasticizers were also unsuccessful. The SWCNT were, however, found to be readily dispersable [Makar and Chan 2009, Makar et al 2005] in Disal at a concentration of 8 g/L. SWCNT sonicated in Disal and water at that concentration were found to be evenly and stably dispersed, forming a black liquid that showed increasing viscosity with increasing SWCNT content. A maximum SWCNT content of 3% of the mass of the mix water was found to be possible before the dispersion was too stiff to be readily mixed with cement. The dispersion was found to be quite stable, remaining fluid and well dispersed for over a year from the date of production.

2.3 Procedure

Two methods of dispersing nanomaterials were used. In the first, single walled carbon nanotubes and nanomaterials were dispersed on the OPC by following a procedure originally developed for use in dispersing carbon nanotubes in alumina composites [Zhan et al. 2002]. Mixtures of unhydrated OPC and each of the various nanoscale materials were placed in small glass jars. Each sample was then sonicated using an high intensity ultrasonic processor (model VCX600, Sonics and Materials, Newtown Connecticut) in isoproponal for 4 hours to evenly disperse the nanoscale material. Sonication took place in an ice bath to avoid overheating the sample. The jars were sealed around the sonicator probe during the sonication process to avoid release of the nanomaterials. All work was done in a fume hood to provide additional environmental protection. Past work has shown that the sonication in isopropanol procedure used here disperses the SWCNT in bundles of a few SWCNT [Makar and Chan 2009].

Sonicated samples were vacuum dried in a specially designed glassware system in order to prevent carbonation of the samples and nanoparticulate release. Samples were then checked by scanning electron microscopy to ensure that the nanomaterials had been adequately dispersed on the surface of the OPC grains. Between 2 and 5 grams of the resulting materials were then hydrated for 7 days at 0.5 w/c ratio for seven days using ultrahigh purity (>1 ppm contaminant) water (Anachemia, Inc.). Hydration characteristics were measured using isothermal conduction calorimetry (TA Instruments, New Castle, Delaware).

Hydration of the samples was stopped by washing in isopropanol. Samples were then vacuum dried and stored under vacuum until needed. They were broken up in a mortar and pestle by hand and the fracture surfaces imaged using high resolution scanning electron microscopy (Hitachi S-4800). Typical imaging conditions were a 1.2 keV accelerating voltage, a 2.5-3 mm working distance and 7 μA emission current. Imaging was done under high vacuum conditions in order to obtain the maximum possible resolution. A discussion of the advantages and disadvantages of using this approach to imaging hydrated OPC samples is given elsewhere [Makar and Chan 2008]. Samples were typically imaged at 40-50,000x magnification, but magnifications as high as 180,000x were used as appropriate. Although the theoretical limit of the resolution of the instrument is ~1 nm, experience has shown that the practical resolution of the instrument in cement samples is on the order of ~ 3 nm. Individual SWCNT were therefore unlikely to be seen, but it was possible to image bundles of SWCNT with widths of 3 or more SWCNT. The higher diameter nanoparticles used in the research were easily resolved.

In the second method, SWCNT were sonicated in a solution of 8 g/L Disal in ultrahigh purity water, a commercial poly-naphthalene sulfonate sodium salt superplasticizer. Concentrations of SWCNT ranged from 0.5% by mass of water to 3%. The resulting dispersion was then mixed with both OPC and commercially produced tricalcium silicate (C_3S) at 0.5 w/c, resulting in SWCNT/cementitious material ratios of 0.25% to 1.5% by mass. Samples were again hydrated for 7 days in an isothermal conduction calorimeter.

3 Effects of SWCNT and Other Nanomaterials on the Hydration of C_3S and OPC

3.1 Comparisons between the Hydration Effects of SWCNT and Other Nanomaterials in Sonicated Systems

One route to improved understanding of the nucleating effects of SWCNT is by comparison to the effects produced by other nanomaterials. Isothermal conduction calorimetry measurements on OPC sonicated with three different nanomaterials are shown in Fig. 1. The OPC was from the same source as the previous work with SWCNT [Makar and Chan 2009]. The three nanomaterials were chosen both for high surface areas and because they were not expected to form reaction products with the hydrating OPC. There were, however, several significant differences between the nanoparticles and SWCNT. All of the comparison nanomaterials were non-conducting, had lower surface areas than the SWCNT and had much lower aspect ratios. The nano-alumina and nano-titania were also in the form of agglomerated particles with total sizes on the order of 5 μm.

A detailed discussion of the mechanisms behind the results in Fig. 1 will be presented elsewhere [Makar and Chan, in preparation]. In comparison to Figure 1 in Makar and Chan [2009], however, examination of Fig. 1 above shows that the extent of the differences between effect of the SWCNT on the hydration process and that observed for other nano-materials. As would be expected from previous work [Sato and Beaudoin 2006, Sato and Beaudoin, in submission], the nano-calcium carbonate additions produce an acceleration of the hydration process. The development of the heat flow was otherwise similar to that produced by the hydration of the OPC sonicated alone. The nano-alumina produced more complex behaviour, with a higher maximum heat flow than any of the other sonicated samples. In addition, the early heat generated by the hydration of the OPC sonicated with the nano-alumina suggests that the addition of that nanoparticle nucleates C_3A hydration reactions. Finally, the sonication of nano-titania with the OPC both retards the hydration reactions and reduces the maximum heat flow generated during hydration.

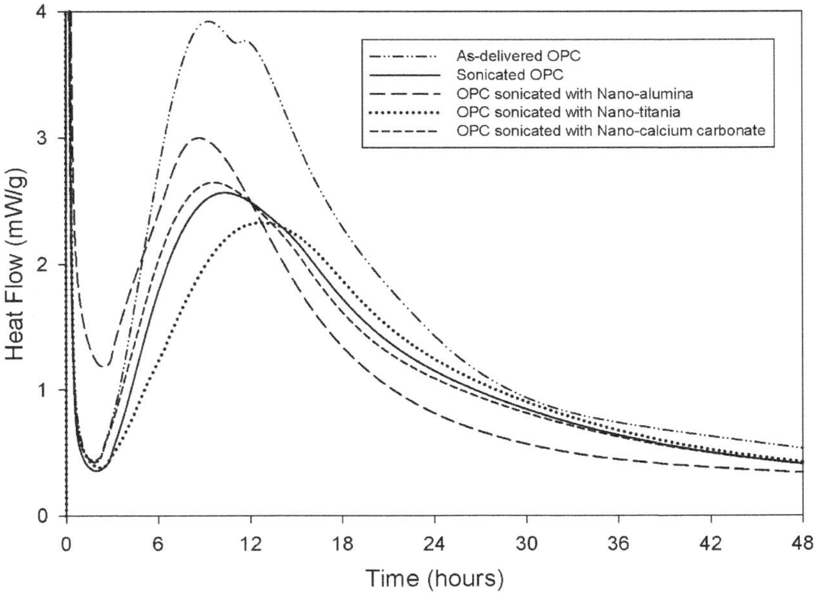

Fig. 1 Heat flow generated by OPC sonicated for 2 hours with and without 2% nanoparticle additions

The effects on OPC hydration produced by the presence of the nanomaterials differed from those reported previously for SWCNT [Makar and Chan 2009] both in terms of magnitude and quality. None of the nanomaterials accelerated the hydration reactions compared to the as-delivered control OPC, nor did they produce higher heat flows. Both characteristics were observed for the OPC sonicated with SWCNT [Makar and Chan 2009]. In addition, the pattern of the heat flow generation by the samples sonicated with nanomaterials was, in general, similar to that of the OPC sonicated alone, with the exception being the early heat flow of the OPC sonicated with nano-alumina. The OPC sonicated with SWCNT had the same general behaviour as the control, unsonicated OPC.

3.2 SWCNT Nucleation Mechanisms

The heat flow measurements in Fig. 1 tend to argue against adsorption as the dominant mechanism behind the observed nucleation effects produced by SWCNT during OPC hydration processes. All of the nanomaterials in Fig. 1 were non-conductive, eliminating electrostatic effects as a possible nucleation mechanism in their case. Adsorption, however, would remain a possible factor in the hydration of the OPC in the presence of the nanomaterials. In fact, it is possible that the retarding effect of the nano-titania was due to adsorption of the ions in the solution on to the surface of the titania, preventing them from taking part in the hydration reactions. While one would expect different degrees of effect from the different nanomaterials if adsorption is the predominant nucleating mechanism, one would still expect to see the same general hydration behaviour in each case. The differences between the hydration behaviour of the samples sonicated with SWCNT and those sonicated with the nanomaterials instead suggests that nucleation effects in the former be attributed to the electrostatic mechanisms that rely on the metallic properties of the SWCNT.

A second line of evidence against the adsorption of metallic ions as a nucleation mechanism is the relative disparity between the nucleation of C-S-H and $Ca(OH)_2$ on the SWCNT bundles. Only one example of $Ca(OH)_2$ nucleating on SWCNT was observed (Fig. 2), with all other episodes of crack bridging occuring in C-S-H. In addition, examination of SWCNT bundles on OPC surfaces during the early stages of hydration did not show evidence of $Ca(OH)_2$ forming around the SWCNT. Although $Ca(OH)_2$ appeared to be less common on the surfaces of OPC during the early stages of hydration than is C-S-H [Makar and Chan 2008, Makar et al. 2007], it was readily identified when it did form. Investigation of the effect of SWCNT on the formation of $Ca(OH)_2$, however, has shown that the same amount of Ca(OH)2 was present in samples with and without SWCNT when the same amount of hydration had occurred [Makar and Chan 2009]. The three observations together suggest the SWCNT may be preferentially nucleating C-S-H, with the $Ca(OH)_2$ forming elsewhere in the system.

It is this relative imbalance between formation of C-S-H around SWCNT bundles and $Ca(OH)_2$ that supports the prevalence of electrostatic effects as a nucleation mechanism. Electrostatic effects would be expected to cause both Ca^{2+} and

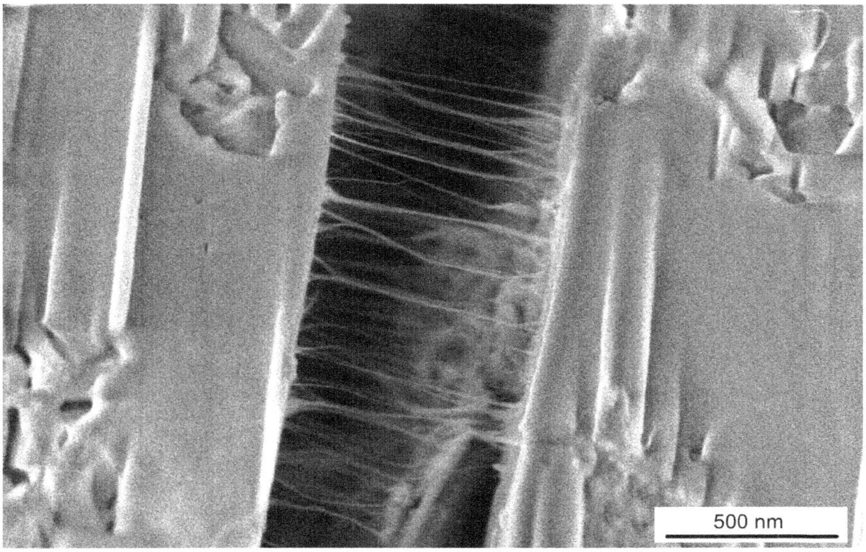

Fig. 2 Crack bridging in $Ca(OH)_2$ crystal (2% SWCNT, 7 days hydration)

Si^{4+} ions to nucleate on the SWCNT surfaces. On the other hand, adsorption of metallic ions would be expected to primarily nucleate Ca^{2+}. In that case, a preferential formation of $Ca(OH)_2$ would be the most likely outcome, not of C-S-H.

3.3 Effect of Distribution of SWCNT in Disal on the Hydration of C_3S

While the effect of SWCNT on the hydration of sonicated OPC was found to be due to electrostatic nucleation effects, other factors came into play when the hydration behaviour of SWCNT/Disal/cement systems were considered. Fig. 3 shows isothermal calorimetry data for C_3S hydrated with and without the presence of Disal and SWCNT. SWCNT content is shown as a percentage of the mass of the C_3S in the mix (i.e. half the percentage of SWCNT by mass of mix water).

As would be expected in a pure C_3S system, the concentration of Disal used in mix water greatly delayed the hydration process. Adding increasing amounts of SWCNT to the mix appears to counteract the retarding effect of the Disal. At 1.5% SWCNT content the hydration process has been accelerated to the point where it is only marginally delayed compared to the control sample without Disal. The apparent acceleration compared to sonicated Disal alone that was produced by the presence of the SWCNT in Fig. 4 was far larger than that seen for samples where the SWCNT were dispersed by sonication [Makar and Chan 2009].

The concentration of SWCNT in the mix to the time of the maximum heat flow of the sample is compared in Fig. 4. The individual data points can be readily fitted to a decaying exponential curve with the empirical equation:

$$t_{max} = 7.73 + 29.7e^{-1.55C} \quad (1)$$

where t_{max} is the time of the maxima in heat flow in hours and C is the concentration of SWCNT as a percentage of the mass of C_3S. The fitted curve has an r^2= 0.994.

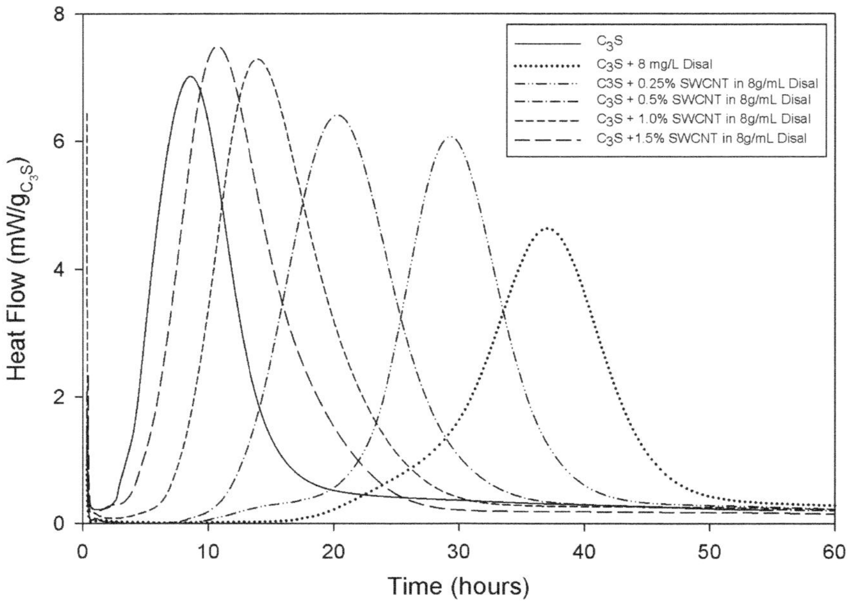

Fig. 3 Effect of SWCNT dispersed in sonicated Disal on C_3S hydration reactions

It is interesting to note that the asymptotic value of the fitted curve is below the time of maximum of the heat flow of the sample without Disal. As shown in Fig. 4, a concentration of ~2.25% by mass SWCNT should be sufficient to eliminate the delay induced by the presence of the Disal. This estimate could not be confirmed experimentally due to the inability to mix C_3S with mix water containing higher concentrations of SWCNT. Fig. 4 also suggests that further additions of SWCNT beyond 2.25% by weight would produce a slight accelerating effect, with a maximum acceleration of ~1 hour at the asymptotic value compared to the C_3S alone. This result compares to the ~2 hour acceleration due to enhanced nucleation effects seen for OPC sonicated with 1% SWCNT [Makar and Chan 2009].

Comparison between the previously published results on OPC sonicated with SWCNT [Makar and Chan 2009] and those here suggests that the major effect on the hydration of the C_3S/sonicated Disal system was not due to SWCNT acting as nucleating sites for the hydration reactions. Instead, the acceleration would appear to have been due to the interaction of the Disal with the SWCNT inhibiting the Disal/C_3S interaction. A separate experiment (not shown) indicates that increasing

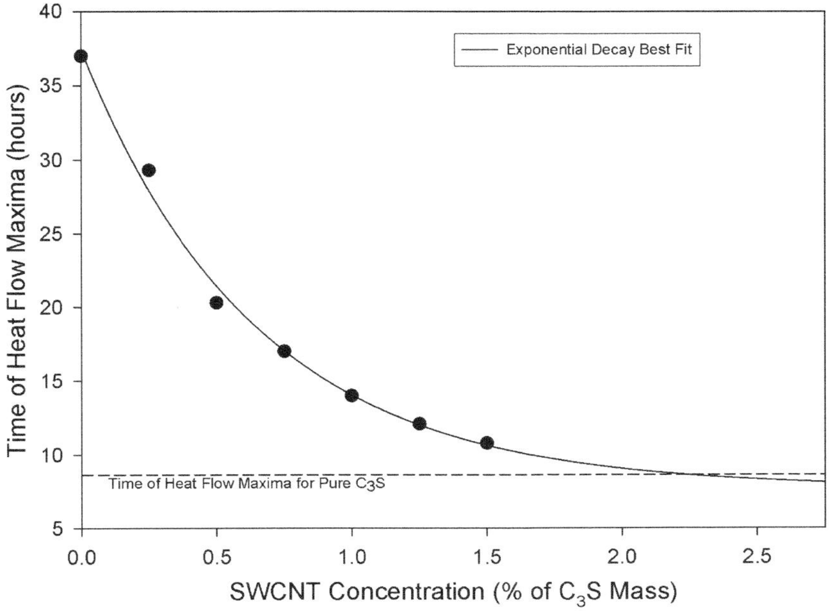

Fig. 4 Change in timing of maxima in heat flow due to the presence of SWCNT dispersed by sonication in 8 g/L Disal.

Disal content in a pure C_3S system results in an exponential growth of the timing of the heat flow maximum. An exponential decay in the timing of the heat flow maximum would therefore be expected if the SWCNT were inhibiting the interaction of the C_3S and Disal.

The loss of interaction with the C_3S in turn suggests that the Disal may be dispersing the SWCNT by wrapping individual nanotubes or bundles. If the Disal molecules were taken up in this form, they would no longer be able to interact with the C_3S. Examples of other molecules wrapping SWCNT can be found in the literature [Narimatsu et al. 2006, Stranos 2006].

3.4 Effect of Distribution of SWCNT in Disal on the Hydration of OPC

Although the effect of SWCNT/Disal combination was pronounced in C_3S systems, Fig. 5 shows that it was significantly reduced in OPC mixes. In addition, the monotonic progression in effect seen in Fig. 4 was not apparent. Instead, SWCNTs between 0.25 and 1.5% by mass all had approximately the same effect on the time of the OPC's major C_3S hydration reaction. The presence of the SWCNT also appeared to accelerate and possibly enhance the secondary C_3A and

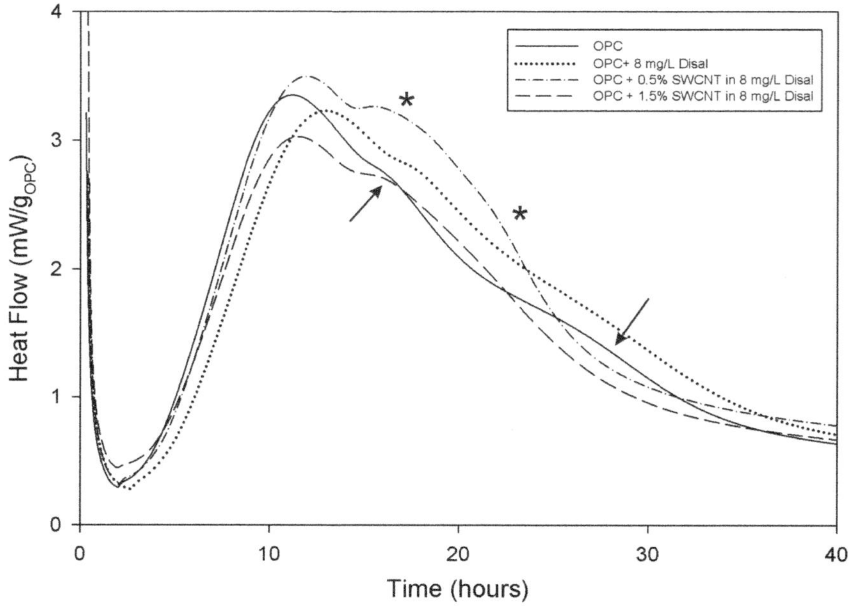

Fig. 5 Effect of SWCNT dispersed in Disal on OPC hydration reactions

C_4AF reactions [Makar and Chan 2008, Pratt and Ghose 1983]. These reactions are marked by arrows in Fig. 5 for the control samples without SWCNT and by stars for the samples with SWCNT. The heat flow reached a maximum value for both the main C_3S reaction and the secondary reactions at 0.5% by mass SWCNT concentration and fell off with increasing concentration.

The acceleration of the secondary reactions in the OPC/SWCNT/Disal system was in direct contrast to the results of the previous work using isopropanol dispersion. There, evidence was provided that the presence of the SWCNT altered the aspect ratio of ettringite formed by very early hydration reactions. No evidence for nucleation of ettringite or other aluminate hydration products was seen and the secondary reaction peak did not appear to be significantly altered from that of the control sample without sonication.

The results of Fig. 5 in comparison to Fig. 3 and the earlier work [Makar and Chan 2009] suggest that although the Disal was taken up by the SWCNT, with a resulting reduction on the effect of the Disal on C_3S hydration, the superplasticizer was still adsorbed on the C_3A surfaces despite the presence of the nanotubes. As a result, the aluminate reactions were enhanced.

The results in Fig. 3 to 5 have a number of implications for the mechanical performance of SWCNT/cement composite materials. As will be discussed in more detail below, the nucleation of C-S-H on SWCNT in sonicated samples appears to produce a high degree of mechanical bonding between the two materials. This bonding would be expected to contribute to the development of classical reinforcing behaviour in the composite materials. If the Disal was wrapping the SWCNT, as suggested by Fig. 3 and 4, nucleation of C-S-H on the SWCNT would be inhibited, potentially reducing the extent of bonding between the two materials. Moreover, if the Disal/SWCNT mixture was preferentially adsorbed on the calcium aluminate surfaces, the results may have been an uneven distribution of SWCNT throughout the matrix. This effect would be expected to be deterimental to the performance of the SWCNT as reinforcement. Not only would areas of minimal reinforcement exist in the matrix, but a sufficiently high concentration of SWCNT in a small number of locations might even act as a crack former in the matrix.

4 Reinforcing Behavior in SWCNT Composites

4.1 Evidence for Classical Reinforcing Behavior

4.1.1 Classical Reinforcing Behaviour

Non-continuous fibers reinforce matrices by absorbing some of the energy associated with crack formation and propagation. Classical microscope evidence for reinforcing behavior by fibers is generally considered to include three different types of behaviour:

- crack bridging (Fig. 6a), where cracks that would otherwise propagate through a matrix are shown to be crossed by fibers;
- fiber pull-out (Fig. 6b) where larger cracks have caused fibers to leave the matrix; and
- crack deflection (Fig. 6c), where a crack that does not cross the fibers is still deflected around them, rather than propagating in an essentially straight line.

Crack bridging and fiber pull out absorb energy as the bond between the fiber and the matrix is broken and the two materials pull away. Crack deflection absorbs energy due to the increased path length of the crack as it goes around the reinforcements. It is also possible for cracks to be deflected to an extent that they no longer propagate perpendicularly to the direction of loading on the composite material, requiring additional energy to continue the failure process.

4.1.2 Reinforcing Behaviour in SWCNT/OPC Composites

All three classical reinforcing mechanisms were observed in SWCNT/OPC composites. Crack bridging was particularly common, being observed in almost all

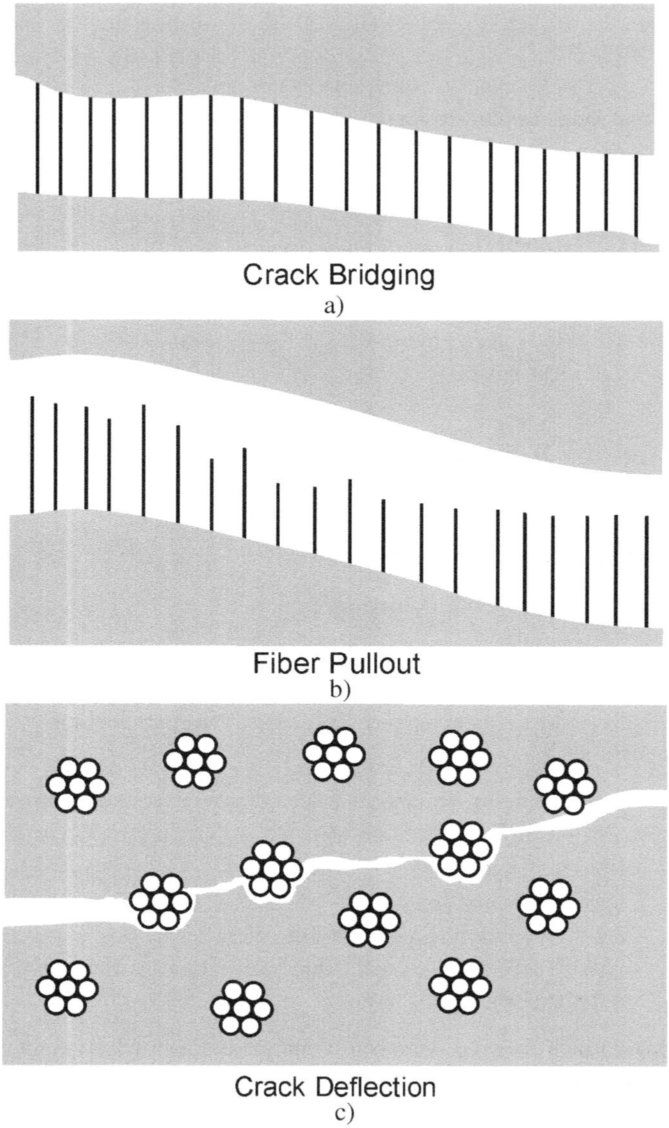

Fig. 6 Schematic Representation of Fiber Reinforcement Mechanisms

samples hydrated for a day or more. A small section of a particularly striking example of crack bridging can be seen in Fig. 7. Here a crack had propagated in a zig-zag pattern across a 1 mm wide particle of cement paste. Typical widths of the crack were less than 1 μm and additional cracks branched away from it where it

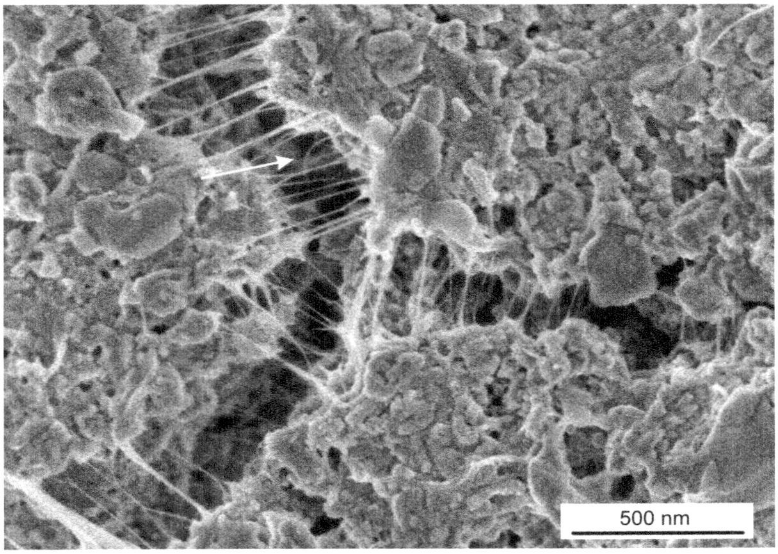

Fig. 7 Crack bridging and fiber pull out (white arrow) in SWCNT/OPC composite (2% SWCNT, 7 days hydration)

changed direction. SWCNT bridged the crack throughout its length and had clearly held the two pieces of the grain together, preventing a complete fracture.

An individual example of fiber pull out is also indicated by the arrow in Fig. 7. Fiber pull out was relatively easy to detect during the early stages of hydration [Makar and Chan 2009], but was much less readily observed in sample hydrated for longer periods such as are shown here. It is possible that the lack of observed SWCNT pull outs was due to the adhesion of the pulled out fibers to the matrix along the fracture surface, making observation by SEM difficult. However, other explanations lie in the distribution of the SWCNT on the surface of the OPC grains and the behaviour of the SWCNT bundles. These points will be discussed further below.

As noted earlier, while almost all incidents of crack bridging were seen in C-S-H, cracking bridging was also seen in a fractured calcium hydroxide crystal (Fig. 2), which was large enough for positive identification by X-ray dispersive spectroscopy in the SEM. This image was particularly significant as it demonstrates that the ability of SWCNT to nucleate hydration reactions was not limited to the formation of C-S-H alone.

The two different examples of crack bridging also show differences in the condition of the SWCNT. In Fig. 7 the nanotube bundles were clearly under tension, with a very linear appearance. In Fig. 2 some of the bundles are also linear, but more appear relaxed, suggesting that the two parts of the $Ca(OH)_2$ crystal had moved together somewhat after the fracture had taken place.

One feature both images have in common is that the ends of many of the bundles branched apart where they were attached to the fracture surfaces. This effect may be due to the way in which the SWCNT bundles were distributed over the surface of the OPC grains. SWCNT bundles distributed by sonication often intersect or overlay each other (Fig. 8). The appearance of the SWCNT in Fig. 7 and Fig. 2 suggests that these contacts are not affected by the hydration process.

Crack deflection (Fig. 9, solid arrows) was also seen in the samples, although again less commonly than crack bridging. It was more obvious in those samples with lower quality of dispersion, where the bundles were thicker and more closely located together. The adhesion of the bundles to the surface of the C-S-H is indicative of a deflection process. As mentioned earlier, SWCNT have elastic moduli of around 1 TPa and, as a result, tend to stand away from the C-S-H once pulled out of the matrix. The dashed arrow in Fig. 9 indicates an example of such a pulled out bundle.

4.1.3 Comparisons to Fracture Surfaces in Control and Other Samples

Comparisons between the OPC-SWCNT composite fracture surfaces shown in section 4.1.2 and those created when other hydrated nanomaterial/OPC blends were fractured help to highlight the importance of the SWCNT in modifying the structure of the C-S-H and in creating reinforcing behaviour. A typical fracture surface for a control sample of OPC sonicated alone and then hydrated for 7 days is shown in Fig. 10, while Fig. 11 to 14 show the effects of addition of the nano-alumina, nano-calcium carbonate and nano-titania.

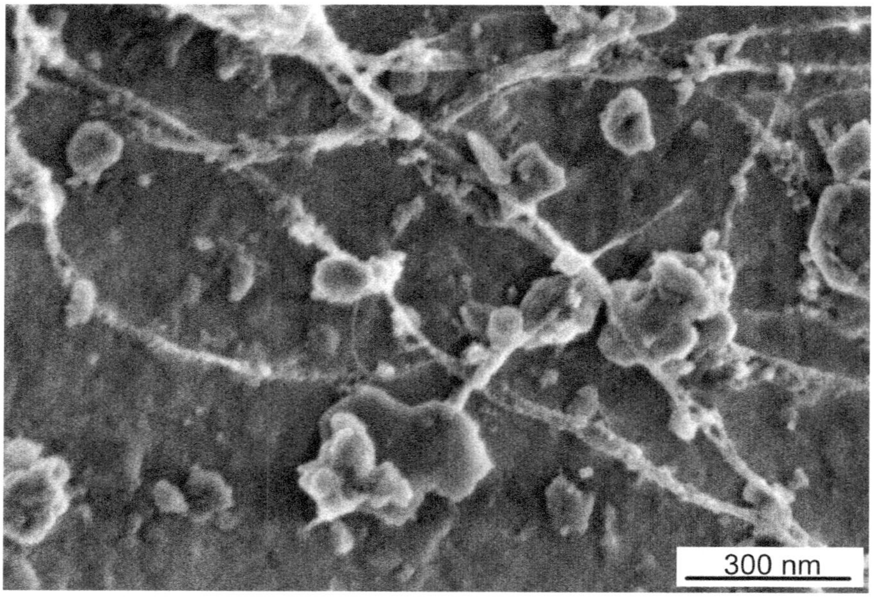

Fig. 8 SWCNT distributed by sonication on OPC surface (1% SWCNT by mass), showing overlaps and intersections of nanotube bundles

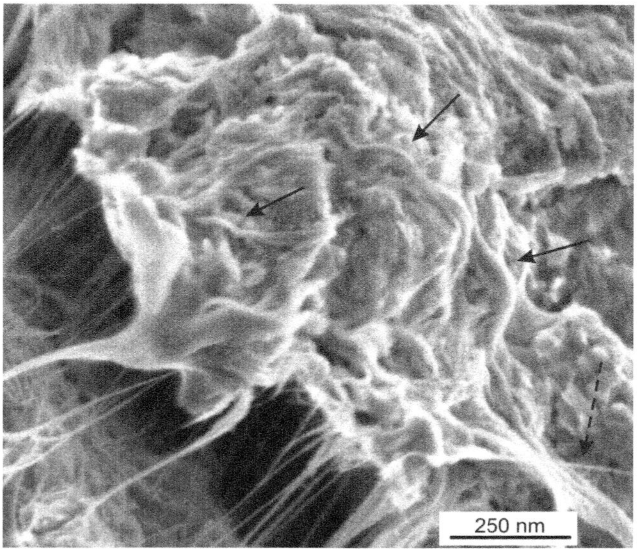

Fig. 9 Crack deflection (solid arrows) and fiber pullout (dashed arrow) in SWCNT/OPC composite

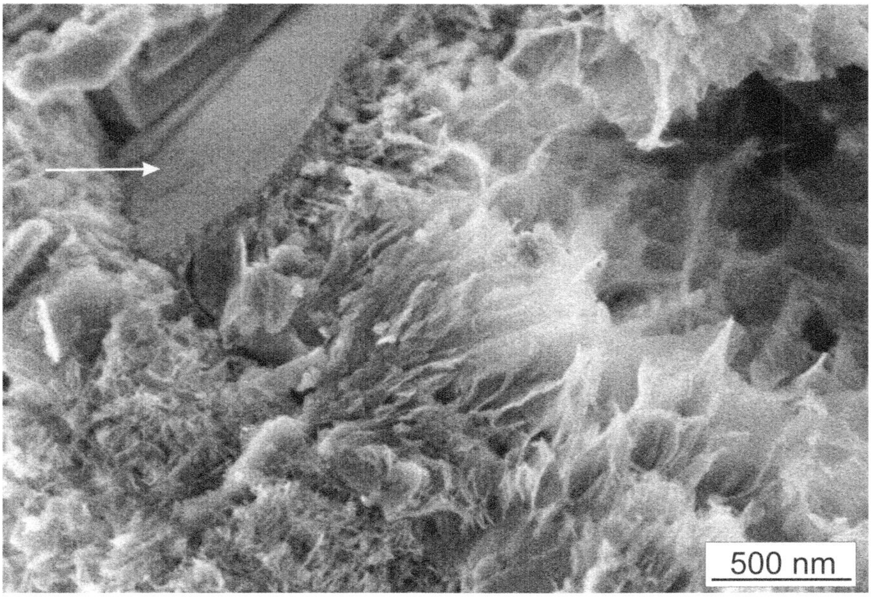

Fig. 10 Typical fracture surface for OPC sonicated in isopropanol (7 days hydration), showing $Ca(OH)_2$ crystal (arrow)

Details about the fracture samples and their analysis will be given elsewhere [Makar and Chan, in preparation]. In general, however, the fracture surfaces at 7 days of hydration of OPC sonicated alone showed a mixture of $Ca(OH)_2$ and C-S-H, with C-S-H predominating. In the case of Fig. 10, a small amount of calcium hydroxide was seen (arrow, identified by morphology), with most of the image taken up by the C-S-H. Other images showed more massive $Ca(OH)_2$ structures. Some images showed ettringite crystals with aspect ratios of 10 or more, which appear to have formed in pores in the structure. The overall structure appeared to be dense at the scale examined.

In contrast, the structure of the fracture surfaces of the OPC sonicated with 2% nano-alumina (Fig. 11) after 7 days of hydration appeared to be less dense, with larger pores visible on the surface. The fracture surfaces themselves were rougher, with a number of images having multiple features similar to that indicated by the arrow in Fig. 11. Both ettringite and Ca(OH)2 were less visible along the fracture surfaces.

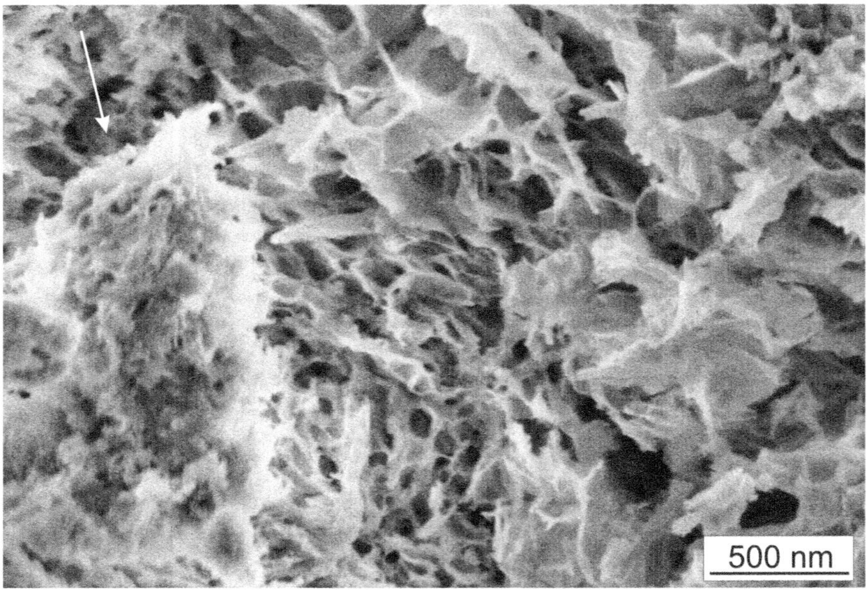

Fig. 11 Typical fracture surface for OPC – 2% nano-alumina composite (7 days hydration), showing typical surface texture (arrow)

The images of the fracture surfaces of the OPC sonicated with 2% nanocalcium carbonate (Fig. 12) had a similar density to that for the OPC sonicated alone. However, the overall surfaces were rougher than the OPC sonicated alone and there was a much higher prevalence of ettringite on the fracture surfaces (82% and 21% of the images of samples with and without nanocalcium carbonate, respectively). None of the images from the nanocalcium carbonate samples show fractures exposing $Ca(OH)_2$.

Fig. 12 Typical fracture surface for OPC – 2% nano calcium carbonate composite (7 days hydration) showing ettringite (rods) and C-S-H

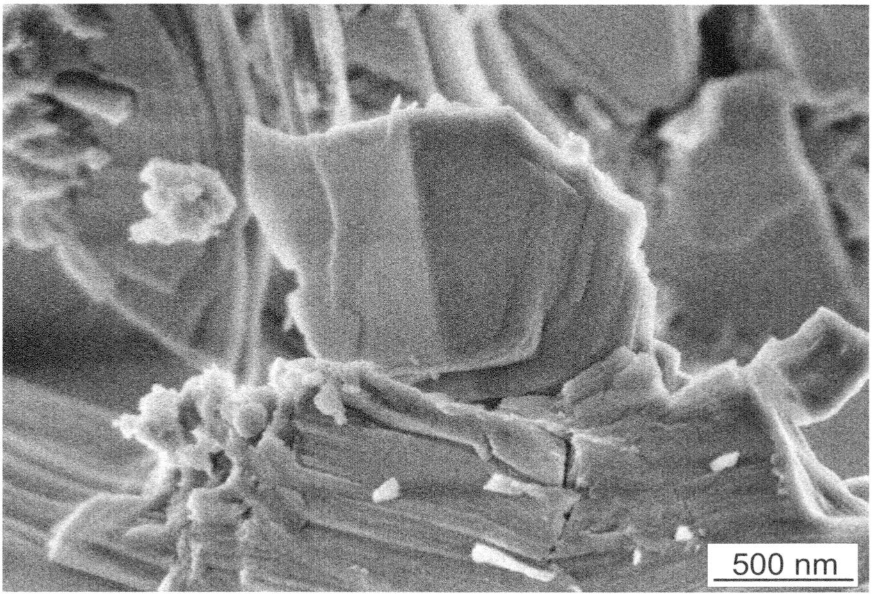

Fig. 13 OPC – 2% nano-titania composite fracture surface with massive $Ca(OH)_2$ (7 days hydration)

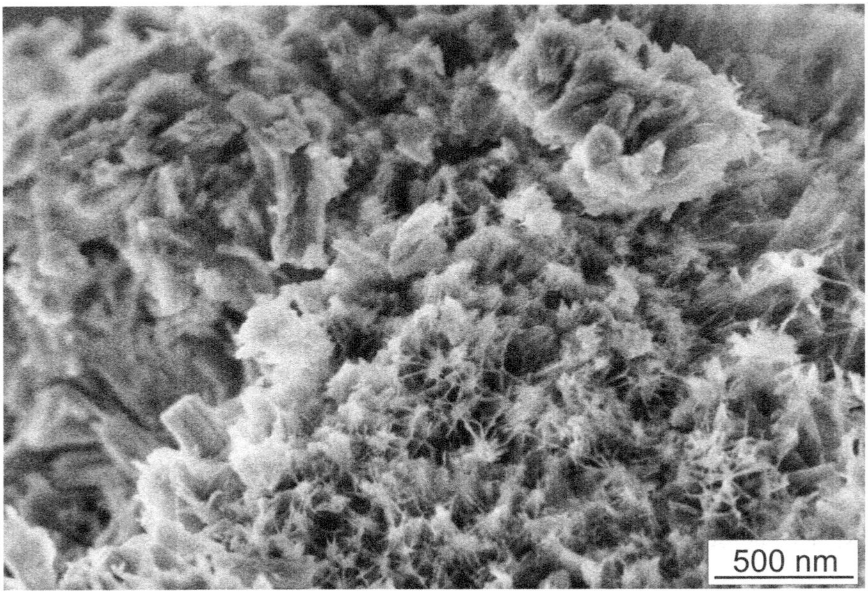

Fig. 14 OPC – 2% nano-titania composite fracture surface with C-S-H (7 days hydration)

Finally, the images of the fracture surfaces from the 7 day hydration samples of OPC sonicated with 2% nano-titania showed either large quantities of $Ca(OH)_2$ (Fig. 13) or relatively dense C-S-H (Fig. 14). The latter surfaces were not dissimilar to those seen for hydrated OPC that was sonicated alone. The prevalence of massive $Ca(OH)_2$ was, however, higher than in any of the other samples.

The observed prevalence of the hydration products discussed above should not be taken to represent the actual distribution of C-S-H, $Ca(OH)_2$ or ettringite in the samples. Instead, it indicated only what is present along the line of fracture of the samples. With that condition in mind, an analysis of the images suggests that the individual grains of cement in the samples hydrated in the presence of the nano calcium carbonate and the nano-alumina were less well bonded together than the control sample of OPC hydrated alone and the sample hydrated in the presence of nano-titania. In addition to the differences in structure and surface texture described above, fracture of the latter two samples produced larger particles, indicating the presence of a tougher matrix.

Despite the differences between the different fracture surfaces described above, all of the four samples had greater similarities to each other than they had to the surfaces of the OPC sonicated with 2% SWCNT samples. As shown in Fig. 7 and 9, the presence of the SWCNT creates a dense matrix with a smoother fracture surface than the other samples. The broken pieces of the matrix were typically much larger than those of the other samples, being in some cases millimeters, rather than micrometers across. Most significantly, no evidence was seen in the other samples of significant cracking such as shown in Fig. 2 and 7. Once cracks had formed in those samples they appear to have propagated through the matrix, causing complete fracture. It was only in the samples containing SWCNT that the crack growth was arrested.

4.2 Evidence for Strength of Bond

Two types of qualitative evidence were observed in SEM images of SWCNT/OPC composite fracture surfaces. In some cases two pieces of the composite material had fractured entirely and remained connected by bridging SWCNT. Subsequently the distance between the pieces of matrix decreased, causing the SWCNT to relax. As a result of the two pieces of the sample moving together, the SWCNT bundles develop curves or bends. SWCNT have elastic moduli of 1 TPa [Salvetat et al. 1999] and a tight bend is therefore indicative of considerable stored energy in the bundle. An equivalent amount of energy would need to be available in the bond between the SWCNT bundle and the matrix in order to prevent the bundle from detaching.

Examples of this process can be seen in Fig. 15, where a pair of SWCNT bundles are shown in the center of image. Both bundles were highly curved and it was not obvious which sections of the bundles are continuous with each other. Closer examination suggested the connections shown by the arrows. The two ends of the bundle marked by the solid arrows and the lower end of the bundle marked by the dashed arrows were attached to C-S-H structures that are in the same approximate focal plane as the bundles themselves, suggesting that the tight curvature seen in the former bundle was real and not an artifact of the angle of

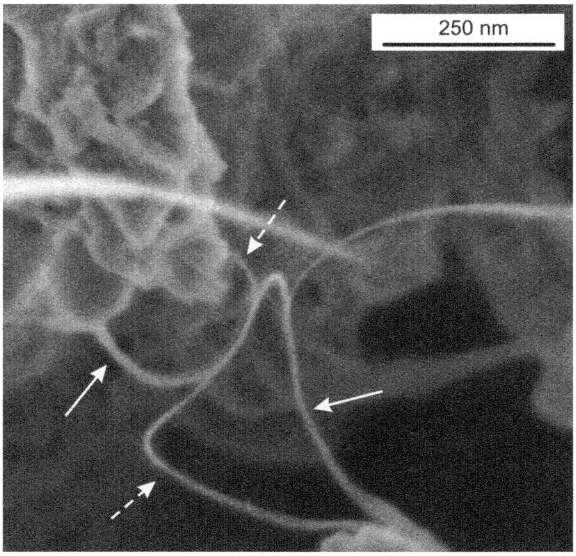

Fig. 15 Adhesion of SWCNT ropes to OPC surface (arrows of same type indicate continuous SWCNT bundles)

imaging. The bundle marked by the dashed arrows, however, appeared to become narrower as it moves towards its upper contact with the C-S-H, suggesting that its curvature may have been less than would be estimated by direct measurement on the image.

Two other types of bending experienced by SWCNT bundles in cement matrices are shown in Fig. 16. Here the bundles were again bridging a gap between two pieces of C-S-H. These bundles were thicker and had not been as well distributed as those discussed in Makar and Chan [2009] and it was possible that they were present in sufficient concentration to act as a crack former, rather than a reinforcement. They do, however, clearly illustrate several important points about the interaction between the C-S-H and the SWCNTs.

The appearance of the gap and the SWCNT between the pieces of C-S-H suggests that they may have rotated slightly as well as being pulled apart. Bundle 1 in Fig. 16 appeared to be curved or buckled in the middle, while the remaining bundles were either straight or pulled completely apart. In addition to the curvature in the bundle 1, all of the bundles showed curvature towards the point of contact with the C-S-H. This effect appeared to be due to each bundle being anchored at a number of places on or in the C-S-H, but held together by van der Wahl's forces in the centre of the bundle. The bending of the SWCNT bundles also suggests a considerable amount of energy was stored in the bundles, again indicating a strong bonding situation.

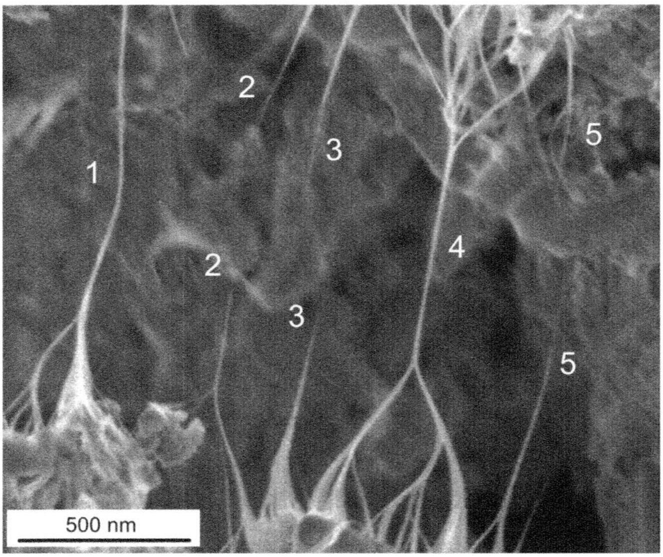

Fig. 16 SWCNT bundles in the process of being pulled apart

The bundles in Fig. 16 also provide another indication of the strength of bond between the SWCNT and the C-S-H. While bundle 1 was curved and bundle 4 was straight and appears to be under some degree of tension, bundles 2, 3 and 5 appear to have been pulled apart at the time of the separation of the C-S-H. In each case corresponding bundle ends, appearing to thin to a lower number of SWCNT, can be seen at the top and bottom of the image. The ends of the separated bundles did not precisely align with each other, likely due to the bundles deflecting to minimize their energy once the two parts of the bundle hand separated. Fig. 17 shows a schematic of the separation process.

Individual SWCNT are held in bundles by van der Waal's forces, which have been estimated as being as high as 11 GPa. While it is impossible to determine from the images the degree to which the individual SWCNT contacted each other, the adhesion of the bundles to the C-S-H when fracture occurs, rather than to other SWCNT, again suggests that the two materials were strongly bonded together.

The apparent degree of bundle to matrix adhesion, as opposed to SWCNT to SWCNT adhesion, was particularly significant when the effect of the size of the bundles was taken into consideration. The strength of bond between the matrix and a SWCNT bundle, as opposed to a single SWCNT, is dependent on the amount of surface contact between the bundles and the matrix. A thicker bundle, with more SWCNT, will have proportionately less contact than a thinner bundle. Bundles of more than 6 SWCNT would be likely to have more SWCNT to SWCNT contact by surface area than SWCNT to matrix. The fact that the SWCNT remained attached to the matrix even when in the form of the relatively thick bundles seen in Fig. 16 also helps to emphasize the strength of the SWCNT/matrix bond.

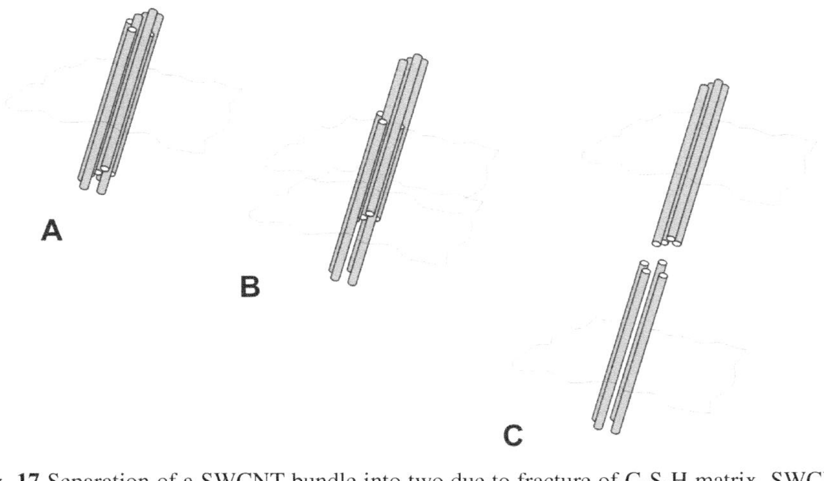

Fig. 17 Separation of a SWCNT bundle into two due to fracture of C-S-H matrix. SWCNT bonded to each side of the fracture (A) slide apart as the crack grows (B), eventually separating entirely (C).

It is worth noting that the branched bundles seen in Fig. 16 may be advantageous with respect to bonding and reinforcement. A branched system would be expected to have greater contact with the matrix than a single linear bundle with the same total number of SWCNT. Secondly, the branches mean that the applied forces producing the crack may not necessarily be aligned perpendicularly to the SWCNT in the matrix even if they are aligned perpendicularly to the crack, reducing the effective load applied to the SWCNT and their bond with the matrix.

While examination of the SEM images provides qualitative information on the strength of the C-S-H/SWCNT bond, direct measurement of the strength of bond between a matrix material and SWCNT is a difficult undertaking. Examples in the literature have relied on attaching one end of an individual nanotube to an AFM probe or other structure and immersing it in unhardened matrix material [ref]. This method can work well when bonds are expected to be weak, but in the case of a well bonded material the results may be affected by changes in the attachment between the sample support structure and the SWCNT as well as the matrix.

Another approach would be to calculate the apparent strength of the bond based on SWCNT or SWCNT bundle deflections. This method uses the known 1 TPa value of the elastic modulus of the SWCNT [Salvetat et al. 1999] and the estimated curvature of bent SWCNT to estimate the energy associated with the interaction between the matrix and the nanotubes. In principle, this calculation could produce quantitative results for bond energy, but there several issues limit its utility. First, determination of degree of curvature is limited by the quality and nature of the images involved. This issue has been discussed above with respect to Fig. 15. Secondly, precise estimates of the number of SWCNT in a bundle can not be obtained from this form of SEM imaging. Finally, it is impossible to determine the extent to which the SWCNT bundles penetrate the surface of the C-S-H. As a result, there is no way to determine the precise length of SWCNT bonded to the C-S-H. In the case of the sliding behaviour shown in Fig. 16, the second and third limitations also apply. In particular, it is certainly possible that there are more SWCNT attached to C-S-H than were present in the center of the bundle.

Given these limitations the best estimate that can be made of the bond strength is that it is likely within an order of magnitude of the van der Waals forces holding the SWCNT bundles together. This estimate arises directly from the behaviour shown in Fig. 16 and 17, where the ends of the bundle remain attached to the matrix, rather than one end completely detaching from the matrix. This estimate in turn suggests that the bonding between the matrix and the SWCNT bundles may itself arise from van der Waals forces. If this is indeed the case, the C-S-H immediately around the SWCNT may be significantly more ordered than the bulk material. Further work, particularly using high resolution transmission electron microscopy, is needed to investigate this possibility.

4.3 Implications of Methods of Dispersion for Composite Performance

The apparent high degree of bonding between the SWCNT bundles and the matrix suggests a corresponding degree of reinforcing capability. Fully taking advantage of that capability requires dispersing the SWCNT as evenly as possible throughout the C-S-H matrix. The cement hydration process, where outer product grows outward from the surface of the grain and inner product forms underneath it limits the locations where reinforcement can take place. Reinforcement is likely to be limited to the outer product unless the reinforcing material can remain in solution and is sufficiently small that it can be carried into the inner product zone before diffusion of ions through the existing hydration products becomes the predominant hydration mechanism [Taylor, 1997].

Neither of the two methods of dispersing SWCNT discussed here is likely to provide a means of reinforcing the inner product. Depositing the SWCNT on the grain surface accelerates the onset of the hydration reactions and provides localized reinforcing behaviour, but once the hydration products have covered them, any remaining product that forms between the cement grains will not be directly reinforced. Similarly, dispersing SWCNT or nanoparticles in superplasticizer appears to be likely to cause concentrations of the nanomaterial on or near C_3A in the unhydrated OPC. The remainder of the OPC is likely to experience a reduced degree of reinforcement, with bonding potentially hampered by the superplasticizer itself.

It is certainly possible that other methods of dispersion of carbon nanotubes and/or other nanomaterials could be developed to position them so that they would stand perpendicularly away from the surface (in the case of nanotubes or nanorods) or remain suspended in the hydration water. Such processes would still leave the inner product region unreinforced, but would improve the extent of outer product reinforcement.

5 Conclusions

At the microscopic level, SWCNT/OPC composites showed evidence of classical reinforcing behaviour in the form crack bridging, fiber pullout and crack deflection. This reinforcing behaviour occured even when the SWCNT were present in the form of bundles, rather than individual nanotubes. The high degree of adhesion between SWCNT in a bundle means that separation of a bundle during crack formation acts as a fourth form of reinforcing mechanism, with energy being absorbed as the van der Waal's forces holding the bundle together are overcome.

The ability of the hydrated OPC matrix to remain attached to the ends of the SWCNT bundles during crack growth suggested that a high degree of bonding existed between the two materials. A qualitative estimate of the strength of bond is that it was within an order of magnitude of the van der Waals forces in the bundle, but no numerical estimate was possible with the data examined here. The high strength of bond appeared to be due to the nucleation of the C-S-H by the SWCNTs and its formation along the SWCNT bundles. Comparisons between the peformance of the SWCNT as a nucleating agent and that of other nano-materials, as well as the preferential nucleation of C-S-H instead of $Ca(OH)_2$ on the SWCNT, suggested that the predominant nucleation mechanism was electrostatic in nature, rather than adsorption of metallic ions.

Dispersion remains a challenge in the use of SWCNT in cement systems. While SWCNT can be successfully dispersed in Disal, the hydration behaviour of those systems suggests that the SWCNT were being wrapped by the admixture, which appeared to inhibit the nucleation behaviour produced by SWCNT dispersed by sonication. In addition, the presence of the admixture appeared to cause the

SWCNT to nucleate C_3A and C_4AF reaction products, in addition to C-S-H. Further work is needed on dispersion issues to reach the full potential of SWCNT as a reinforcing material.

Despite the need for improved dispersion techniques, SWCNT remain extremely promising as nanometric reinforcing materials. As with other nanomaterials, the real challenges to use in the field are not the development of the material. Instead, they involve the safe and appropriate use of the material in construction. Scaling laboratory procedures to industrial production, producing the material in an manner safe for construction construction workers (i.e. without release of nanomaterials in the cement or ready mix plant), and following appropriate environmental and health guidelines through the product's life cycle from placement to demolition all present significant, long term challenges that need to be overcome.

Acknowledgements

Sonicated OPC samples and those made with nano-alumina, nano-calcium carbonate and nano-titania were prepared and imaged by Gordon W. Chan. The sample shown in Fig. 8 was also prepared and imaged by Gordon W. Chan. The remaining SWCNT samples shown in Section 4 were prepared by Jeanne Luh. Kaela Essegehaier prepared the SWCNT samples for isothermal conduction calorimetry in section 3.

References

Ausman, K.D., Piner, R., Lourie, O., Ruoff, R.S., Korobov, M.: Organic Solvent Dispersions of Single-Walled Carbon Nanotubes: Toward Solutions of Pristine Nanotubes. J. Phys. Chem. B 104(38), 8911–8915 (2000)

Campillo, I., Dolado, J.S., Porro, A.: High performance nanostructured materials for construction. In: Bartos, P.J.M., Hughes, J.J., Trtik, P., Zhu, W. (eds.) Nanotechnology in Construction, Proceedings of the First International Symposium on Nanotechnology and Construction, June 23-25, Paisley, Scotland, Roy. Soc. Chem., London, pp. 215–226 (2003)

Chiang, I.W., Brinson, B.E., Huang, A.Y., Willis, P.A., Bronikowski, M.J., Margrave, J.L., Smalley, R.E., Hauge, R.H.: Purification and characterization of single-wall carbon nanotubes (SWNTs) obtained from the gas-phase decomposition of CO (HiPco process). J. Phys. Chem. B. 105(35), 8297–8301 (2001)

Cwirzen, A., Habermehl-Cwirzen, K., Penttala, V.: Surface decoration of carbon nanotubes and mechanical properties of cement/carbon nanotube composites. Ad. Cem. Res. 20, 65–73 (2008)

De Ibarra, Y.S., Gaitero, J.J., Campillo, I.: Atomic force microscopy and nanoindentation of cement pastes with nanotube dispersions. Phys. Stat. Sol. (a) 203, 1076–1081 (2006)

Kingston, C.T., Homenick, C.M., Guan, J., Simard, B.: Nondestructive purification of single-walled carbon nanotubes using simple solvents (in preparation)

Kingston, C.T., Jakubek, Z.J., Denommee, S., Simard, B.: Efficient laser synthesis of single-walled carbon nanotubes through laser heating of the condensing vaporization plume. Carbon 42, 1657–1663 (2004)

Konsta-Gdoutos, M.S., Metaxa, Z.S., Shah, S.P.: Highly dispersed carbon nanotube reinforced cement based materials. Cem. Conc. Res. 40(7), 1052–1059 (2010)

Li, G.Y., Wang, P.M., Zhao, X.: Mechanical behaviour and microstructure of cement composites incorporating surface-treated multi-walled carbon nanotubes. Carbon 43, 1239–1245 (2005)

Li, G.Y., Wang, P.M., Zhao, X.: Pressure-sensitive properties and microstructures of carbon nanotube reinforced cement composites. Cem. Conc. Comp. 29, 377–382 (2007)

Makar, J.M.: Carbon Nanotube / Cement Composite Materials. In: Somani, P.R., Umeno, M. (eds.) Carbon Nanotubes: Multifunctional Materials, Applied Science Innovations Pvt. Ltd., Pune, Maharashtra, India (2009)

Makar, J.M., Chan, G.W.: End of the induction period in ordinary Portland cement as examined by high resolution scanning electron microscopy. J. Amer. Ceram. Soc. 91, 1292–1299 (2008)

Makar, J.M., Chan, G.W.: Growth of cement hydration products on single walled carbon nanotubes. J. Am. Ceram. Soc. 92, 1303–1310 (2009)

Makar, J.M., Chan, G.W.: Effects of nanoparticles and ultrahigh surface area materials on growth of cement hydration products (in preparation)

Makar, J.M., Chan, G.W., Esseghaier, K.Y.: An Additional Hydration Reaction at the End of the Cement Induction Period. J. Mater. Sci. 42, 1388–1392 (2007)

Makar, J.M., Margeson, J., Luh, J.: Carbon nanotube/cement composites – early results and potential applications. In: Banthia, N., Uomoto, T., Bentur, A., Shah, S.P. (eds.) Proceedings of ConMat 2005 and Mindess Symposium, Construction Materials, August 22-24, vol. 32, UBC Press, Vancouver (2005)

Narimatsu, K., Niidome, Y., Nakashima, N.: Pulsed-laser induced flocculation of carbon nanotubes solubilized by an anthracene-carrying polymer. Chem. Phys. Lett. 429(4-6), 488–491 (2006)

Raki, L., Beaudoin, J., Alizadeh, R., Makar, J., Sato, T.: Cement and Concrete Nanoscience and Nanotechnology. Materials 3, 918–942 (2010)

Pratt, L., Ghose, A.: Electron microscope studies of Portland cement microstructures during setting and hardening. Phil. Trans. R. Soc. Lond. A 310, 93–103 (1983)

Salvetat, J.-P., Bonard, J.-M., Thomson, N.H., Kulik, A.J., Forró, L., Benoit, W., Zuppiroli, L.: Mechanical Properties of Carbon Nanotubes. Appl. Phys. A 69, 255–260 (1999)

Sato, T., Beaudoin, J.J.B.: Effect of Nano-CaCO3 on Hydration of Cement Containing Supplementary Cementitious Materials. Adv. Cem. Res. (in submission)

Sato, T., Beaudoin, J.J.B.: The Effect of nano-sized CaCO3 addition on the hydration of OPC containing high volumes of ground granulated blast-furnace slag. In: 2nd International RILEM Symposium on Advances in Concrete Through Science and Engineering, Québec City, September 11, pp. 355–366 (2006)

Strano, M.S.: Polymer-wrapped nanotubes. Nat. Matr. 5(6), 433–434 (2006)

Taylor, H.F.W.: Cement Chemistry, 2nd edn. Thomas Telford, London (1997)

Vodenitcharova, T., Mylvaganam, K., Zhang, L.C.: Mechanical interaction between single-walled carbon nanotubes during the formation of a bundle. J. Matr. Sci. 42(13), 4935–4941 (2007)

Walters, D.A., Ericson, L.M., Casavant, M.J., Liu, J., Colbert, D.T., Smith, K.A., Smalley, R.E.: Elastic strain of freely suspended single-wall carbon nanotube ropes. App. Phys. Let. 74, 3803–3805 (1999)

Wansom, S., Kidner, N.J., Woo, L.Y., Mason, T.O.: A.C. impedance response of multi-walled carbon nanotube/cement composites. Cem. Conc. Comp. 28, 509–519 (2006)

Xiang, X.J., Torwald, T.L., Staedler, T., Trettin, R.H.F.: Carbon nanotubes as a new reinforcement material for modern cement-based binders. In: NICOM2 Proceedings of the Second International Symposium on Nanotechnology and Construction, Bilbao, Spain, November 13-16, pp. 209–213 (2005)

Yu, M.-F., Lourie, O., Dyer, M.J., Moloni, K., Kelly, T.F., Ruoff, R.S.: Strength and Breaking Mechanism of Multiwalled Carbon Nanotubes Under Tensile Load. Science 287, 637–640 (2000)

Zhan, G.-D., Kuntz, J.D., Wan, J., Mukherjee, A.K.: Single-wall carbon nanotubes as attractive toughening agents in alumina-based nanocomposites. Nat. Mater. 2, 38–42 (2002)

Zheng, L.X., O'Connell, M.J., Doorn, S.K., Liao, X.Z., Zhao, Y.H., Akhadov, E.A., Hoffbauer, M.A., Roop, B.J., Jia, Q.X., Dye, R.C., Peterson, D.E., Huang, S.M., Liu, J., Zhu, Y.T.: Ultralong single-wall carbon nanotubes. Nat. Matr. 3, 673–676 (2004)

Nanomaterials-Enabled Multifunctional Concrete and Structures

Hui Li, Jinping Ou, Huigang Xiao, Xinchun Guan, and Baoguo Han

Abstract. Nanomaterials-enabled multifunctional concrete that has self-sensing ability and high mechanical properties is attractive for guaranteeing the safety of infrastructure. Self-sensing ability of nanoconcrete which is based on its piezoresistivity is obtained by adding appropriate concentration of nano-carbon black into concrete. Effect of various loading states on pieroresistivity of nanoconcrete was studied experimentally, and a theoretical model was proposed to predict and modify the strain gauge factor of nanoconcrete under various loading or environmental conditions. Effect of moisture on resistance of nanoconcrete was studied and a water-proof method was proposed to eliminate the unfavorable effect of polarization on resistance measurement. Finally, a cement-based strain sensor was fabricated and used in monitoring the strain of concrete column. Other benefits from inclusion of nanomaterials in concrete include enhancement in mechanical properties, including strength, abrasion resistance and fatigue properties. Microstructures of nanoconcrete was studied with help of SEM pictures, which showed that the hydration product of nanoconcrete was more uniform and compact than that of normal concrete.

1 Introduction

Civil infrastructures usually suffer from fatigue load, environmental corrosion or/and natural disasters. To improve the safety of infrastructures, structural health monitoring (SHM) is getting more and more important. SHM is normally performed by measuring the strain/stress of a structure's critical zone with sensors; the health of the structure can be evaluated based on the measured information, which can be used when deciding whether or not to repair the structure [1].

Hui Li · Jinping Ou · Huigang Xiao · Xinchun Guan · Baoguo Han
School of Civil Engineering, Harbin Institute of Technology, Harbin
150090, China

Jinping Ou
School of Civil and Hydraulic Engineering, Dalian University of Technology, Dalian
116024, China

K. Gopalakrishnan et al. (Eds.): Nanotechnology in Civil Infrastructure, pp. 131–173.
springerlink.com

Traditional strain sensors (such as optical fiber sensors, electrical resistance strain gauges, and piezoelectric ceramics) have been widely used in SHM [2-5]. However, the cost and durability of traditional sensors, and the comparability between traditional sensors and concrete are limited, motivating us to develop new generation of sensors that is more preferable for concrete structures.

The recently developed multifunctional nanomaterials enabled concrete provides an efficient way to improve the safety of concrete structures. By incorporating some conductive nanomaterials (such as carbon nanofiber or carbon black) with concrete, the concrete can be conductive and piezoresistive [6, 7], the so-called smart concrete. Smart concrete is a new generation of structure materials that developed from carbon fiber reinforced concrete (CFRC) in the past decade by utilizing the piezoresistivity for sensing strain [8-13] or by utilizing the relation between damage and resistance for sensing damage [14-15]. To improve the application of smart concrete in health monitoring, some approaches such as using smart concrete as surface coating [8] or as structural materials [10;14] have been proposed. Recently, an advanced application approach using smart concrete as an embedded strain sensor was proposed. In such approach, a standard cubic strain sensor was fabricated with smart concrete [11,12]. CBCC sensor is just this kind of embedment cement-based sensor[16]. The strain-sensing property, humidity insulation method and piezoresistivity model of CBCC sensor has been studied to promote the application of CBCC sensors [17,18]. Besides the self-sensing ability, another merit for concrete by incorporating nanomaterials is the improvement of microstructure and mechanical properties[19,20].

Due to an ultrafine size, nano-particles show unique physical and chemical properties different from those of the conventional materials. Because of their unique properties, nano-particles have been gaining increasing attention and been applied in many fields to fabricate new materials with novelty functions[21-23]. If nano-particles are integrated with traditional building materials, the new materials might possess outstanding or smart properties for the construction of super high-rise, long-span or intelligent civil infrastructure systems. This paper will introduce some progresses on utilizing nano-SiO_2, nano-TiO_2 to improve the strength, wear resistance and fatigue properties of concrete [24-27].

2 Self-sensing Nano-concret and Structure

2.1 Materials and Test Methods

Carbon black (CB) of 120nm came from Liaoning Tianbao Energy Co., Ltd (Liaoning, China). The specific gravity of CB was 1.98g/cm^3. CB in the amount of 5%, 10%, 12%, 15%, 20% and 25% by weight of cement (i.e., 3.11%, 6.04%, 7.22%, 8.79%, 11.39%, and 13.85% by volume of composite, respectively) were used and the corresponding mixture types were called A-5, A-10, A-12, A-15, A-20 and A-25 respectively. The cement used was Portland cement (P.O42.5) from Harbin Cement Company (Harbin, China). The water-cement ratio was 0.4 for all specimens. A water-reducing agent UNF (one kind of –naphthalene sulfonic acid and formaldehyde condensates) was used in the amount of 1.5% by weight of cement. The water-reducing agent could increase the dispersion of CB particles and facilitate the workability of the mixture. The defoamer, tributyl phosphate (made in China), was used in the amount of 0.13 vol.% to decrease the number of air bubbles.

Defoamer and UNF water-reducing agent were dissolved in water, then CB was added and stirred at high speed in a mortar mixer for 3 minutes. This mixture and the cement were mixed at high speed for 2 min. After this, the mix was poured into oiled molds to form prisms of 30×40×50mm for compressive testing. After pouring, an external vibrator was used to facilitate compaction and decrease the number of air bubbles. The samples were demolded after 24h and then cured in a moist room (relative humidity 100%) for 28 days. Afterwards, the specimens were dried in an oven at 60 ℃ for 2 days to extract redundant water to eliminate the polarization effect on resistance measurement. The dried specimens were then tested at ambient temperature.

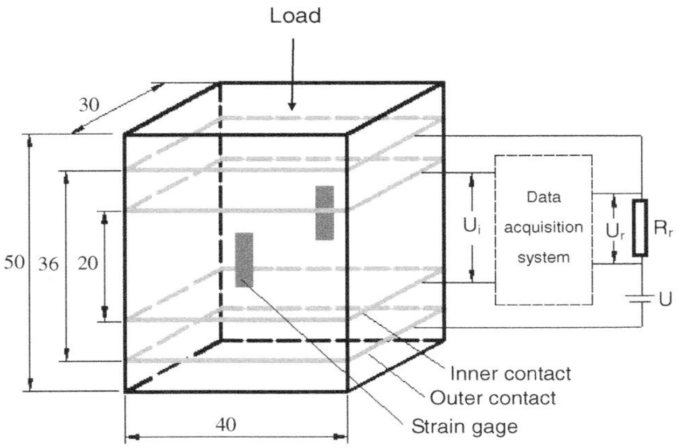

Fig. 1 Schematic of the experimental set-up (mm).

DC electrical resistance measurement was made in the longitudinal axis, using the four-probe method, in which copper nets served as electrical contacts. The copper nets were placed into the specimen when pouring the mix into molds. Four contacts were placed across the whole cross-section of 30×40mm of the specimen, these were all perpendicular to the longitudinal axis and symmetrically positioned with respect to the mid-point along the height of the specimen (i.e., two contacts were in planes above the mid-point and two contacts were in planes below the mid-point). The outer two contacts (36mm apart) were for passing current. The inner two contacts (20mm apart) were for measuring the voltage (see Fig. 1).

A DC circuit, developed by Han [28] was used to measure the resistance of the specimen as follows:

$$R_i = U_i \cdot R_r / U_r \tag{1}$$

where R_r is the standard reference resistor, U_r is the voltage applied on the reference resistor, and U_i and R_i are the voltage and resistance between the two inner contacts of the specimen, respectively. U_i and U_r were collected using the data acquisition board.

2.2 Piezoresistivity of Nano-concrete and the Modeling

2.2.1 Piezoresistivity Behavior of Nano-concrete

Compressive testing was performed on a 30×40mm side of each specimen. The strain was measured by using strain gauges attached to the middle of the opposite sides of a specimen and parallel to the stress axis. Compressive testing under force control was conducted using a hydraulic mechanical testing system with 120-kN maximum loading capacity. The scheme of monotonically static loading up to specimen failure was arranged. During the loading process, DC electrical resistance measurement was simultaneously made in the stress axis, using the four-probe method as described earlier. Three specimens of each type mixture were tested.

Fig. 2 shows the resistivity (ρ) as a function of CB volume content (V) of CB-filled composites. It can be observed from Fig.2 that the resistivity of the composites decreased dramatically with increasing CB content from 7.22 vol.% to 11.39 vol.%, i.e. from A-12 to A-20. The resistivity of the composites varied slightly outside the above range. The content range over which the resistivity varied precipitously was called percolation threshold. Therefore, in this study, the percolation threshold zone of the composites was CB in the amount of 7.22 vol.%~11.39 vol.%.

According to tunneling effect theory, the natural logarithm of resistivity of composites $\ln(\rho)$ is a linear function of potential barrier width (s), i.e. the distance between CB particles. Furthermore, it is well accepted that distribution of CB

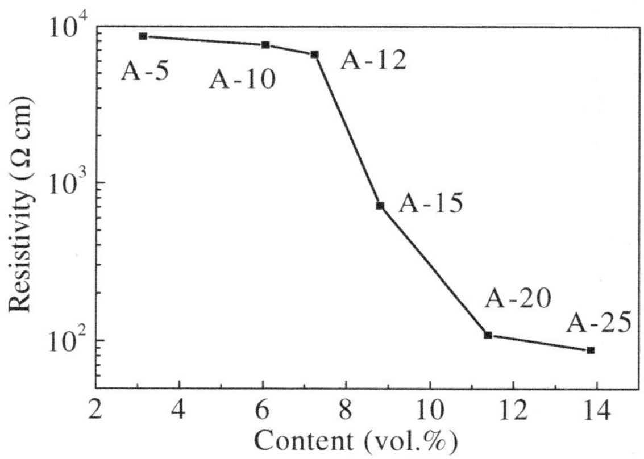

Fig. 2 Logarithm of resistivity as a function of volume content of CB.

particles is random and the average distance between particles is proportional to $V^{-1/3}$. Thus, $\ln(\rho)$ is a linear function of $V^{-1/3}$. For clear observation, the $V^{-1/3}$ and natural logarithm of resistivity were respectively nominated as abscissa and ordinate and the curve in Fig.2 was redrawn in Fig.3. The curve can be divided into three stages according to the variation of slope. The curve from 7.22 vol.% to 11.39 vol.% in Fig. 3 was just linear, so the tunneling effect dominated conductivity and electromechanical properties of the composites in stage II.

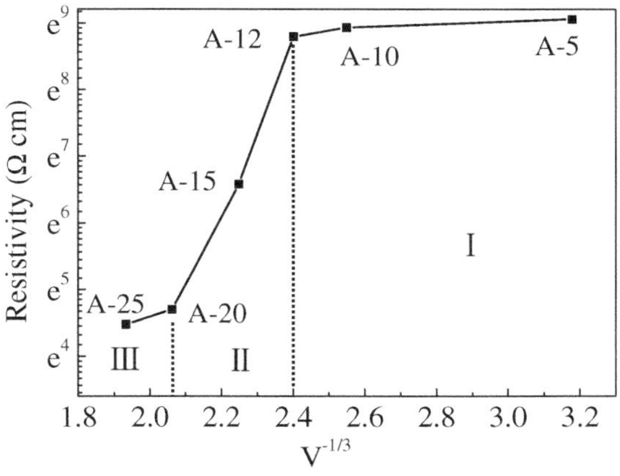

Fig. 3 Natural logarithm of resistivity as a function of V-1/3.

Tunneling current is an exponential function of barrier width, implying a precipitous change of resistivity upon distance between CB particles. The distance between CB particles is shortened with increasing compressive strain on specimens. As a consequence, the resistivity of composites decreases with increasing compressive strain according to tunneling effect theory.

Figs. 4 show the fractional change in resistivity versus the compressive strain curves of A-15. Additionally, the distance between probes was shortened during loading, which induced change in electrical resistance. However, it could be observed from Fig. 4 that the change in resistance induced by this factor was very small and negligible. The resistance was essentially proportional to the volume resistivity that was selected as a measurement in this study. For A-15, the resistivity decreased linearly with increasing compressive strain up to failure of the specimens except for a small perturbation over the strain range of [0.003~0.004] which indicated the occurrence of microcracks. The three curves for the three specimens of this mixture were almost the same, indicating that the results were repeatable. Linear fit of the experimental data showed that the relationship between the fractional change in resistivity and compressive strain was nearly linear. The fractional change in resistivity per unit strain (i.e., the strain gauge factor) was 55.28 as shown in Fig. 4.

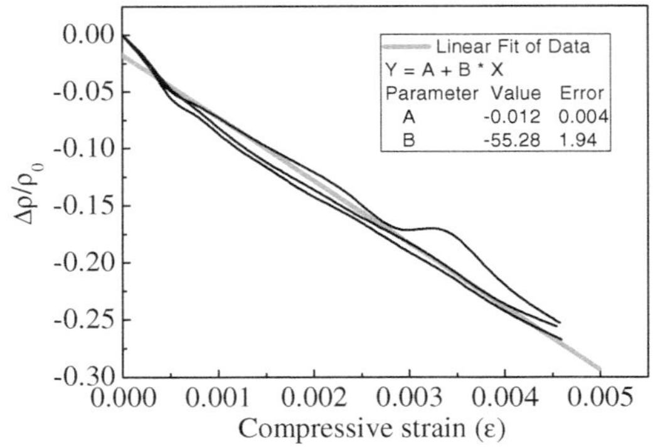

Fig. 4 Fractional change in resistivity of A-15 as a function of compressive strain.

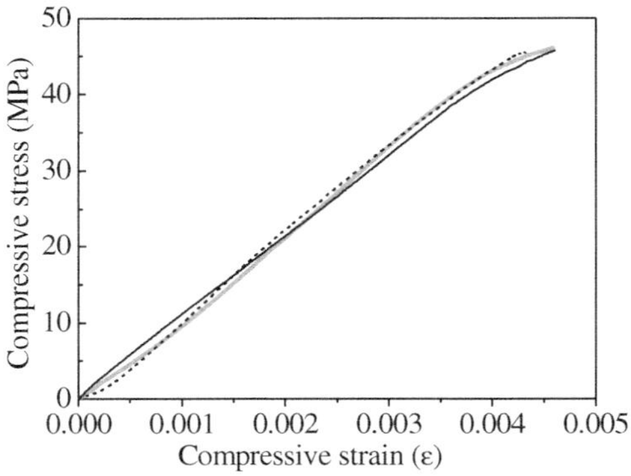

Fig. 5 Stress-strain curves of A-15.

Fig. 4 presents the strain-stress curves of A-15. It can be seen from Fig.4 that the strength of the CB-filled composite could reach more than 40MPa. Therefore, it can be used as a structural material like common concrete. Similar results have been obtained for other mixtures. In the following text, CBCC denotes the cement-based composites filled with 15% carbon black.

2.2.2 Modeling on Piezoresistivity of CBCC

According to tunneling effect theory [30], current density in a tunneling resistor formed by two adjacent CB particles at a low voltage region can be depicted by the following equation:

$$J = [3(2m\varphi)^{1/2} / 2S_0](e/h)^2 U \cdot \exp[-(4\pi S_0 / h)(2m\varphi)^{1/2}] \quad (2)$$

where J is current density, m, e and h are the electron mass, charge on an electron and Planck's constant, respectively, φ, S_0 and U are the height of tunneling potential barrier, tunneling width and voltage applied across tunneling resistor, respectively. R_{t0} denotes the tunneling resistor and A and S_0 denote the section area and length of the resistor, respectively. Then the resistance of the resistor can be obtained as

$$R_{t0} = k_1 S_0 \exp(k_2 S_0) \tag{3}$$

where $k_1 = (2/3)(2m\varphi)^{-1/2}(e/h)^{-2}A^{-1}$ and $k_2 = (4\pi/h)(2m\varphi)^{1/2}$.

For a given composite, k_1 and k_2 are constants. S_0 can be obtained as follows:

$$S_0 = D\left[(\pi/6)^{1/3}V_c^{-1/3} - 1\right] \text{[28]}, \tag{4}$$

where V_c and D are the volume concentration and diameter of CB particle, respectively. Equation (2) indicates that even a slight change of S_0 may cause a large change of resistance. Conductive networks in CBCC are composed of a large number of R_{t0} and the change in resistance of CBCC is the integrated result of the change of each R_{t0}. Therefore, the piezoresistivity model of CBCC will be established based on the resistance behavior of each R_{t0}. The resistance of each R_{t0} under strain can be quantified with equation (3); hence, the key task of the modeling is to obtain the deformation of each R_{t0} under external strain.

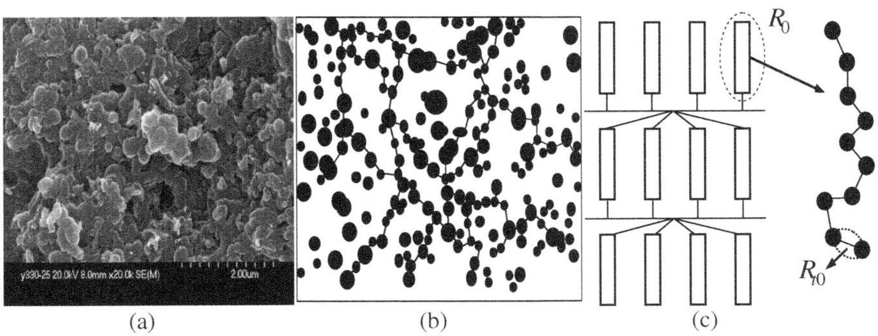

Fig. 6 Schematic of conductive network in CBCC. (a) SEM picture, (b) Schematic picture, (c) conductive network model.

The microstructure of CBCC was observed with SEM to study the characteristics of the conductive network of CBCC. Fig. 6 (a) shows the microstructure of CBCC, in which the bright spherical objects and rounded hollowness denotes CB particles. To see the pattern of conductive network clearly, CB particles are emphasized in a white background, as shown in Fig. 6 (b). The conductive path is formed by adjacent CB particles and is presented in Fig. 6 (b) by joining the CB particles. Based on the schematic characteristics of a conductive network, a conductive model is proposed, as shown in Fig. 6 (c). First, R_{t0} is connected in series to form a resistor element R_0. Then, R_0 forms the conductive network by connections in parallel and then series. It can be easily derived that the fractional change of resistance of CBCC is equal to that of R_0. Therefore, piezoresistivity modeling of CBCC is focused on the behavior of R_0 under external loading.

The resistance value of each R_{t0} can be calculated based on equations (2) and (3). The orientation direction of each R_{t0} is assumed to be uniformly distributed by considering the infinite and randomly distributed CB particles in CBCC. Fig. 7 (a) and (b) show the schematics of orientation and distribution of tunnel resistor, respectively. Assume that there are a total $4\times3(M+1)^2$ of R_{t0} in the space and each $4(M+1)^2$ of R_{t0} surround a coordinator. For example, $4(M+1)^2$ of R_{t0} is first defined by θ_z and φ_z, as shown in Fig. 7 (a). Both θ_z and φ_z should be over the range of $[0,2\pi]$; however, considering the symmetrical characteristic of R_{t0} in each quadrant, $\theta_z, \varphi_z \in [0,\pi/2]$ is sufficient to represent the orientation character of R_{t0} and is adopted in this paper. Therefore, θ_z and φ_z are in

$$\theta_z, \varphi_z = [0, \pi/2M, \pi/M, 3\pi/2M, \cdots\cdots, \pi/2(M-1), \pi/2], \tag{5}$$

where θ_x, φ_x, and θ_y, φ_y are defined in the same way as that of θ_z, φ_z. Hence, the resistance behavior of CBCC can be described by the 3(M+1)2 of R_{t0}. As shown in Fig. 7 (b), the distribution of CB particles defined by above method is not uniform in space that the distribution density is higher near each coordinate, but it is uniform in calculating the effect of multi-axial strain on resistance.

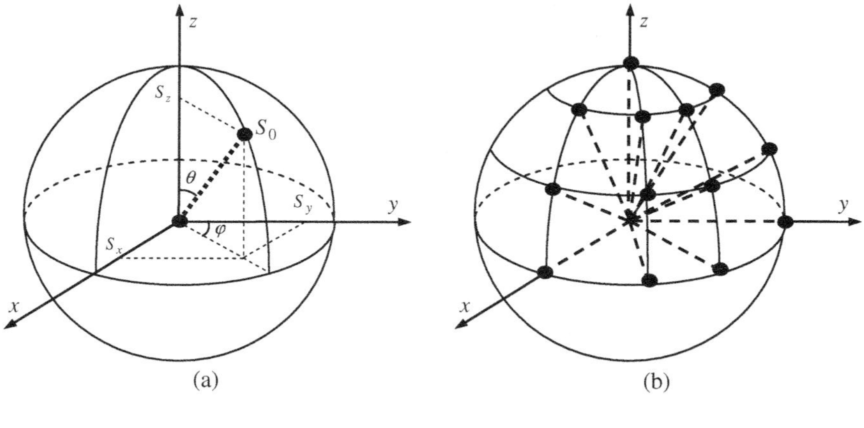

Fig. 7 Schematic of (a) orientation and (b) distribution of R_{t0}.

When CBCC is load free, the original length of R_{t0} is S_0, and R_0 is the sum of 3 $(M+1)^2$ of R_{t0} as

$$R_0 = 3(M+1)^2 k_1 S_0 \exp(k_2 S_0), \tag{6}$$

Then, assuming that CBCC is subjected to multi-axis strain of ε_{mx}, ε_{my} and ε_{mz} in each axis, the corresponding strain of R_{t0} in each axis direction can be calculated as

$$\varepsilon_{x,y,z} = \frac{\varepsilon_{mx,my,mz}}{V_c(E_m / E_c) + V_m}, \tag{7}$$

where V_c and V_m are the ratio of volume occupied by CB and matrix to the whole volume of composite, respectively. E_c and E_m are the modulus of CB and matrix, respectively. The length of R_{t0} under multi-axial strain changes from S_0 to either $S_{\varepsilon x}$, or $S_{\varepsilon y}$, or $S_{\varepsilon z}$ as

$$\begin{aligned} S_{\varepsilon x} &= S_0\sqrt{\cos^2\theta_x(1+\varepsilon_x)^2 + \sin^2\theta_x \sin^2\varphi_x(1+\varepsilon_y)^2 + \sin^2\theta_x\cos^2\varphi_x(1+\varepsilon_z)^2} \\ S_{\varepsilon y} &= S_0\sqrt{\cos^2\theta_y(1+\varepsilon_y)^2 + \sin^2\theta_y \sin^2\varphi_y(1+\varepsilon_x)^2 + \sin^2\theta_y\cos^2\varphi_y(1+\varepsilon_z)^2} \\ S_{\varepsilon z} &= S_0\sqrt{\cos^2\theta_z(1+\varepsilon_z)^2 + \sin^2\theta_z \sin^2\varphi_z(1+\varepsilon_x)^2 + \sin^2\theta_z\cos^2\varphi_z(1+\varepsilon_y)^2} \end{aligned}. \tag{8}$$

Equation (8) indicates that even under the same external strain status, the strain of each R_{t0} with different orientation angle is different. Therefore, based on equations (6) and (8), the resistance of CBCC under multi-axial strain changes from R_0 to R as

$$R = \sum_{i=1}^{M+1}\sum_{j=1}^{M+1} k_1 S_0[F_x \exp(k_2 S_0 F_x) + F_y \exp(k_2 S_0 F_y) + F_z \exp(k_2 S_0 F_z)] \tag{9}$$

where

$$F_x = \sqrt{\cos^2(\frac{(i-1)\pi}{2M})(1+\varepsilon_x)^2 + \sin^2(\frac{(i-1)\pi}{2M})\sin^2(\frac{(j-1)\pi}{2M})(1+\varepsilon_y)^2 + \sin^2(\frac{(i-1)\pi}{2M})\cos^2(\frac{(j-1)\pi}{2M})(1+\varepsilon_z)^2}$$

$$F_y = \sqrt{\cos^2(\frac{(i-1)\pi}{2M})(1+\varepsilon_y)^2 + \sin^2(\frac{(i-1)\pi}{2M})\sin^2(\frac{(j-1)\pi}{2M})(1+\varepsilon_x)^2 + \sin^2(\frac{(i-1)\pi}{2M})\cos^2(\frac{(j-1)\pi}{2M})(1+\varepsilon_z)^2}$$

$$F_z = \sqrt{\cos^2(\frac{(i-1)\pi}{2M})(1+\varepsilon_z)^2 + \sin^2(\frac{(i-1)\pi}{2M})\sin^2(\frac{(j-1)\pi}{2M})(1+\varepsilon_x)^2 + \sin^2(\frac{(i-1)\pi}{2M})\cos^2(\frac{(j-1)\pi}{2M})(1+\varepsilon_y)^2}$$

Hence, the change in resistance of CBCC can be calculated as

$$\frac{R}{R_0}=\frac{1}{3(M+1)^2}\sum_{i=1}^{M+1}\sum_{j=1}^{M+1}\{F_x\exp[k_2S_0(F_x-1)]+F_y\exp[k_2S_0(F_y-1)]+F_z\exp[k_2S_0(F_z-1)]\} \tag{10}$$

Conducting series expansion on the exponent function in equation (10), the following equation can be obtained:

$$\frac{R}{R_0}=\frac{1}{3(M+1)^2}\sum_{i=1}^{M+1}\sum_{j=1}^{M+1}(F_x+F_y+F_z)+\frac{k_2S_0}{3(M+1)^2}\sum_{i=1}^{M+1}\sum_{j=1}^{M+1}(F_x{}^2-F_x+F_y{}^2-F_y+F_z{}^2-F_z)$$
$$+\frac{(k_2S_0)^2}{6(M+1)^2}\sum_{i=1}^{M+1}\sum_{j=1}^{M+1}[F_x(F_x-1)^2+F_y(F_y-1)^2+F_z(F_z-1)^2]+\cdots\cdots \tag{11}$$

The first item of the right hand side in equation (11) is the inherent behavior of the integrated change in length of each R_{t0} under external strain and is denoted as $\overline{S}/S_0$, which represents the resistance change trend of CBCC under multi-axial strain. k_2S_0 is denoted as β, which represents the sensitivity factor of CBCC and is dependent on the intrinsic property of CBCC. Therefore, $\overline{S}/S_0$ and β represent the piezoresistivity characteristics (trend and sensitivity factor) of a composite. Considering the uncertainty in CBCC synthesis, β is best obtained from a simple uniaxial compressive test and can then be utilized to predict the resistance behavior of CBCC under complex loading and environmental conditions. The applicability of obtaining β from a simple compressive test will be validated by utilizing it in predicting resistance behavior under other effects (CB concentration and temperature). Orientation uniformity of R_{t0} is dependent on the value of M. $M\geq 30$ can meet the precision requirement of calculation.

For CBCC, Based on the previously measured strain gage factor of 55.5 [6], coefficient β can be calculated as 337, and k_2 is then obtained as 3.45 nm^{-1}. Utilizing these parameters, the model can predict the strain gauge factors of a composite under various complex strain statuses. Besides predicting the resistance behavior of CBCC under various strain states, the model can be used to predict the resistance behavior of CBCC under various ambient conditions by transforming such conditions change into strain.

Thermal expansion will lead to a change in S_0 that may cause a change in strain gauge factor. By substituting thermal expansion strain into the proposed model, the strain gauge factors under various temperatures can be predicted. Experiments were conducted for CBCC. The thermal expansion coefficient (TEC) of CBCC in longitudinal direction and transverse direction was measured as 16.5 and 13.9, respectively. The slight discrepancy of TEC in different directions was caused by the embedded electrical probes. Taking the distance S_0 under 20℃ as the baseline, the original distances between each two adjacent CB particles under various temperatures can be calculated based on the measured TEC. Then, based on the proposed model, strain gauge factors of CBCC under various temperatures were obtained, as shown in Fig. 8. The theoretical strain gauge factors change slightly from 55.197 (-10 ℃) to 55.249 (50 ℃), indicating that the effect of temperature on strain gauge factor is slight. The experimental measured strain gauge factors, 54.95 (-10℃), 55.23(20℃) and 55.64 (50℃), respectively, as shown in Fig. 8, also show a small dependence on temperature.

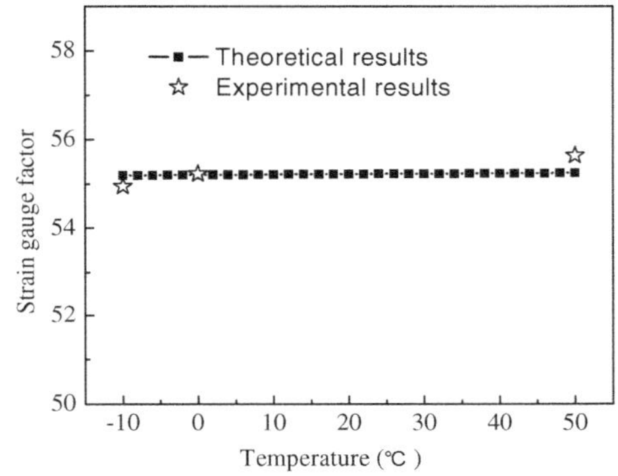

Fig. 8 Effect of temperature on the strain gauge factors.

2.3 Effect of Water Content on Electrical Property of CBCC and the Water-Proofing Method

2.3.1 Effect of Water Content on Resistance Measurement of CBCC

Water content is defined as the ratio of the absorbed water mass to the dried specimen mass and is measured after the specimen is submerged in water and the surface dried. Previous research indicated that when an electric field was present during electrical resistance measurements, a polarization-induced electric field in the material in opposite direction occurred and ions were transported to the contact [31]. To investigate the influence of water content on the polarization and electrical properties of CBCC, the resistances of CBCC with various water contents were continuously collected for five minutes for each specimen and the first value referred to as the initial resistance of the specimen.

Fig.9 (a) shows the fractional change in initial resistance of CBCC with water content. Fig.9 (b) shows the relative fractional change in resistance of CBCC with measurement time. Three specimens were tested. The fractional change in initial resistance η is given as follows:

$$\eta = (R_0^{'} - R_0) / R_0 , \tag{12}$$

where $R_0^{'}$ and R_0 show the initial resistance of a specimen with a given water content and with water content of 0%, respectively.

The relative fractional change in resistance β is given as follows:

$$\beta = (R - R_0^{'}) / R_0^{'}, \tag{13}$$

where R and $R_0^{'}$ are resistances of the specimen with a given water content at measurement time t and at the beginning of the measurement interval (t = 0), respectively.

It can be seen from Fig.9 (a) that the initial resistance of CBCC increased slightly when the water content was lower than 3%. Next, the initial resistance of specimens increased dramatically. The initial resistance increased by about 86% for the specimen with the highest water content (i.e. 12.9%). As shown in Fig.9 (b), the resistance of the dried specimen (i.e. water content 0%) stayed almost constant during measurement. However, the resistance of specimens with substantial water content increased over time. Also, the higher the water content, the more dramatic the increase of resistance was with measurement time.

Because the mobility of ions in the cement paste increased with water content, the polarization phenomena was also related to water content. When the water content was very small, it was difficult for ions to move to the contacts due to the absence of solution. As a consequence, the resistance of the specimen stayed constant during measurement. The results indicate that it was necessary to keep the composite dry and to be insulated from ambient moisture. The waterproofing method was then proposed.

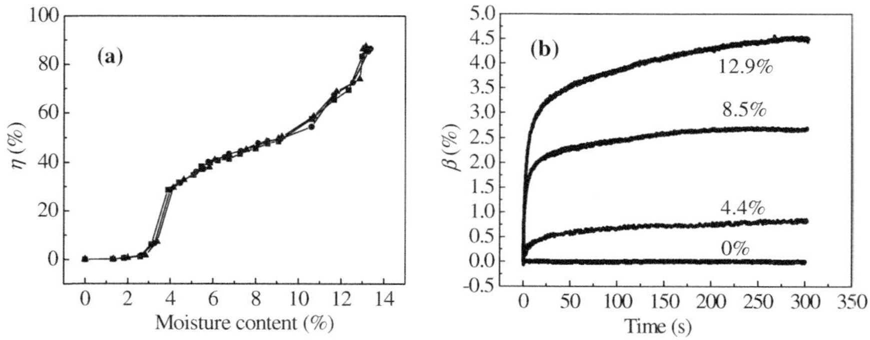

Fig. 9 Fractional change in initial resistance of CBCC as a function of water contents (a), Relative fractional change in resistance of CBCC as a function of measurement time (b)

2.3.2 Waterproof Method of CBCC Encapsulated with Epoxy

Specimens were cast with the same materials and methods as described before. Cured specimens were dried at 60℃ for 48 hours to remove water and then encapsulated with epoxy with a thickness of about 0.9mm. The epoxy was a kind of polyamide resin with the tension strength of 43.6MPa, tension modulus of 3,

680MPa, elongation percentage of 1.2% and a density of 1.254g/cm^3. Before encapsulation, two strain gauges were attached on the two opposite sides of each specimen in the longitudinal direction.

Six encapsulated specimens were prepared for this test. The specimens were dipped in a container filled with water and were controlled to be 50℃ in an oven to simulate the environment of fresh concrete during the first three days. Resistances of specimens were measured each day. Once the first three days' measurement was completed, the container was moved out of the oven to the ambient air and the resistance of the specimens was measured at room temperature every 3 days for 15 days. Ambient temperature was synchronously measured using a thermometer for temperature compensation (with a temperature gauge factor of 0.0007409/℃).

After the two-week unloaded test, the specimens were taken out of the water and enclosed with a sponge full of water. The loading with creep was then applied to the specimens. Three specimens were applied with loading of stress ratio of 0.2 and the other three specimens were applied with loading of stress ratio of 0.6. The tests lasted for one month. The creep strain of specimens was measured using a micrometer gauge. Resistance, strain and temperature were measured every three days.

The variation in resistance could be an indicator of the waterproof effectiveness of epoxy. The initial resistance (R_I) and temperature of six specimens were first measured as references. During the water dipping test, the resistance of specimens (R) and the ambient temperature were continuously measured. The compensated resistance (R_C) was obtained after temperature compensation from R. The fractional change in resistance (percent ratio of change in resistance to initial resistance ($\Delta R / R_I$) and to compensated resistance ($\Delta R / R_C$), respectively) with time of two specimens are shown in Fig. 10; the results of the other specimens were similar.

It can be seen from Fig. 10 that the resistance of specimens decreased with temperature for the first three days (which was attributed to the tunnel effect being strengthened at high temperature) and then was completely recovered to the initial resistance when the temperature returned to the ambient temperature. The resistance was almost constant except for a slight difference in the next 11 days, which implies that no water immerged into the specimen, and the slight difference was attributed to the small scatter of temperature compensation factor. Therefore, the waterproof effectiveness of epoxy was validated for unloaded condition.

The effect of the creep strain and the temperature on resistance should be removed in order to see clearly the effect of moisture. Therefore, the strain gauge factor (fractional change in resistance per unit strain) of each specimen was first measured as 58.2, 52.7, 56.4, 54.5, 59.6 and 58.5, respectively. The results in reference [32] indicated that the creep strain gauge factor of composites was the same as that measured in the elastic strain regime.

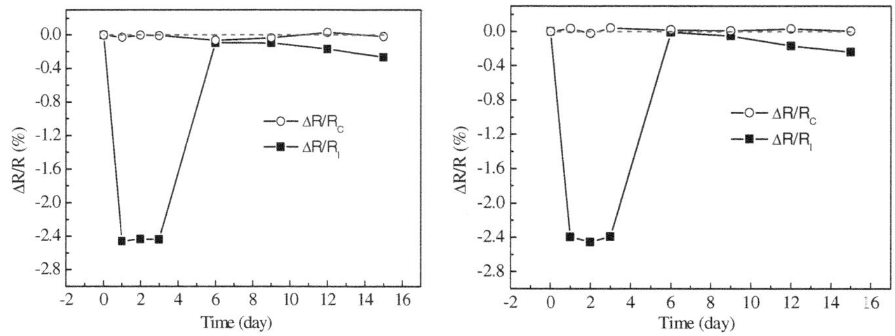

Fig. 10 Fractional change in resistance of CBCC encapsulated with epoxy under unloaded condition in moist environment

Fig.11 (a) shows the fractional change in resistance during long-term loading with stress ratio of 0.2. The instantaneous elastic strain of three specimens was 725, 714 and 739 $\mu\varepsilon$, respectively. Accordingly, the change in resistance was 4.556%, 3.852% and 4.457%, respectively. Based on the strain gauge factor obtained earlier, the change in resistance of specimens under this loading level should be 4.22%, 3.763% and 4.168%, respectively, which agrees well with the measured resistance. Therefore, the epoxy was not cracked at this applied strain. The resistance of specimens decreased simultaneously with increasing creep strain as shown in Fig.11 (a) as $\Delta R / R_I$. Based on the measured temperature and strain gauge factor, R_C was obtained after temperature and creep strain compensation; therefore, $\Delta R / R_C$ was obtained to indicate if the epoxy was cracked. $\Delta R / R_C$ was very slight, which indicated that water content in the CBCC did not change.

Fig.11 (b) shows the fractional change in resistance of three specimens during long-term loading with stress ratio of 0.6. The instantaneous elastic strain of three specimens was 2556, 2519 and 2591 $\mu\varepsilon$, respectively. $\Delta R / R_C$ was slightly

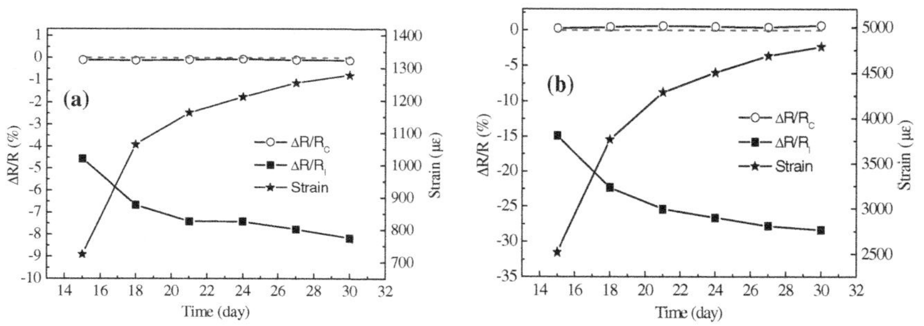

Fig. 11 Fractional change in resistance and creep behavior of CBCC encapsulated with epoxy during long-term loading with stress ratio of 0.2 (a) and with stress ratio of 0.6 (b) in moist environment.

larger than that shown in Fig.11 (a). The largest difference between R and R_C of three specimens was 1.0% (occurring at the 24th day), 0.68% (occurring at the 30th day) and 0.71% (occurring at the 21th day), respectively. However, fractional changes in resistance of these specimens remained very small during the measurement process, indicating that the difference between R and R_C was not caused by water content change in CBCC. The difference was mainly attributed to the slight strain gauge factor change under various loading levels; creep strain compensation was conducted with an obtained strain gauge factor in the elastic regime, but micro-cracks would occur with stress ratio over 0.6 resulting in a slightly decreased strain gauge factor.

In addition to the water absorbing capacity discussed above, the deformation ability of epoxy was also important to ensure the waterproof effectiveness under loading. When a specimen was applied compressive loading, the specimen should expand in transverse direction, which would lead to a tension deformation of side epoxy layer. The Poisson ratio of composite was about 0.17. Therefore, for specimens under long-term loading of 0.2 stress ratio as shown in Fig.11(a), the tension strain of epoxy was less than 221$u\varepsilon$, for specimens under long-term loading of 0.6 stress ratio as shown in Fig. 11(b), The tension strain of epoxy was less than 800 $u\varepsilon$. The tension capacity of epoxy was more than 5000 $u\varepsilon$, so the epoxy should not crack under the loading condition.

The experimental results indicate that waterproof measure, i.e. the specimen encapsulated with epoxy, could effectively insulate specimens from moisture and provided a potential way of using CBCC as a strain sensor material. The full water absorbing capability of epoxy was the most important factor for waterproof efficiency, for CBCC, the full water absorbing capability of epoxy should be less than 1.83. During long-term loading, creep behavior was observed for CBCC, but waterproof efficiency of epoxy remains well, which should be attributed to the high deformation ability. The resistance of CBCC encapsulated with epoxy exposure to moist environment was only influenced by temperature and strain, which could be conveniently compensated.

2.4 *Self-sensing Concrete Structures*

2.4.1 Preparation of Concrete Column with Embedded CBCC Sensor

Epoxy encapsulated CBCC sensor with shape of 30×40×50 mm was prepared with the developed methods, and the properties of CBCC sensor were shown in table 1.

Table 1 Properties of CBCC sensor

Strain sensing properties		Mechanical properties			
Resistivity (Ωcm)	Gage factor	Strength (MPa)	Peak strain ($\mu\varepsilon$)	Modulus (×10^4 MPa)	Poisson ratio
719.4±53.7	55.5±2.87	44.7±0.32	4400±136	1.44±0.06	-0.17±0.015

Concrete columns were made with C40 and C80 concrete respectively for the aim of studying the strain sensing property of CBCC sensors in different strength grades of concrete matrix. The cement used for C40 and C80 concrete were Portland cement P.O32.5 and P.O42.5, respectively. The other materials used for fabricating C40 and C80 were all the same. The fine aggregate was natural river sand with a fineness modulus of 2.94. The coarse aggregate was crushed diabase with a maximum radius of 20 mm. A water-reducing agent (FDN) and silica fume (average radius is 100nm) were used. The mix proportions are shown in Table 2.

Table 2 Mix proportions of concrete columns (kg/m3).

Strength grade	Cement	Water	Sand	Coarse aggregate	Silica fume	FDN
C40	435	190	535	1240	-	-
C80	480	148	690	1080	60	5.76

Concrete columns were cast with molds of 100×100×300 mm, and a CBCC sensor was placed at the center of each column as shown in Fig. 12. The longitudinal axis (resistance measurement direction) of the CBCC sensors was parallel to the 300mm side of concrete columns. For casting concrete column, FDN was firstly dissolved in half volume of water and mixed with cement, sand and coarse aggregate in a concrete centrifugal blender for 2 min. The remainder of the FDN solution and water were poured into the mixture and mixed for another 2 min to achieve good workability. Finally, the fresh concrete was poured into molds with central located CBCC sensor to form prisms. After pouring, an external vibrator was used to facilitate compaction and decrease the amount of air bubbles. The specimens were demolded at 24 h and then cured in a standard moist room for 28 days.

In addition to the concrete columns embedded with CBCC sensor, three plain columns that without CBCC sensors were fabricated for C40 and C80, respectively. The strength of concrete columns with and without CBCC sensors will be comparatively studied to see the effect of CBCC sensor embedment on the mechanical property of concrete column.

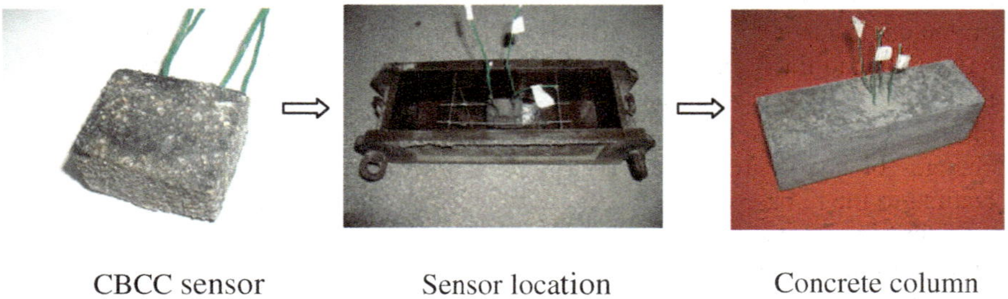

Fig. 12 CBCC sensor and concrete column with embedded CBCC sensor.

2.4.2 Test Methods

Uniaxial compressive test was performed on the 100×100mm side of a concrete column by MTS with 2,500 kN maximum loading capacity. Two displacement transducers were vertically fixed to measure the longitudinal strain, and two additional longitudinal strain gauges were attached on the two other longitudinal sides to help centrally install the column on the MTS (if the concrete column is central installed, the strains measured by two transducers and two strain gauges under a low level preloading will be almost the same). Two strain gauges were horizontally attached on two opposite side surfaces (100×300mm) of the column to measure the transverse strain. During compressive test, the longitudinal strain was synchronously measured with the embedded CBCC sensor by measuring its resistance.

Two loading schedules, i.e. cyclic loading and monotonic loading, were arranged. For cyclic loading, the amplitude of each cycle was 30% of the ultimate strength of concrete column. The loading and unloading rates were all 0.5MPa/s. The load was locked for 60s at the amplitude (maximum and minimum) of each cycle to investigate the sensing stability of CBCC sensor. For monotonic loading, in order to generate data in the descending part of the stress-strain curve, force control at a rate of 0.5 MPa/s was first applied until load reached 70% of ultimate strength and then switched to deformation control at a rate of 0.003 mm/s until the specimen could no longer sustain the load. Three specimens were tested for each loading case.

2.4.3 Measured Strain under Cyclic Loading

Fig. 13 shows the strain of a C40 concrete column (Fig. 13 (a)) and a C80 concrete column (Fig. 13 (b)) under cyclic loading measured by CBCC sensor and displacement transducer, respectively. Strain measured by CBCC sensor was obtained by dividing the fractional change in resistance with the gauge factor 55.5.

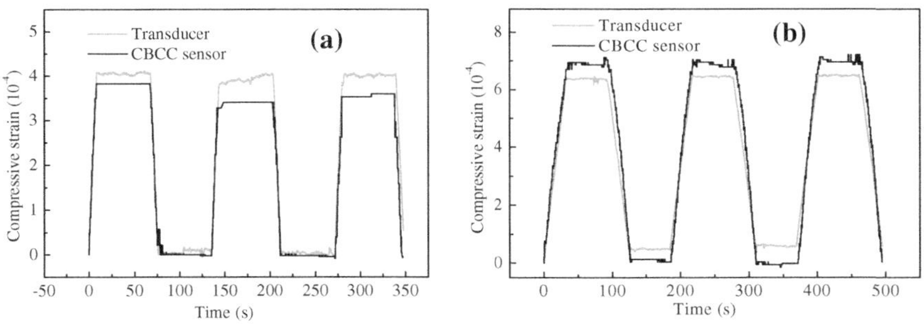

Fig. 13 he strains of (a) C40 concrete column and (b) C80 concrete column under cyclic loading measured by displacement transducer and CBCC sensor. Strain measured by CBCC sensor was obtained by dividing the fractional change in resistance with the calibrated gauge factor 55.5.

It can be seen from Fig. 13 that the strain amplitude of C40 concrete column measured by displacement transducer, and CBCC sensor is about 400 $\mu\varepsilon$ and 380 $\mu\varepsilon$, respectively, and the strain amplitude of the C80 concrete column is about 700 $\mu\varepsilon$ by displacement transducer and 630 $\mu\varepsilon$ by CBCC sensor. During the interval of load locking, the measured strain by displacement transducer and CBCC sensor both keep stable. The little discrepancy between strain measured by CBCC sensor and that by displacement transducer is caused by the accuracy of the calibrated gauge factor used here as 55.5. Considering the deformation capacity of CBCC sensor as 4400 $\mu\varepsilon$, the deformation of the embedded CBCC sensor under cyclic loading is in elastic regime, which can be verified by the zero residual strain after unloading. The experimental results indicate that the repeatability and stability of embedded CBCC sensor under cyclic elastic loading are reliable.

2.4.4 Measured Strain under Monotonic Loading

Fig. 14 shows the strain-stress curves of a C40 concrete column and a C80 concrete column under monotonic loading measured by CBCC sensor and displacement transducer, respectively. Strain measured by CBCC sensor was obtained by dividing the fractional change in resistance with the gauge factor 55.5. The fractional change in resistance of CBCC sensor upon strain measured by displacement transducer (considered as the baseline strain) is also shown in Fig. 14. The stress-strain curves measured by displacement transducer are the typical curves of low strength concrete (Fig. 14(a)) and high strength concrete (Fig. 14(b)), respectively. The strains measured by transducer are regarded as the baseline (real strain) to study the sensing property of embedded CBCC sensors. For C40 concrete column, the strain measured by CBCC sensor agrees well with that measured by displacement transducer when the strain is smaller than 1200 $\mu\varepsilon$ (concrete column is in elastic regime). But with the further increase in loading, the strain measured by

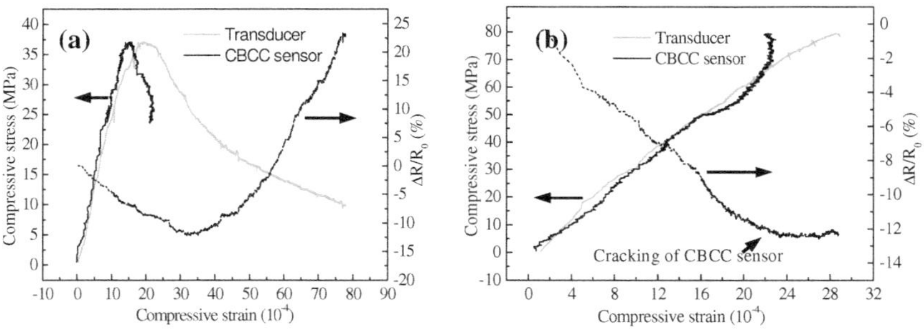

Fig. 14 The strains of (a) C40 concrete column and (b) C80 concrete column under monotonic loading measured by displacement transducer and CBCC sensor, and the fractional change in resistance of CBCC sensors during loading. Strain measured by CBCC sensor was obtained by dividing the fractional change in resistance with the calibrated gauge factor 55.5.

CBCC sensor becomes lower and lower than that measured by displacement transducer and at last turns to be minus. Corresponding to the measured strain-stress curve, the resistance of CBCC sensor decreases at a constant rate at the first stage, then the decrease rate becomes slow and slow and at last turns to increase. For the C80 concrete column, similar phenomenon to that of C40 concrete column is observed from Fig. 14(b), i.e. the strain measured by CBCC sensor agrees well with that measured by displacement transducer when the strain is smaller than 1800 $\mu\varepsilon$ (concrete column in elastic regime), then the strain measured by CBCC sensor becomes smaller than that measured by displacement transducer.

The changing of CBCC sensor's accuracy in different loading stages is caused by the change of CBCC sensor's gauge factor which is related with the transverse deformation characteristics of concrete column in different loading stages. The ratios of transverse strain to the longitudinal strain (RTL) of concrete columns are calculated and shown in Fig. 15. The rapid increase of RTL means the quick development of cracks in concrete column. For C40 concrete column, RTL keeps at about -0.19 when the concrete column is in elastic stage as shown in Fig. 15 (a). On the other hand, the Poisson ratio of CBCC sensor is also in the order of -0.19[32], therefore, the gauge factor of CBCC sensor can be directly used for calculating the strain. However, when the strain of concrete column exceeds 1200 $\mu\varepsilon$, the absolute value of RTL becomes larger and larger than 0.19, indicating that the concrete column steps into plastic stage. Moreover, the increasing RTL causes a different transverse deformation condition for the embedded CBCC sensor from that of the free CBCC sensor and results in the discrepancy between real and using gauge factor. In other words, CBCC sensor's gauge factor at concrete column's plastic stage is different from that calibrated at free condition, however, the gauge factor used for calculating strain was still the calibrated gauge factor 55.5. Hence, the strain measured by CBCC sensor becomes lower and lower than that measured by displacement transducer. The relationship between RTL behavior and the strain measured by CBCC sensors of C80 concrete columns is similar to that of C40 concrete column. The experimental results indicate that the transverse deformation of concrete column affects the gauge factor of CBCC

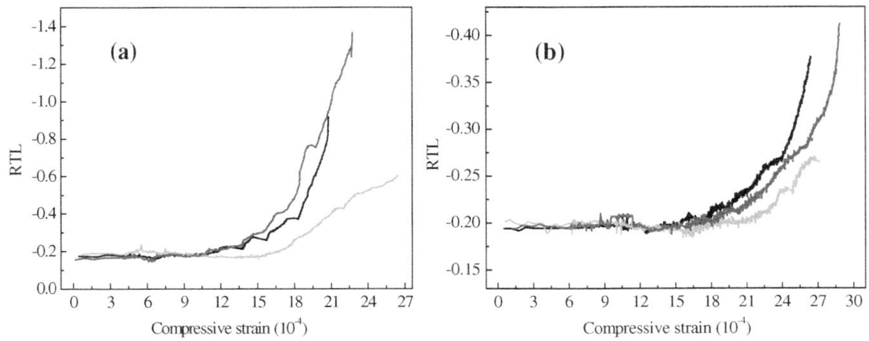

Fig. 15 The RTL during compressive loading of concrete column (a) C40 and (b) C80.

sensor. The calibrated gauge factor of CBCC sensor should be modified based on the proposed piezoresistivity model (see section 2.2) when used in different loading conditions. The experimental results indicate that the CBCC sensor has a good strain sensing ability especially when the concrete column in elastic regime.

2.4.5 Damaged Characteristics of CBCC Sensors Embedded in Various Concrete Columns

The strength of CBCC sensor should match with that of the concrete structure to avoid breaking earlier than the monitored object. The damage of CBCC sensor is defined as the point of its resistance turns to increasing from decreasing.

It can be seen from Fig. 14 (a) that the fractional change in resistance of CBCC sensor embedded in C40 concrete column decreases with strain untill 3230 $\mu\varepsilon$ (the other two CBCC sensors are 3250 $\mu\varepsilon$ and 3013 $\mu\varepsilon$,respectively). On the other hand, the peak strains of the C40 concrete columns are 1919, 2024and 2003 $\mu\varepsilon$, respectively, indicating that the CBCC sensors damge later than the C40 concrete columns. Fig.16 (a) shows an "integral" CBCC sensor in the broken C40 concrete column, supporting the conclusion that CBCC sensor doest not synchronously break with C40 concrete column, but the interface between CBCC sensor and matrix may damage first because of the well known reason that interface in low strength concrete is a weak zone. For CBCC sensor embedded in C80 column, It can be seen from Fig. 14(b) that the fractional change in resistance of the CBCC sensor turns to increase from decrease after the strain of concrete column exceeds 2768 $\mu\varepsilon$ (the other two embedded CBCC sensors are 2607and 2562 $\mu\varepsilon$,respectively), On the other hand, the peak strains of C80 concrete columns are 2849 $\mu\varepsilon$, indicating that the CBCC sensors fail a little earlier than the C80 concrete columns, which can be supported by Fig. 16 (b) that the CBCC sensor is fully fracture, but the interface seems in a good condition.

Generally, the modulus and Poisson ratio between CBCC sensor and concrete are different. Therefore, when a concrete column embedded with CBCC sensor is under loading, the embedded CBCC sensor will be in a multi-axial loading condition and the distributed stress on the embedded CBCC sensor may be different from that on the concrete matrix. Hence, the strength matching is not between the uniaxial compressive strength of a free CBCC sensor and the strength of the concrete matrix, but between the triaxial loading strength (depending on the Poisson ratio relationship between the CBCC sensor and the concrete matrix) of the embedded CBCC sensor and the distributed stress (depending on the modulus relationship between the CBCC sensor and the concrete matrix) on the embedded CBCC sensor at the time of concrete column break. The uniaxial compressive strength of CBCC sensor is much lower than that of C80 concrete. However, they were damaged almost simultaneously, indicating that the strength matching is not between the uniaxial compressive strength of CBCC sensor and the strength of concrete column.

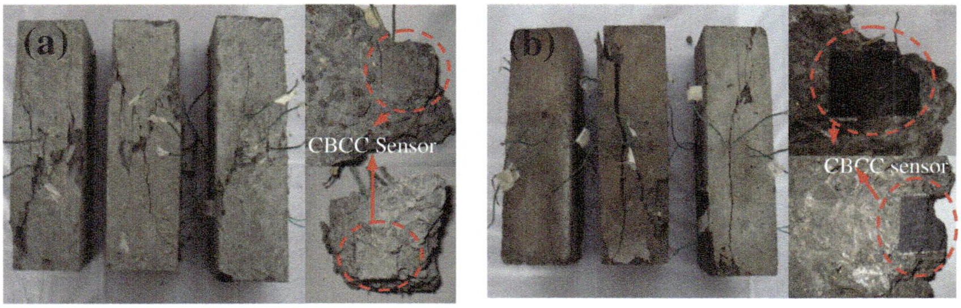

Fig. 16 Failure modes of the embedded CBCC sensors (a) in C40 concrete column and (b) in C80 concrete column.

3 Mechanical Properties of Nano-concret

3.1 Microstructure of Nano-concrete

3.1.1 Materials

A cement paste is composed of small grains of hydrated calcium silicate gels, nanosized individual pores, capillary pores (structural defects), and large crystals of hydrated products. There should be rooms for nanophase materials to improve the properties of pure cement paste. However, as nano-particles are easy to aggregate due to their great surface energy, large quantity of these particles cannot be uniformly dispersed. In this pilot study, the nano-particle contents in the mortar specimens were 3%, 5% and 10% by weight of cement.

Table 3 Mix proportion of the specimens

	Mix proportion of the specimens (per 768 cm^3)							
Mixture no	W/b	Water (ml)	Cement (g)	Sand (g)	Nano-SiO_2 (g)	Nano-Fe_2O_3 (g)	UNF (g)	Silica fume (g)
A	0.5	225	450.0	1350	–	–	–	–
B_1	0.5	225	436.5	1350	–	13.5	3.4	–
B_2	0.5	225	427.5	1350	–	22.5	6.5	–
B_3	0.5	225	405.0	1350	–	45.0	11.2	–
C_1	0.5	225	436.5	1350	13.5	–	6.8	–
C_2	0.5	225	427.5	1350	22.5	–	11.2	–
C_3	0.5	225	405.0	1350	45.0	–	22.5	–
D	0.5	225	427.5	1350	9.0	13.5	7.9	–
E_1	0.5	225	427.5	1350	–	9.0	3.0	13.5
E_2	0.5	225	405.0	1350	–	18.0	6.0	27.0
F	0.5	225	382.5	1350	–	–	3.4	67.5

Although there are various nanophase materials supplied by some companies in China, nano-SiO_2 and nano-Fe_2O_3 only were used in this study. The nano-SiO_2 with a mean grain diameter of 15-nm was purchased from Zhoushan Mingri

Nanophase Material Company, Zhejiang province, China. The nano-Fe_2O_3 with a mean grain diameter of 30-nm and α phase was purchased from Fangyuan Nanophase Material Institute, China. The cement used was Portland cement (P.O32.5). UNF-water reducing agent (UNF) was added to disperse the nano-particles, and defoamer was also used to decrease the amount of air bubbles.

Eleven mixtures were cast with different mix proportions (see Table 3). The water/binder (w/b) ratio for all mixtures was 0.5, where the binder weight is the total weight of cement, nanophase materials and silica fume.

3.1.2 Methods

A rotary mixer with a flat beater was used for mixing. Defoamer and dispersant agent (if applicable) were dissolved in water and then the nano-particles were added and stirred at high speed for about 2 min. Then the cement and silica fume (if applicable) were added to the mixer and stirred for another 1 min. Afterwards, sand was added into the mixture and auto-stirred for about 1.5 min. The well-mixed mortar was poured into molds to form the cubes of size 4×4×4cm for all mixing proportion for compressive testing and prisms of size 4×4×16cm for mixtures A, B_1, B_2, C_1, C_2 and D only for flexural testing. An external vibrator was used to facilitate compaction and decrease the amount of air bubbles. The samples were demolded after 24 hours and then cured in air at room temperature for 7days and 28 days, respectively.

Six cubic specimens were made from each mixture. Three cubes were tested at the 7th day and the other three were tested at the 28th day to observe the influence of different age strengths of mortars with nano-particles. However, three prism specimens for mixtures B_1, C_1, and D were cast for the flexural strength test at the 28th day.

Compressive testing for the strength at the 7th and 28th day was performed using a hydraulic mechanical testing system (MTS) under load control. Bending testing for flexural strength at the 7th day and at the 28th day was carried out on the long surface of prism specimens using a bend tester under load control. In addition, the compressive testing for strength of the specimens after the bending test for mixtures A, C_1, C_2 and D was also conducted at the 28th day, and the results so obtained were compared with those of the cube specimens. After the mechanical tests, the crushed specimens were selected for scanning electronic microscope (SEM) tests.

3.1.3 Strength

Table 4 shows the compressive strength of eleven mortar mixtures. It can be seen that the compressive strengths of specimens with mixtures B_1, B_2 and B_3 at the 7th day and 28th day were all higher than that of plain cement mortar with the same w/b, as for the strength at the 28th day. The effectiveness of the nano-Fe_2O_3 in increasing strength increased in the order: B_1> B_2>B_3 (with the decrease on nano-Fe_2O_3 volume fraction). Furthermore, that the strength enhancement for B_3 at the 7th day is evidently higher than that at the 28th day. These results indicated that the optimal content of nano-Fe_2O_3 for reinforcing concrete purposes should be less than 10% (by weight of cement) under the present dispersion condition.

The compressive strengths of the specimens with mixtures C_1, C_2 and C_3 at the 7th day and 28th day were higher than that of the plain cement mortar with the same w/b too. The effectiveness of the nano-SiO_2 in increasing strength increased in the order: $C_{3>}$ $C_{2>}C_1$ (with the increase on nano-SiO_2 volume fraction), which is opposite to those for the nano-Fe_2O_3 test series.

Comparison of the strength of the cement mortars with and without silica fume are listed in Table 4, it indicates that the nano-particles are more valuable in providing strengthening than silica fume.

Table 5 shows the flexural strength at the 7th day and 28th day. It is increased by the addition of nano-SiO2 or nano-Fe2O3. The effectiveness of the nano-particles in increasing the flexural strength increased in the order: B1> B2 and C1> C2, respectively.

The strength of the mortar mixed with nano-SiO2 and nano-Fe2O3 together was lower than that of the mortar mixed with only nano-SiO2 or nano-Fe2O3.

Table 4 Compressive strength of mixtures

Mixture no	Compressive strength at the 7th day		Compressive strength at the 28th day	
	Target (MPa)	Enhanced extent (%)	Target (MPa)	Enhanced extent (%)
A	17.6	0	28.9 (29.7)	0
B_1	21.4	22.7	36.4	26.0
B_2	20.6	16.7	33.1	14.5
B_3	21.1	20.0	30.0	3.7
C_1	18.6	5.7	32.9 (33.4)	13.8 (12.6)
C_2	21.3	20.1	33.8 (34.1)	17.0 (14.7)
C_3	21.3	20.1	36.4	26.0
D	22.4	27.0	35.4 (34.8)	22.0 (17.1)
E_1	19.4	10.0	29.8	3.0
E_2	23.2	32.0	34.3	18.6
F	18.9	7.4	31.8	10.0

Note that the data in bracket were obtained from the corresponding mixtures of 4×4×8 cm (by-specimens after bending test of specimens with dimension 4×4×16 cm) at the 28th day.

Table 5 Flexural strength of mixtures at the 7th day and 28th day, respectively

Mix-ture no	Flexural strength at the 7th day		Flexural strength at the 28th day	
	Target (Mpa)	Enhanced extent (%)	Target (MPa)	Enhanced extent (%)
A	3.28	0	4.9	0
B_1	–	–	5.8	17.8
B_2	4.3	30	6.0	23.0
C_1	–	–	5.8	19.2
C_2	4.2	28	6.2	27.0
D	–	–	6.0	21.8

3.1.4 Microstructure and Discussion

Figs 17~20 show the microstructure of cement pastes with and without nano-SiO_2 and nano-Fe_2O_3. It was found that in Fig 17 (the microstructure photograph of the plain cement paste) that C-S-H gel existed in the form of 'stand-alone' clusters, lapped and jointed together by many needle hydrates. At the same time, deposit $CaOH_2$ crystals were distributed among the cement paste. The Figs 18~20 show the microstructures of mixtures B_1, C_1 and D, which are of higher strength. They were different from that of the plain cement paste, i.e., the texture of hydrate products was more dense and compact. Big crystals such as $Ca(OH)_2$ were absent. Although the cement paste pattern of these three mixtures showed some differences, their microstructures were uniform and compact.

The mechanism that the nano-particles could improve the microstructure and strength of cement paste can be illustrated as follows. When a small quantity of the nano-particles were uniformly dispersed in the cement paste, the hydrate products of cement will deposit on the nano-particles due to their great surface energy during hydration and grow to form conglomeration containing the nano-particles as 'nucleus'. The nano-particles located in the cement paste as nucleus will further promote and accelerate cement hydration due to their high activity. In the consideration of the nano-particles uniformly disperse situation, a good microstructure could be formed with the uniformly distributed conglomeration. At the same time, according to Wu's 'centroplasm' hypothesis, the aggregates, sands and other particles are considered as centroplasm that acts as skeleton, and gel as transmitter substance. The binding force between centroplasm and transmitter substance has an important effect on the strength of concrete [33]. Innumerable nano-particles distributing in cement paste as 'sub-centroplasm' can tightly bond with the hydrated products around the transition zone between the nano-particle and hydrate products. On the other hand, the nano-particles among the hydrate products will prevent the crystal from growing, such as $CaOH_2$ and AFm, and such fine crystals are favorable for the strength of cement paste [34-36]. Also, the nano-particles will fill pores to increase the strength as silica fume does. However, when the nano-particles cannot be well dispersed, as the case of extensive nano-particles content, the aggregating nano-particles will create weak zone, in form of voids. Consequently, the homogeneous hydrate microstructure could not be formed, and low strength will be expected. On the other hand, as the nano-SiO_2 can participate in the hydration process to generate C-S-H through reacting with $CaOH_2$, the small quantity of aggregating nano-SiO_2 will not be a weak zone, so the strength increases with the content of nano-SiO_2 increases even when small quantity of nano-SiO_2 is not very well dispersed.

The strength of the cement mortars with nano-particles has a preferably improvement, as demonstrated in this study. Furthermore, it can be predicted that the strengthening effect of nano-particles would be further enhanced in concrete because the nano-particles improve not only the cement paste, but also the interface between paste and aggregates.

Fig. 17 SEM photograph of mixture A

Fig. 18 SEM photograph of mixture B1

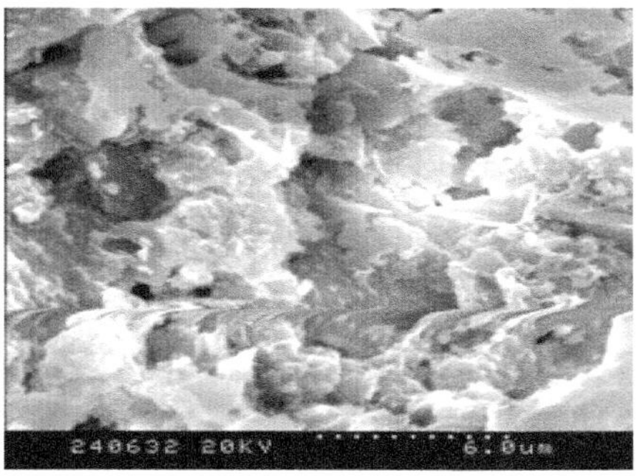

Fig. 19 SEM photograph of mixture C1

Fig. 20 SEM photograph of mixture D

3.2 Abrasion Resistance of Concrete Containing Nano-particles

3.2.1 Materials and Mixture Proportions

The cement used is Portland cement (P.O42.5). Fine aggregate is natural river sand with a fineness modulus of 2.4. The coarse aggregate used is crushed diabase with diameter of 5~25mm. UNF water-reducing agent (one kind of β–naphthalene sulfonic acid and formaldehyde condensates, China) is employed to aid the dispersion of nano-particles in concrete and achieve good workability of concrete. The defoamer, tributyl phosphate (made in China) is used to decrease the amount of air bubbles.

The nano-particles are purchased from Zhoushan Mingri Nano-phase Material Co. (Zhejiang, China) and their properties are shown in Table 6. The modified PP fibers are obtained from Zhangjiagang Synthetic Fiber Co. (Jiangsu, China) and their properties are shown in Table 7.

Table 6 The properties of nano-particles

Item	Diameter (nm)	Specific surface area (m^2/g)	Density (g/cm^3)	Purity (%)	Phase
SiO_2	10 ± 5	640 ± 50	<0.12	99.9	-
TiO_2	15	240 ± 50	0.04~0.06	99.7	Anatase

Table 7 The properties of PP fibers

Item	Elongation (%)	Fiber number (D)	Diameter (μm)	Length (mm)
Target	40 ± 3	11 ± 0.5	84~92	15 ± 1

The water-to-binder (the sum of cement and nano-particles) ratio used for all mixtures is 0.42. Sand ratio is 34%. The mixture proportions for cubic meter of concrete are given in Table 8. Herein, PC denotes plain concrete. PPC1and PPC2 denote the concrete containing PP fibers in the content of 0.6 kg/m^3 and 0.9 kg/m^3, respectively. NSC1 and NSC3 denote the concrete containing nano-SiO_2 in the amount of 1% and 3% by weight of binder, respectively. And NTC1, NTC3 and NTC5 denote the concrete containing nano-TiO_2 in the amount of 1%, 3% and 5% by weight of binder, respectively.

Table 8 Mix proportions of specimens (kg/m^3)

Mixture no.	Water	Cement	Sand	Coarse aggregate	PP fiber	Nano-SiO_2	Nano-TiO_2	UNF	Defoamer	Slump (cm)
PC	151	360	650	1260	-	-	-	5.4	-	5~6
PPC1	151	360	650	1260	0.6	-	-	5.4	-	3~4
PPC2	151	360	650	1260	0.9	-	-	5.4	-	2~3
NSC1	151	356.4	650	1260	-	3.6	-	5.4	0.216	2~3
NSC3	151	349.2	650	1260	-	10.8	-	7.2	0.288	1~2
NTC1	151	356.4	650	1260	-	-	3.6	5.4	0.216	2~3
NTC3	151	349.2	650	1260	-	-	10.8	7.2	0.288	2~3
NTC5	151	342	650	1260	-	-	18	7.2	0.288	1~2

3.2.2 Specimen Fabrication

To fabricate the concrete containing nano-particles, water-reducing agent is firstly mixed into water in a mortar mixer, and then nano-particles are added and stirred at a high speed for 5 min. Defoamer is added as stirring. Cement, sand and coarse aggregate are mixed at a low speed for 2 min in a concrete centrifugal blender, and then the mixture of water, water-reducing agent, nano-particles and defoamer is slowly poured in and stirred at a low speed for another 2 min to achieve good workability.

To fabricate plain concrete and the concrete containing PP fibers, water-reducing agent is firstly dissolved in water. After cement, sand, coarse aggregate and PP fibers (if used) are mixed uniformly in a concrete centrifugal blender, the mixture of water and water-reducing agent is poured in and stirred for several minutes.

Finally, the fresh concrete is poured into oiled molds to form cubes of size 100×100×100 mm that are used for compressive testing, prisms of size 100×100×400 mm for flexural testing and cubes of size 150×150×150 mm for abrasion testing. After pouring, an external vibrator is used to facilitate compaction and decrease the amount of air bubbles. The specimens are de-molded at 24h and then cured in a standard moist room at a temperature of 20℃.

3.2.3 Test Methods

Both compressive and flexural tests are performed in accordance with JTJ 053-94 (Testing methods of concrete for highway engineering, China). Abrasion testing is conducted according to GB/T16925-1997 (Test method for abrasion resistance of concrete and its products, China), testing equipment is ball bearing abrasion machine.

Firstly, the abraded face of specimen is put upward. The abrasion head is put on the abraded face. The machine rotates the abrasion head rubbing the abraded face. The abraded face is flushed by water in order to clean the dust from the abraded face as the machine rotates. The abrasion head abrades a circular groove with diameter of 75mm on the abraded face of specimen. The depth of abrasion groove is measured to be as the initial depth when abrading 30 revolutions. Henceforth, the machine stops once time when rotating per 1000 revolutions, and the depth of abrasion groove is measured. When the revolutions of abrasion head reach 5000 revolutions or the depth of abrasion groove exceeds 1.5mm, the testing can be stopped. The revolutions of abrasion head are recorded and the final depth of abrasion groove is measured.

Then the index of abrasion resistance is defined as

$$I_a = \sqrt{R}/P \tag{14}$$

where, I_a is the index of abrasion resistance; R is the revolutions of abrasion head (kilo revolution); P is the depth of abrasion groove (mm), which is the difference between the final depth and initial depth of abrasion groove.

The index obtained from the test that takes the molded face of specimen as the abraded face is called surface index of abrasion resistance, and that taking the side of specimen as the abraded face is called side index of abrasion resistance. The surface index is slightly discreter than the side index of abrasion resistance, but it is more in agreement with the actual circumstance of pavement concrete.

The larger the value of the index of abrasion resistance is, the stronger the abrasion resistance of concrete is. Generally, when surface index of abrasion resistance is larger than 1.60, the abrasion resistance of concrete pavement can be ensured.

3.2.4 Compressive and Flexural Strengths

Table 9 shows the compressive and flexural strengths of all specimens at the 28th day. It can be seen that, when nano-particles in a small amount are added, both the compressive and flexural strengths of concrete can be enhanced. However, when nano-particles in a large amount are added, the flexural strength of concrete is lower than plain concrete although the compressive strength is still a little enhanced. The effectiveness of nano-TiO_2 in increasing the compressive and

Table 9 Compressive and flexural strengths of specimens

Mixture no.	Flexural strength		Compressive strength	
	Target (MPa)	Enhanced extent (%)	Target (MPa)	Enhanced extent (%)
PC	5.46	0	59.08	0
PPC1	5.99	9.81	61.02	3.28
PPC2	6.60	20.87	63.29	7.12
NSC1	5.69	4.21	66.36	12.31
NSC3	5.36	-1.87	61.16	3.51
NTC1	6.02	10.28	69.73	18.03
NTC3	5.62	3.04	66.62	12.76
NTC5	5.28	-3.27	60.00	1.55

flexure strengths increases in the order: NTC5<NTC3<NTC1 (with the decrease on nano-TiO_2 content) and the similar results can be observed in Table 4 for the concrete containing nano-SiO_2. When the content of PP fibers increases from 0.6kg/m^3 to 0.9kg/m^3, compressive strength of the concrete containing PP fibers enhances only a little, but its flexural strength improves markedly.

If concrete is perfect and has no flaws, its compressive and flexural strengths should be enhanced synchronously. But in practice, the enhanced extent of compressive strength of concrete is greatly larger than that of flexural strength. This is primarily attributed to the presence of micro-cracks with different scales in concrete, and the effect of micro-cracks on flexural strength of concrete is greater than on compressive strength. However, with the addition of PP fibers, the formation and propagation of micro-cracks are inhibited, and the scales of micro-cracks are reduced, so the enhanced extent of flexural strength of the concrete containing PP fibers is higher than that of compressive strength. When the content of nano-particles is larger, such as the concretes containing 3% nano-SiO_2 and 5% nano-TiO_2, the workability of concrete is worse, and the number of micro-cracks in concrete increases, which results in the decrease of flexural strength of concrete. In addition, because nano-particles are more difficult to uniformly disperse when the contents are large, and the weak zone in concrete increases, which also causes the decrease of flexural strength of concrete.

3.2.5 Abrasion Resistance

Table 10 shows the results of abrasion resistance of all specimens at the 28th day. It can be seen that the abrasion resistance of concretes containing nano-particles and PP fibers is remarkably improved, in particular the abrasion resistance of concrete containing nano-particles. The enhanced extent of the abrasion resistance of concrete containing nano-particles is much higher than that of concrete containing PP fibers. The side indices of abrasion resistance of all concretes are larger than their surface indices of abrasion resistance.

Table 10 The results of abrasion resistance of specimens

Mixture no.	Surface index of abrasion resistance		Side index of abrasion resistance	
	Target	Enhancedextent (%)	Target	Enhancedextent (%)
PC	1.19	0	1.55	0
PPC1	1.42	19.1	2.42	55.9
PPC2	1.60	34.4	2.62	69.2
NSC1	3.06	157.0	3.71	139.4
NSC3	2.39	100.8	2.93	89.0
NTC1	3.34	180.7	4.24	173.3
NTC3	2.95	147.7	3.72	140.2
NTC5	2.27	90.4	2.88	86.0

The effectiveness of nano-TiO_2 in enhancing abrasion resistance increases in the order: NTC5<NTC3<NTC1 (with the decrease on nano-TiO_2 content). The abrasion resistance of concrete containing nano-TiO_2 in the amount of 1% by weight of binder increases by 180.7% for the surface index and 173.3% for the side index. Even for the concrete containing nano-TiO_2 in the amount of 5% by

weight of binder, the abrasion resistance increases by 90.4% for the surface index and 86% for the side index. The similar results can be found for the concrete containing nano-SiO_2.

The index of abrasion resistance of concrete containing PP fibers increases with increasing fibers content. The enhanced extent is almost the same as that presented in the related literatures.

The addition of nano-particles is much more favorable to the abrasion resistance of concrete than that of PP fibers.

The mechanism of nano-particles improving the abrasion resistance of concrete can be interpreted as follows. Supposed that nano-particles are uniformly dispersed and each particle is contained in a cube pattern, the distance between nano-particles can be specified. After hydration begins, hydrate products diffuse and envelop nano-particles as kernel. If the content of nano-particles and the distance between them are appropriate, the crystallization will be controlled to be a suitable state through restricting the growth of $Ca(OH)_2$ crystal by nano-particles. This makes the cement matrix more homogeneous and compact. As a consequence, the abrasion resistance and strength are improved evidently such as the concrete containing 1% nano-TiO_2. With increasing content of nano-particles, the distance between nano-particles decreases, $Ca(OH)_2$ crystal can not grow up enough, which leads to the ratio of crystal to C-S-H gel small and the microstructure of cement matrix is loose. The abrasion resistance and strength of concrete decrease relatively.

In addition, because cement content and water content in this study are small, the slump of fresh concrete, especially the concrete containing nano-particles is less than 60mm, which leads to a thin mortar layer on the surface of concrete forming. The thin mortar layer is favorable to the improvement of abrasion resistance of concrete.

However, the abrasion resistance and strength of concrete are decreased with increasing content of nano-particles. Studies have shown that the content of nano-particles has a strong influence on the water demand of cement paste. The larger the content of nano-particles is, the more the water demand of cement paste is. If the mixture proportions of concrete are the same, the workability of concrete will decrease remarkably with increasing content of nano-particles, which consequently decreases the abrasion resistance and strength of concrete. In addition, the uniform dispersion of nano-particles in cement paste is difficult with increasing content of nano-particles, which also decreases the strength and abrasion resistance of concrete.

The abrasion resistance of concrete containing modified PP fibers is improved, which is mostly attributed to the crack-arresting effect and crack-thinning effect of PP fibers, and the bridge effect of PP fibers on cracks and diversion effect of PP fibers on separated cement blocks. The larger the content of PP fibers is, the stronger these effects are. In addition, friction work is consumed when PP fibers are pulled out from concrete.

3.2.6 The Relationship between Abrasion Resistance and Compressive Strength of Concrete

As mentioned in introduction, there are different views on the relationship between abrasion resistance and compressive strength of concrete at present. The relationship between abrasion resistance and compressive strength of concrete is obtained from this test, and shown in Fig. 21 and 22, respectively.

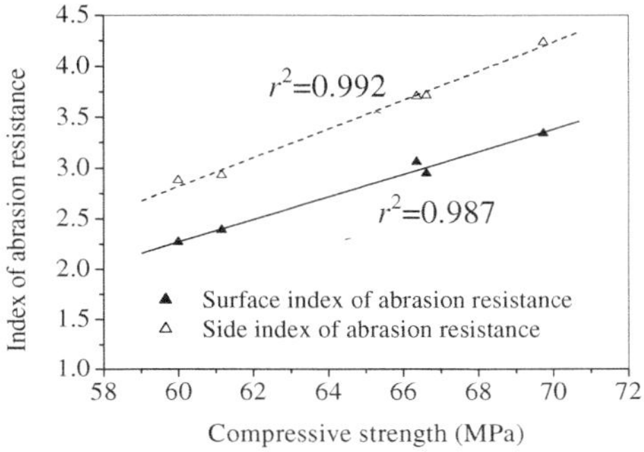

Fig. 21 The relationship between the index of abrasion resistance and compressive strength of concrete containing nano-particles

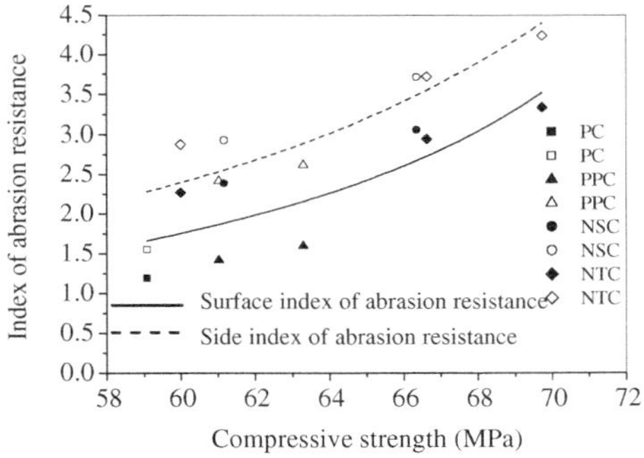

Fig. 22 The relationship between the index of abrasion resistance and compressive strength for all mixtures

It can be seen from Fig.21 that the indices of abrasion resistance of concrete containing nano-particles increase with increasing compressive strength. The relationship appears to be linear, and the correlation coefficient is close to 1.0. The correlation of side index of abrasion resistance is slightly larger than that of surface index.

Fig. 22 shows the total relationship between the indices of abrasion resistance and compressive strength for plain concrete, the concrete containing nano-particles and the concrete containing PP fibers. It can be seen that the curves approach hyperbola basically, which confirms that compressive strength is a critical factor affecting the abrasion resistance of concrete.

The relationship between the index of abrasion resistance and compressive strength can be expressed by

$$I_a = \frac{f_{cu}}{\theta_1 f_{cu} + \theta_2} \tag{15}$$

where, I_a is the index of abrasion resistance of concrete; f_{cu} is the compressive strength (MPa); θ_1 and θ_2 are constants that can be obtained by curve fitting techniques and reported in the Table 11.

Table 11 Regression coefficient in the Eq. (15)

Index of abrasion resistance	θ_1	θ_2	Correlation coefficient r^2
Surface index of abrasion resistance	-1.47819	122.867	0.67625
Side index of abrasion resistance	-0.94309	81.5936	0.78734

3.3 Flexural Fatigue Performance of Concrete Containing Nano-particles for Pavement

3.3.1 Materials and Mixture Proportions

The materials and fabrication methods were same to that shown in section 2.2. Six kinds of mixtures are cast in this study. The water/binder ratio used for all mixtures is 0.42, where the binder weight is the total weight of cement and nano-particles. Sand ratio is 34%. The mixture proportions for cubic meter of concrete are given in Table 12. Herein, PC denotes plain concrete. PPC denotes the concrete containing PP fibers in the content of 0.9 kg/m^3. NSC1 denotes the concrete containing nano-SiO_2 in the amount of 1% by weight of binder. NTC1 and NTC3 denote the concrete containing nano-TiO_2 in the amount of 1% and 3% by weight of binder, respectively. And NTPC denotes the concrete containing both nano-TiO_2 in the amount of 1% by weight of binder and PP fibers in the content of 0.9 kg/m^3.

Table 12 Mix proportions of specimens (kg/m^3)

Mixture type	Water	Cement	Fine aggregate	Coarse aggregate	PP fiber	Nano-SiO_2	Nano-TiO_2	UNF	Defoamer	Slump (cm)
PC	151	360	650	1260	-	-	-	5.4	-	5~6
PPC	151	360	650	1260	0.9	-	-	5.4	-	2~3
NSC1	151	356.4	650	1260	-	3.6	-	5.4	0.216	2~3
NTC1	151	356.4	650	1260	-	-	3.6	5.4	0.216	2~3
NTC3	151	349.2	650	1260	-	-	10.8	7.2	0.288	2~3
NTPC	151	356.4	650	1260	0.9	-	3.6	7.2	0.288	-

3.3.2 Test Methods

Flexural strength testing is performed in accordance with JTJ 053-94 (Testing Methods of Concrete for Highway Engineering, China).

For flexural fatigue testing, a four-point bending test method is applied with an effective span of 300 mm in a 100 kN Material Testing System (MTS). The test is carried out in load control using a continuous sinusoidal waveform with a loading frequency of 10 Hz.

The load cycle characteristic value R is defined as follows: $R = P_{\min}/P_{\max}$, where $P_{\min}$ and $P_{\max}$ refer to the minimum and maximum load of sinusoidal wave in each cycle. R is taken as 0.1 in this test.

The stress level S is defined as: $S = \sigma_p/\sigma_f$, where σ_p and σ_f are the flexural fatigue strength and the flexural strength, respectively. The following four stress levels are selected: 0.70, 0.75, 0.80 and 0.85 for all specimens referred in this study.

The input data for the test include the waveform, maximum and minimum load amplitude, loading frequency, maximum number of cycles.

The reference for the loading stress levels is the average ultimate static flexural strength of specimens measured just before the fatigue testing. The preload of 100~200 N is put on the specimen to eliminate the error caused by poor contact. The fatigue failure numbers of specimens are recorded.

3.3.3 Fatigue Equation and Weibull Distribution

3.3.3.1 The Type of Fatigue Equation and the Physical Meaning of the Parameters in Fatigue Equation

There are two types for fatigue equation. One is single-logarithm fatigue equation, i.e. $S = \lg\alpha - \beta\lg N$; the other is double-logarithm fatigue equation, i.e. $\lg S = \lg\alpha - \beta\lg N$. Where, S is the stress level, N is the fatigue life of concrete.

The fatigue performance of concrete is dependent on the two important parameters (α and β) in fatigue equation. The parameter α reflects the height of fatigue curve. The larger the parameter α is, the higher the fatigue curve is, and the better the fatigue performance of concrete is. The parameter β reflects the steep degree of fatigue curve. The larger the parameter β is, the steeper the fatigue curve is, and the fatigue life of concrete is more sensitive to the change of stress.

At present, the single-logarithm fatigue equation is extensively used. It is available in the common range ($0.55<S<0.85$) of fatigue life N, but it cannot extend. The double-logarithm fatigue equation is a perfect type. It can not only agree well with the test results but also extend suitably. Both two types of fatigue equations are used to analyze the fatigue performance of concrete in this study.

3.3.3.2 Weibull Distribution

Because concrete belongs to the heterogeneous material, the data of flexural fatigue test are very scattered, even though the stress level S is the same. Therefore, it is necessary to take an appropriate method to process the data.

In 1939, Weibull, an engineer of Sweden put forward a probability density distribution function to process the data of fatigue test [37,38]. The cumulative distribution function $P_f(N)$ of double-parameters Weibull distribution can be expressed by

$$P_f(N) = 1 - \exp\left[-\left(\frac{N}{u}\right)^{\lambda}\right] \tag{16}$$

The function $P_f(N)$ corresponds to the failure probability. Where, λ is the shape parameter or the Weibull slope at the stress level S; u is the scale parameter; N is the position parameter or the fatigue life of concrete at the stress level S.

Taking twice natural logarithm for both sides of Eq. (16) gives

$$\ln\ln\frac{1}{1-P_f(N)} = \lambda\ln(N) - \lambda\ln(u) \tag{17}$$

Setting $Y = \ln\ln\frac{1}{1-P_f(N)}$, $X = \ln(N), \eta = \lambda\ln(u)$, then

$$Y = \lambda X - \eta \tag{18}$$

It can be seen from Eq. (18) that, if the relationship between Y and X is linear (when the correlation coefficient r is larger), the data of fatigue test follow the Weibull distribution. On the contrary, it will be not true.

According to Eq. (17), it can be figured out

$$N = \ln^{-1}\left\{\left\{\ln\left[\ln\frac{1}{1-P_f(N)}\right] + \lambda\ln u\right\}\Big/\lambda\right\} = e^{\frac{\ln\left[\ln\frac{1}{1-P_f(N)}\right]+\eta}{\lambda}} = e^{\frac{Y+\eta}{\lambda}} \tag{19}$$

The corresponding relationship between the fatigue life and failure probability can be found from Eq. (19). When a failure probability p is given, the corresponding fatigue life N at certain stress level S can be calculated.

3.3.4 Flexural Strengths

Table 13 shows the flexural strengths of all specimens at the 28th day. It can be seen that the flexural strengths of the specimens with additives NSC1, NTC1 and NTC3 are all higher than that of plain concrete with the same water/binder ratio. The effectiveness of nano-particles in enhancing flexural strength increases in the order: NTC3 <NSC1< NTC1.

Table 13 Flexural strengths of specimens

Mixture type	PC	PPC	NSC1	NTC1	NTC3	NTPC
Target (MPa)	5.46	6.60	5.69	6.02	5.62	4.85
Variation coefficient (%)	7.56	8.35	4.85	7.77	7.84	8.55
Enhanced extent (%)	0	20.87	4.21	10.28	3.04	-11.21

Generally, nano-particles are more difficult to uniformly disperse when the content is large, and thus the workability of concrete is worse, the number of microcracks and weak zone in concrete increase, which leads to smaller enhancement in flexural strength, e.g. the flexural strength of NTC3 is smaller than that of NTC1.

The flexural strength of NSC1 is smaller than that of NTC1 even though the contents of the two kinds of nano-paticles are the same. It may be attributed to the fact that the specific surface area of nano-SiO_2 is much larger than that of nano-TiO_2, so nano-SiO_2 is more difficult to uniformly disperse than nano-TiO_2 in concrete.

The flexural strength of PPC improves markedly. Due to the addition of PP fibers, the formation and propagation of microcracks in concrete are inhibited, and the scales of microcracks are reduced.

However, the flexural strength of NTPC is lower than that of plain concrete, which indicates that the interaction between PP fibers and nano-particles is negative. When nano-TiO_2 and PP fibers are respectively added into concrete, the workability of fresh concretes (NTC1 and PPC) becomes bad compared with plain concrete (shown in Table 13); when nano-TiO_2 and PP fibers are added into concrete together, the workability of fresh concrete (NTPC) becomes worse, which leads to the low density of hardened concrete, more porosity in concrete and low flexural strength.

3.3.5 Weibull Distribution Verification

According to the probability theory of Weibull distribution, the failure probability p corresponding to the failure life N can be expressed by

$$p = \frac{i}{k+1} \tag{20}$$

where, k is the total number of the fatigue test data at certain stress level; i is the sequence number of failure specimens at certain stress level, i=1, 2, …, k.

Then the linear regression is carried out for the fatigue test data according to Eq. (18) when the stress level S is respectively equal to 0.70, 0.75, 0.80 and 0.85. Table 14 shows the results of Weibull regression analysis for all concretes referred in this study.

Table 14 The results of Weibull regression analysis

Stress level *S*	PC			PPC			NSC1		
	λ	η	r	λ	η	r	λ	η	r
0.85	0.5372	3.1176	0.9648	0.5084	3.7853	0.9776	0.5555	3.5917	0.9897
0.80	0.4552	3.6308	0.9947	0.7748	5.8196	0.9838	0.7047	5.0214	0.9916
0.75	0.5516	4.8485	0.9950	0.6405	5.7947	0.9948	0.6991	6.0511	0.9554
0.70	0.5639	5.7642	0.9636	0.7062	7.2156	0.9952	0.6125	6.2229	0.9925

Table 15 The results of Weibull regression analysis (continued)

Stress level *S*	NTC1			NTC3			NTPC		
	λ	η	r	λ	η	r	λ	η	r
0.85	0.7687	4.8185	0.9609	0.5402	3.6665	0.9928	0.6178	4.0804	0.9539
0.80	0.6269	5.1252	0.9903	0.4725	3.8831	0.9947	0.4628	4.0289	0.9906
0.75	0.7951	7.4281	0.9995	0.6216	5.7011	0.9960	0.7052	6.4582	0.9929
0.70	0.7913	8.0021	0.9982	0.6277	6.2475	0.9922	0.6252	6.3174	0.9974

It can be seen from Table 15 that all correlation coefficients are larger than 0.95, which indicates that the relationship between $\ln\ln\frac{1}{1-P_f(N)}$ and $\ln(N)$ is linear for four stress levels. This validates that the fatigue lives of all concretes referred in this study follow the double-parameters Weibull distribution.

In this study, the test data processed by Weibull distribution is used to analyze the flexural fatigue performance of various concretes.

3.3.6 Flexural Fatigue Performance of Concretes under the Same Failure Probability

The test data that have 10% failure probability after being processed by Weibull distribution are used to compare the flexural fatigue performance of various concretes.

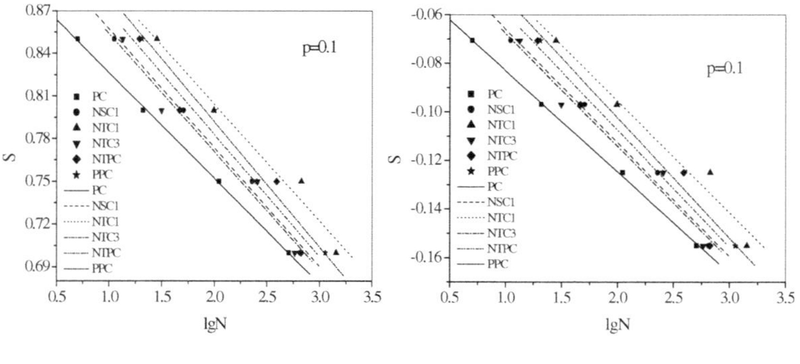

Fig. 23 Fatigue curves of concretes when the failure probability p is 10%

Fig. 23 shows the fatigue curves of various concretes when the failure probability p is 10%. Although the slopes of fatigue curves are little different for various concretes, on the whole the flexural fatigue performance of concretes increases in the order: PC< NTC3<NSC1<NTPC<PPC<NTC1. The results of single-logarithm fatigue equation are consistent with that of double-logarithm fatigue equation.

Table 16 Fatigue equations of concretes (failure probability p=0.1)

Mixture type	Single-logarithm fatigue equation		Double-logarithm fatigue equation	
	$S = \alpha - \beta \lg N$	Correlation coefficient r	$\lg S = \lg \alpha - \beta \lg N$	Correlation coefficient r
PC	$S = 0.9005 - 0.0741 \lg N$	0.9995	$\lg S = -0.0413 - 0.0417 \lg N$	0.9995
PPC	$S = 0.9678 - 0.0880 \lg N$	0.9954	$\lg S = -0.0035 - 0.0494 \lg N$	0.9946
NSC1	$S = 0.9403 - 0.0834 \lg N$	0.9966	$\lg S = -0.0191 - 0.0467 \lg N$	0.9939
NTC1	$S = 0.9682 - 0.0820 \lg N$	0.9880	$\lg S = -0.0035 - 0.0460 \lg N$	0.9851
NTC3	$S = 0.9365 - 0.0828 \lg N$	0.9826	$\lg S = -0.0211 - 0.0465 \lg N$	0.9818
NTPC	$S = 0.9545 - 0.0857 \lg N$	0.9746	$\lg S = -0.0112 - 0.0481 \lg N$	0.9723

Table 17 Regression parameters of fatigue equations (failure probability p=0.1)

Mixture type	Single-logarithm fatigue equation				Double-logarithm fatigue equation			
	α	Enhanced extent (%)	β	Enhanced extent (%)	α	Enhanced extent (%)	β	Enhanced extent (%)
PC	0.9005	0.00	0.0741	0.00	0.9093	0.00	0.0417	0.00
PPC	0.9678	7.47	0.0880	18.76	0.9920	9.09	0.0494	18.47
NSC1	0.9403	4.42	0.0834	12.55	0.9570	5.24	0.0467	11.99
NTC1	0.9682	7.52	0.0820	10.66	0.9920	9.09	0.0460	10.31
NTC3	0.9365	4.00	0.0828	11.74	0.9526	4.76	0.0465	11.51
NTPC	0.9545	6.00	0.0857	15.65	0.9745	7.18	0.0481	15.35

Tables 16 and 17 show the fatigue equations of various concretes and their regression parameters (α and β) when the failure probability p is 10%. It can be seen that, the regression parameters of fatigue equations of concretes containing nano-particles and PP fibers are all increased in different extent, and all correlation coefficients are larger than 0.97. The increase of α indicates that the flexural fatigue performance of concretes is significantly improved. The increase of β indicates that the fatigue curves of concretes become steep, and the sensitivity of their fatigue lives to the change of stress is increased.

For the concrete containing PP fibers, when the single-logarithm fatigue equation is used, the values of α and β increase by 7.47% and 18.76%, respectively; and when the double-logarithm fatigue equation is used, the values of α and β increase by 9.09% and 18.47%, respectively. The similar conclusion can be found in related literatures [39,40].

For the concrete containing 1% nano-TiO_2, the values of α and β increase by 7.52% and 10.66% (in the single-logarithm fatigue equation), respectively, and 9.09% and 10.31% (in the double -logarithm fatigue equation), respectively. Compared with other concretes, the enhanced extent of α is the largest, whereas, the enhanced extent of β is the lowest. Therefore, the fatigue performance of NTC1 is improved in the largest extent, and the sensitivity of its fatigue life to the change of stress is increased in the smallest extent in this test.

For NSC1 and NTC3, the enhanced extent of α is smaller, whereas, the enhanced extent of β is larger. Therefore, the fatigue performance of NSC1 and NTC3 improves a little, but their fatigue lives are more sensitive to the change of stress. The fatigue performance of NTC3 and NSC1 is close to each other.

For the concrete containing both nano-particles and PP fibers, i.e. NTPC, the enhanced extent of α is smaller than NTC1and PPC, whereas the enhanced extent of β is larger than NTC1 and smaller than PPC. Therefore, the fatigue performance of NTPC is poor than NTC1and PPC, but the sensitivity of its fatigue life to the change of stress is intervenient.

Tables 18 and 19 show the theoretic fatigue lives of various concretes calculated by using single-logarithm fatigue equation and double-logarithm fatigue equation, respectively, at four stress levels. It can be seen that the theoretic fatigue lives of concretes containing nano-particles and PP fibers are all enhanced in different extent. With increasing stress level, the enhanced extent of theoretic fatigue number of the concretes containing nano-particles and PP fibers is increased remarkably, which indicates that the concretes containing nano-particles and PP fibers have excellent flexural fatigue performance at high stress level (corresponding to heavy traffic loads) compared with plain concrete. Additionally, the fatigue lives of concretes containing nano-particles and PP fibers are also decreased with increasing stress level.

Table 18 Theoretic fatigue lives of concretes calculated by single-logarithm fatigue equation

	$S = 0.85$		$S = 0.80$		$S = 0.75$		$S = 0.70$	
Mixture type	Theoretic fatigue number	Enhanced extent (%)	Theoretic fatigue number	Enhanced extent (%)	Theoretic fatigue number	Enhanced extent (%)	Theoretic fatigue number	Enhanced extent (%)
PC	10	0.00	45	0.00	215	0.00	1016	0.00
PPC	44	354.07	161	255.26	597	177.95	2209	117.46
NSC1	24	151.90	96	111.83	383	78.13	1522	49.79
NTC1	55	475.38	225	395.39	916	326.52	3730	267.22
NTC3	22	130.77	89	96.01	358	66.49	1437	41.41
NTPC	33	245.03	127	179.59	487	126.56	1865	83.59

The fatigue lives, as well as their enhanced extent calculated by using single-logarithm fatigue equation are almost the same as that calculated by using double-logarithm fatigue equation. The fatigue lives of various concretes increase in the order: PC<NTC3<NSC1<NTPC<PPC<NTC1.

It can be seen from Table 18 that the theoretic fatigue number of PPC increases by 354.07% when the stress level S is 0.85, which is approximately consistent with the test results in related literature [40]. However, the theoretic fatigue number of NTC1 increases by 475.38% at the same stress level. When the stress level

Table 19 Theoretic fatigue lives of concretes calculated by double-logarithm fatigue equation

Mixture type	$S = 0.85$		$S = 0.80$		$S = 0.75$		$S = 0.70$	
	Theoretic fatigue number	Enhanced extent (%)	Theoretic fatigue number	Enhanced extent (%)	Theoretic fatigue number	Enhanced extent (%)	Theoretic fatigue number	Enhanced extent (%)
PC	10	0.00	43	0.00	203	0.00	1060	0.00
PPC	46	352.62	156	260.84	575	183.50	2322	119.05
NSC1	25	151.30	93	115.08	369	82.23	1618	52.65
NTC1	57	470.30	215	397.83	873	330.77	3912	269.05
NTC3	23	130.10	85	98.04	342	68.80	1508	42.30
NTPC	34	240.68	121	180.76	463	128.51	1944	83.36

S is 0.70, the theoretic fatigue number of PPC increases by 117.46%, however, the theoretic fatigue number of NTC1 increases by 267.22%. Apparently, the concrete containing 1% nano-TiO_2 has much better flexural fatigue performance than the concrete containing PP fibers.

The enhanced extent of fatigue life of the concrete containing 3% nano-TiO_2 is the lowest. Even for this concrete, the theoretic fatigue number increases by 130.77% and 41.41% at the stress levels of 0.85 and 0.70, respectively.

Table 20 presents the theoretic stress level of NTC1 and PC calculated by fatigue equation. The theoretic stress level of NTC1 is enhanced compared with PC when the fatigue failure number *N* is equal to 10^6. The theoretic stress level of NTC1 increases by 4.45% (calculated by single-logarithm fatigue equation) and 2.80% (calculated by double-logarithm fatigue equation). This further validates that the fatigue performance of NTC1 is superior to PC.

Table 20 Theoretic stress level of NTC1 and PC when the fatigue failure number *N* is 10^6

Mixture type	Single-logarithm fatigue equation		Double-logarithm fatigue equation	
	Theoretic stress level	Enhanced extent (%)	Theoretic stress level	Enhanced extent (%)
PC	0.4559	0.00	0.5111	0.00
NTC1	0.4762	4.45	0.5254	2.80

3.3.7 Mechanism of Fatigue Improvement

The addition of PP fibers benefits the fatigue performance of concrete under flexural cyclic loading. This can be attributed to that the fibers are able to bridge microcracks and retard their growth, thereby prolonging the fatigue life of concrete. Otherwise, the addition of PP fibers can increase the ability to absorb energy of concrete [41]. That is to say, in the concrete containing PP fibers, the action of fibers bridging and fibers pullout dissipate energy in the wake of the crack tip, and therefore substantially improve load-bearing capacity and resistance to crack growth. Furthermore, the inclusion of PP fibers produces a more ductile behavior during fatigue loading [42].

The mechanism of nano-particles improving the flexural fatigue performance of concrete can be interpreted as follows. Supposed that nano-particles are uniformly dispersed and each particle is contained in a cube pattern, the distance between nano-particles can be specified. After hydration begins, hydrate products diffuse and envelop nano-particles as kernel. If the content of nano-particles and the distance between them are appropriate, the crystallization will be controlled to be a suitable state through restricting the growth of $Ca(OH)_2$ crystal by nano-particles. Moreover, the nano-particles located in cement paste as kernel can further promote cement hydration due to their high activity. This makes the cement matrix more homogeneous and compact. Additionally, nano-particles can act as fillers to improve the density of concrete, which leads to the porosity of concrete reduced significantly. As a consequence, the flexural fatigue performance and strength of concrete are improved evidently, as presented in previous sections of this article.

Studies have shown that the content and specific surface area of nano-particles have a strong influence on the water demand of cement paste. The larger the content of nano-particles is, the more the water demand of cement paste is. If the water/binder ratio is the same for the concrete containing different contents of nano-particles, the workability of concrete will decrease with increasing content of nano-particles, which consequently has a disadvantageous influence on the flexural fatigue performance and strength of concrete. Additionally, the nano-particles become more difficult to disperse uniformly in cement paste with increasing content of nano-particles. The aggregation of nano-particles will form weak zones in cement paste. As a result, the enhanced extent of flexural fatigue performance and strength of concrete decreases, e.g. NTC3.

NTC1 has better flexural fatigue performance than NSC1 even though they contain the same content of nano-particles. This may be attributed to the larger specific surface area of nano-SiO_2 than that of nano-TiO_2, which leads to nano-SiO_2 being more difficult to disperse uniformly in cement paste, as mentioned early in the article.

The flexural fatigue performance of the concrete containing both nano-particles and PP fibers (NTPC) is inferior to that of the concrete containing the same amount of nano-TiO_2 alone or PP fibers alone. This is primarily attributed to the bad workability of NTPC.

4 Future of Multifunctional Nano-concrete

Concrete structures constitute a large portion of civil infrastructures, but their reliability is relatively low because of wide material discreteness and complex service environment. Consequently, the safety of concrete structures is an important problem being paid attention to at all times in civil engineering field. Therefore, it is necessary to take reasonable measures to monitor the state of concrete structures. In order to monitor the performance and state of concrete structures during their service periods, the information of structural state need be obtained by appropriate monitoring technologies. Now local monitoring for concrete structures is usually achieved by embedding such sensors as electric-resistance strain gauges, optic sensors, piezoelectric ceramic, shape memory alloy and fiber reinforced polymer bar in key structural positions. However, these sensors have such drawbacks as poor durability, low sensitivity, high cost, and unfavorable compatibility with concrete structures. Smart concrete has favorable piezoresistivity, great durability and good compatibility with concrete structures, etc., and it can therefore be used to develop retrofit or new installations, including traffic monitoring, weighing in motion, corrosion monitoring of rebar, strain-sensing coating.

Acknowledgement

This work is financially supported by NSFC grants No. 50525823, 50238040, 50808059, 50538020, 50278029, the Ministry of Science and Technology grant No. 2006BAJ03B05 and 2007AA04Z435, the Outstanding Young Teachers in Harbin Institute of Technology HITQNJS.2007.030.

References

1. Scott, W.D., Charles, R.F., Michael, B.P., Daniel, W.S.: Damage Identification and Health Monitoring of Structural and Mechanical Systems from Changes in Their Vibration Characteristics: A Literature Review (1996); Report of Los Almos Lab
2. Chen, J.C.: Intelligent Monitoring System for Suspension Bridge Damage Detection. In: Proc. of Bridge into the 21st Century Conference, Hongkong (October 25, 1995)
3. Li, H., Ou, J.P., Zhao, X.F.: Structural health monitoring system for the Shandong Binzhou Yellow River Highway Bridge. Computer-Aided Civil and Infrastructure Engineering 21(4), 306–317 (2006)
4. Austin, T., Singh, M., Gregson, P.J., Dakin, J.P., Powell, P.M.: Damage Asseeement in Hybrid Laminates Using an Array of Embedded Fiber Optic Sesnors. In: Proc. The SPIE conference on Smart Systems for Bridges, Structures and highways, Newport Beach, California, SPIE, vol. 3671, pp. 281–287 (March 1999)
5. Claus, R.O., Mckeenman, J.C., May, R.G., et al.: Opitical fiber sensors and signal processing for smart materials and Structures ARO Smart Materials. In: Structures and Mathematical Issues workshop Proceeding Virginia Polytechnic Institute and state University Blacksburg, VA, pp. 15–16 (1988)
6. Li, H., Xiao, H.G., Ou, J.P.: Effect of compressive strain on electrical resistivity of carbon black-filled cement-based composites. Cement and Concrete Composites 28, 824–828 (2006)
7. Xiao, H.G., Li, H., Ou, J.P.: Self-monitoring properties of concrete columns with embedded cement-based strain sensors. Journal of Intelligent Materials Systems and Structures 22, 191–200 (2011)
8. Wen, S.H., Chung, D.D.L.: Carbon fiber-reinforced cement as a strain-sensing coating. Cement and Concrete Research 31, 665–667 (2001)
9. Reza, F., Batson, G.B., Yamamuro, J.A., Lee, J.S.: Resistance changes during compression of carbon fiber cement composites. Journal of Materials in Civil 15, 476–483 (2003)
10. Song, X.H., Zheng, L.X., Li, Z.Q.: Temperature compensation in deformation testing for smart concrete structures. Key Engineering Materials 326-328, 1503–1506 (2006)

11. Han, B.G., Ou, J.P.: Embedded piezoresistive cement-based stress/strain sensor. Sensors and Actuators A 138, 294–298 (2007)
12. Ou, J.P., Han, B.G.: Piezoresistivie cement-based strain sensors and self-sensing concrete components. Journal of Intelligent Material Systems and Structures 20, 329–336 (2009)
13. Yu, X., Kwon, E.: A carbon nanotube/cement composite with piezoresistive properties. Smart Materials and Structures 18, 1–5 (2009)
14. Chen, B., Liu, J.Y.: Damage in carbon fiber-reinforced concrete, monitored by both electrical resistance measurement and acoustic emission analysis. Construction and Building Materials 22, 2196–2201 (2008)
15. Wang, S., Shui, X., Fu, X., Chung, D.D.L.: Early fatigue damage in carbon-fibre composites observed by electrical resistance measurement. Journal of Materials Science 33, 3875–3884 (1998)
16. Xiao, H.G., Li, H.: A study on the application of CB-filled cement-based composites as a strain sensor for concrete structures. In: SPIE Nondestructive Evaluation and Health Monitoring of Aerospace Materials, Composites, and Civil Infrastructure, San Diego, CA, USA (March 2006)
17. Li, H., Xiao, H.G., Ou, J.P.: Electrical property of cement-based composites filled with carbon black under long-term wet and loading condition. Composites Science and Technology 68, 2114–2119 (2008)
18. Xiao, H.G., Li, H., Ou, J.P.: Modeling of piezoresistivity of carbon black filled cement-based composites under multi-axial strain. Sensors and Actuators A: Physical 160, 87–93 (2010)
19. Li, H., Xiao, H.G., Yuan, J., Ou, J.P.: Microstructure of cement mortar with nano-particles. Composites Part B: Engineering 35(2), 185–189 (2004)
20. Li, H., Xiao, H.G., Ou, J.P.: A study on mechanical and pressure-sensitive properties of cement mortar with nanophase materials. Cement and Concrete Research 34(3), 435–438 (2004)
21. Lau, K.T., Hui, D.: The Revolutionary Creation of New Advanced Materials-Carbon Nanotube Composites. Composites: Part B 33, 263–277 (2002)
22. Yihua, Z., Hongjie, Z., Jinyi, H.L.H.: Electric Properties of Ag/Si_3N_4 Nanostructured Composites. Journal of Inorganic Materials 11, 348–352 (1996)
23. Qing, Y.: Research on the Comparison of Pozzolanic Activity between Nano SiO_2 and Silica Fume. Concrete 3, 19–22 (2001) (in Chinese)
24. Li, H., Zhang, M.H., Ou, J.P.: Abrasion resistance of concrete containing nano-particles for pavement. Wear 260(11-12), 1262–1266 (2006)
25. Li, H., Zhang, M.H., Ou, J.P.: Flexural fatigue performance of concrete containing nano-particles for pavement. International Journal of Fatigue (29), 1292–1301 (2007)
26. Zhang, M.H., Li, H.: The resistance to chloride penetration of concrete containing nano-particles for pavement. In: Proc. of SPIE The SPIE Nondestructive Evaluation for Health Monitoring and Diagnostics Symposium-Testing, Reliability and Application of Micro- and Nano-Material Systems IV, San Diego, USA, March 2006, vol. 6175, 61750E (2006)
27. Xiao, H.G., Lan, C.M., Ji, X.Y., Li, H.: Mechanical and sensing properties of structural materials with nanophase materials. Pacific Science Review 5, 122–127 (2003)
28. Han, B.G., Guan, X.C., Ou, J.P.: Electrode design, measuring method and data acquisition system of carbon fiber cement paste piezoresistive sensors. Sensors and Actuators: A Physical 135, 360–369 (2007)

29. Boettger, H., Bryksin, U.V.: Hopping conduction in solids, pp. 108–148. Berlin Verlag Akademie, Berlin (1986)
30. Simmons, J.G.: Generalized formula for the electric tunnel effect between similar electrodes separated by a thin insulating film. Journal Applied Physics 34, 1793–1803 (1963)
31. Wen, S.H., Chung, D.D.L.: Electric polarization in carbon fiber-reinforced cement. Cement and Concrete Research 31, 141–147 (2001)
32. Xiao, H.G.: Piezoresistivity of Cement-based Composites Filled with Nanophase Materials and Self-sensing Smart Structural System. PhD Thesis, Harbin Institute of Technology, School of Civil Engineering (2006)
33. Zhongwei, W., Huizhen, L.: High Performance Concrete, pp. 49–50. China Railway Publishing Company, Beijing (1999)
34. Xin, W., Xunyan, T., Yansheng, Y., Yu, Z.: Analysis on Toughening Mechanisms of Ceramic Nano-Composites. Journal of Ceramics 2, 107–111 (2000) (in Chinese)
35. Xijun, W., Mingwen, Z.: Properties and Interfacial Microstructures for Nanostructured Materials. Chinese Journal of Atomic and Molecular Physics 2, 148–152 (1997) (in Chinese)
36. Colston, S.L., O'Connor, D., Barnes, P.: Functional micro-concrete: The incorporation of zeolites and inorganic nano-particles into cement micro-structures. Journal of Materials Science Letters 19, 1085–1088 (2000)
37. Fang, K., Xu, J.: Statistical Distribution. Science Press, Beijing (1987) (in Chinese)
38. Zhentong, G.: Fatigue application statistics. National Defence Industry Press, Beijing (1986) (in Chinese)
39. Ma, B., Hu, C.S., Lu, X.M., et al.: Experimental Study of Concrete for Ultra-thin Whitetopping Pavement. Journal of Highway and Transportation Research and Development 20(6), 8–12 (2003) (in Chinese)
40. Chen, S.F.: The Concrete Flexural Fatigue Property by Adding Polypropylene Fiber. Journal of Xi'an Highway University 21(2), 18–20 (2001) (in Chinese)
41. Lee, M.K., Barr, B.I.G.: An overview of the fatigue behavior of plain and fibre reinforced concrete. Cement & Concrete Composites 26, 299–305 (2004)
42. Spadea, G., Bencardino, F.: Behavior of fiber-reinforced concrete beams under cyclic loading. J. Struct. Eng. 123(5), 660–668 (1997)

Nano-optimized Construction Materials by Nano-seeding and Crystallization Control

Michael Kutschera, Luc Nicoleau, and Michael Bräu

Abstract. Nanotechnology and nanoscience are rapidly creating new possibilities to control and improve material properties. This trend can be also seen in construction materials used for civil infrastructure applications. Special focus in this area is put on Portland cement and gypsum. Together their annual production is by far larger than for any other material worldwide.

Nano-engineering and nano-modification of these materials can be done between their dissolution and hardening. Especially the nucleation step and the crystallization period are most suitable to change the material properties by adding active supramolecular components or colloidal (particle) nano-seeding-additives. This chapter summarizes existing technologies for analyzing and changing the nano-structure of cement and gypsum construction materials. It also shows first results in homogeneous seeding the precipitation of calcium silicate hydrates within a real Portland cement composition. All these procedures, when done correctly, will result in improved material properties opening up with new possibilities for civil infrastructure applications.

1 Introduction

When talking about nanotechnology and cementitious systems it is not always easy to span the bridge between modern advanced technology and know how which exists since some two thousand years. But it is always good to start from

Michael Kutschera
BASF SE, Polymer Research, Physics for Polymer Systems,
67056 Ludwigshafen, Germany
e-mail: `Michael.Kutschera@BASF.com`

Luc Nicoleau
BASF Construction Chemicals GmbH, Polymer Research, Mineralogy,
83308 Trostberg, Germany
e-mail: `Luc.Nicoleau@BASF.com`

Michael Bräu
BASF Construction Chemicals GmbH, Polymer Research, Mineralogy,
83308 Trostberg, Germany
e-mail: `Michael.Braeu@BASF.com`

K. Gopalakrishnan et al. (Eds.): Nanotechnology in Civil Infrastructure, pp. 175–205.
springerlink.com

the very beginning. Since the dawn of time, the need to create a shelter, a home, alongside acquiring food and partnership remains in the top 3 creature priorities. This inevitably implicates that historically, to create the said shelter or housing, the best developed technologies that were suitable and affordable during those times have been used. Starting from wooden shacks, clay cottages, natural stone dwellings up to brick mansions and concrete buildings there has been a constant and tremendous development. But it is often not recognized as high-tech due to the fact that more or less everybody has access and can use the technology on a very normal basis.

However we will show that modern hydraulic binder systems including cement and gypsum are on the eve of a technological revolution. Modern possibilities of engineering and modification on the nanoscale can influence the existing nanostructure of cementitious systems in a targeted, specific and systematic way.

A start can be made with existing building materials. They can roughly be sorted into three classes:

- Solid mass natural materials (e.g. stone, wood, straw)
- Pre-processed, manufactured materials (e.g. iron, steel, glass, brick)
- On site reactive materials (e.g. clay, lime mortar, gypsum, cement, concrete)

Each class has its own quality and application area and nearly all buildings are made of a combination of those materials. From the engineering side, solid mass natural materials are typically quite cheap but require certain workmanship and are not easily compatible with modern industrial construction methods or procedures. Here lies the advantage of pre-processed, manufactured materials. They come in standardized dimensions and can be delivered on a regular basis with guaranteed quality. Additionally they have some extraordinary material properties

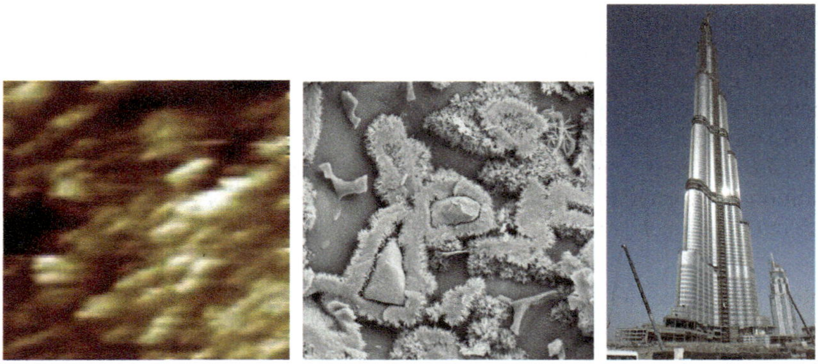

Fig. 1 Bridging the gap of several orders of magnitudes in length scale: Nano-engineering and nano-modification of hydraulic binder systems. Left: nanoscale calcium silicate hydrates on sintered tricalcium silicate surface in a calcium hydroxide solution after 2 hours observed by atomic force microscopy (particles of 30x60 nm²) [Nicoleau 2004], middle: microstructure within reacted cement paste (μm…mm), right: Burdsch Chalifa during construction (m…km). (picture sizes left: 0.4 μm•0.4 μm, middle: 20 μm•20 μm, right: ~500 m•830 m)

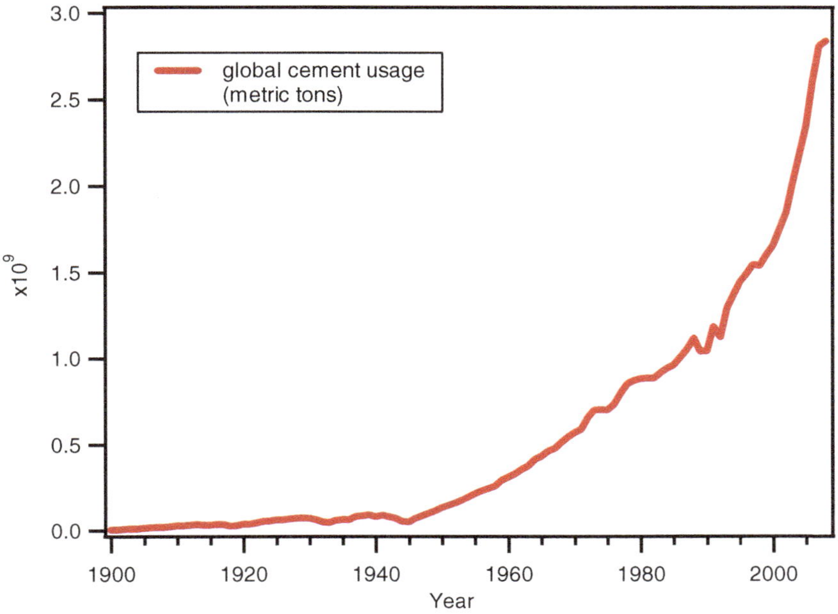

Fig. 2 Dramatic increase of global cement usage during the last century ([U.S. Geological Survey 2010]).

(like steel or glass). Finally on-site reactive materials fulfill double function. They are used to bond together other construction materials (like brick) and on the other hand can be filled into a boarding to form a construction element of any desired form when hardened.

This classification is not fully distinctive. There are lots of materials which belong to multiple categories like e.g. prefabricated concrete parts or clay-straw composites. Nonetheless the classification is still helpful.

1.1 A Short History of Cementitious Systems

When looking back on the history of construction, gypsum is by far the most ancient material. Traces of the use of gypsum go back to 7000 b.c. and there are clear descriptions dating back to 3000 b.c. when Egyptians burned gypsum in open-air fires. To produce gypsum, natural calcium sulfate hydrates are thermally treated (burned) to generate a mixture of calcium sulfate hemi-hydrate and anhydrate. The processing itself as well as reaction times and temperatures determine the resulting composition and subsequent product properties and quality. Modern production methods focus on low prices (flash calciner, high temperatures, short residing times) or high quality and robust reactivity (autoclave-type calciner, lower controlled temperature profiles, longer burning times).

Proven traces of cement production and usage go back to the Roman age where the so called Opus Caementicium was used as a recipe for the inside of brick-faced or solid concrete buildings. The Romans also used aggregates and different hydraulic binder systems (lime, pozzolans, cement) to form lime mortars, structural mortars, underwater mortars and concrete [Vitruvius, 25].

Modern cement development started some 200 years ago. In 1817 Louis Vicat a French engineer invented an artificial cement called white gold. This finding was superseded in 1824 when Joseph Aspdin who was a master craftsman for masonry in Leeds (England) filed a patent called "an improvement in the mode of producing an artificial stone". This so called ordinary Portland cement (OPC) mainly consists of calcium silicates with the most prominent modifications tricalcium silicate (Alite) and dicalcium silicate (Belite) among aluminates, ferrites and many more components and impurities. For Portland cement the production consists of a calcination (de-carbonization) process in a rotary kiln. Raw materials are among others limestone and clays. Due to the high temperatures needed for calcination (approx. 1400°C) optimizing the production with respect to a decreased energy demand, management of the energy source used as well as use of secondary energy carriers (e.g. plastic waste or tires) were important tasks in modern cement production. CO_2 emission is still a major future challenge for cement producers.

From a scientific view it is valid to treat cement and gypsum within one common scope. In a physico-chemical sense both raw materials dissolve in contact with water leading to a very high local supersaturation with respect to calcium-silicate-hydrates or calcium-sulfate-hydrates respectively. This is due to the fact that hydrated species have a lower solubility when compared to the not hydrated ones. The following precipitation process is therefore running highly thermodynamically unbalanced. But also complex kinetics like supply of fresh ions, formation of critical nuclei and self-passivation play an active role during the hardening process. In addition both materials undergo some transitions from amorphous precursors to intermediate and final crystalline structures. A detailed description of these reactions will be given in chapter 2 for cement and gypsum.

In the view of nano-technology (hardened) cement itself without any modifications is clearly a nano-material. It has a hierarchical structure ranging from sub-millimeter dimensions down to nanometer scale. And it is known that a lot of its material properties strongly depend on the structures and the structure development below 100nm [Taylor 1997]. Examples are rheological behavior in liquid state, shrinkage during hardening as well as the development of the final compressive and flexural strength. Things are not so clear for gypsum. Here the $CaSO_4{\cdot}2H_2O$ crystals are well in the µm range. On the other hand the mechanical properties of gypsum depend on the cohesion between these crystallites. This can be easily seen by the decrease of mechanical strength upon humidity or wetting of gypsum specimens [Tesarek et al 2004; McGowan 2007]. Theses cohesion forces again very much depend on Van der Waals forces, inner surface structure matching and roughness on nanometer scale. Therefore cement and gypsum are by nature a nanostructured material with complex hierarchical super and sub-structures.

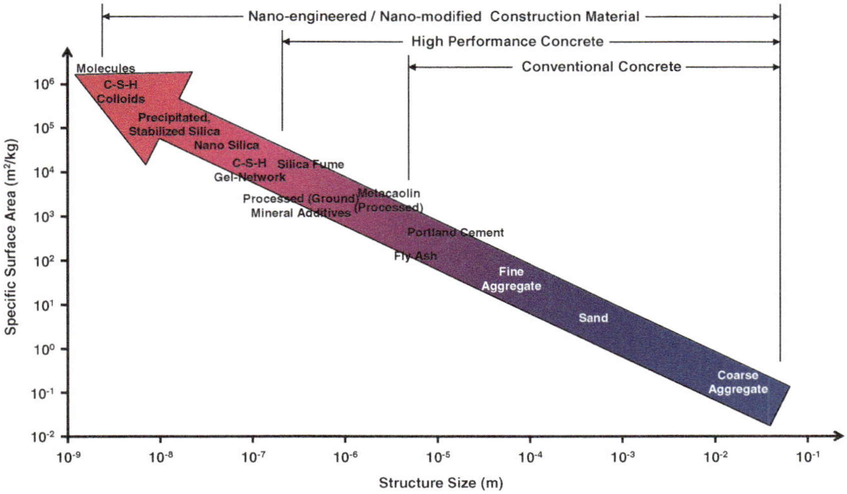

Fig. 3 General view on length scales and surface areas related to construction materials and additives for construction materials [Sobolev 2006].

1.2 Current Trends in Nano-modification of Cementitious Systems

In the last decades there has been a continuously increasing trend to modify and optimize cementitious binders by means of nanotechnology and nanoscale additives. In a roughly chronological order the following modifications evolved:

- Supramolecular additives
- Nanoscale fillers (inert)
- Functional nano additives which influence hydration and/or structure development

Supramolecular additives for cement and concrete are known and used since the mid-1960s. They can act as for example high performance dispersants, rheology modifier or anti shrinkage agents. All of them act in the liquid pore solution before the final hardening of the concrete. They modify and partially cover the surfaces both of the starting materials as well as of the forming hydrate particles. On these surfaces they control rheological properties of the cement paste by electrostatic, hydrophobic or steric interactions. They also modify (lower) the surface energies leading to changes in wetting properties and decreased capillary forces. Typical chemistries used have sulfonate, carboxylate or amine functionalities (among many others…).

The development of improved supramolecular additives allowed incorporation to a certain extent, nanoscale filler particles and materials into cement paste and concrete mixtures. These fillers are mostly inert meaning that they neither interfere with the hydration process nor do they change the hydration products. Typical examples for those fillers are micro-silica, nanoscale pyrogenic SiO_2, $Ca(OH)_2$ or even TiO_2 particles. Their main task is to optimize the grain size distribution leading to a highly filled and compact cement matrix with reduced pores and voids. There has been a noticeable development in the mix design using nanoscale fillers. The resulting ultra high performance concretes (UHPCs) outperform standard concretes with very high compressive strengths and better flexural strength and increased durability. The use of these UHPCs is still limited to high performance civil infrastructure applications like bridges, isolated parts or in ocean underwater structures. The UHPC parts are prefabricated and require novel joining and application techniques (e.g. gluing instead of grouting) [Fehling, Schmidt and Stürwald 2008].

The latest means of nano-modification of cementitious systems are functional nano additives which influence hydration and/or structure development. Known systems comprise of nano-tubes or nano-rods (mainly carbon nanotubes [Akkaya 2003; Trettin and Kowald 2005 & Shah 2009]), nanoscale C-S-H particles and nanoscale gypsum particles. They shall act as internal reinforcement as well as nucleation and crystallization seeds. In the following chapters we will focus on the latter two means of nano-modifications. In chapter 2 we will discuss the chemical reactions occurring during hardening of cement and gypsum followed by a short overview of relevant analytical tools and modeling approaches in chapter 3. Chapter 4 is about nano-modification of the nucleation steps during hydration. Finally chapter 5 demonstrates a means of influencing and modifying the subsequent crystallization processes on the nanometer scale. Some final concluding remarks will summarize the contents. We explicitly note that there are numerous hydration models, modeling techniques as well as characterization methods that are published and proven but cannot be repeated due to size and scope limitations. Our main focus will be the manipulation of the hydration process of cement and gypsum by nano-modification of the nucleation period of those materials.

2 The Hardening of Construction Materials

2.1 Hydration in Ordinary Portland Cement

Concrete is the complex mix of cement powder, sand, gravel, water and admixtures. It is daily used almost everywhere on earth and allows raising sky-creepers of hundredth meters high. The secret for building such tall and at the same time fragile structures originates in a structure transformation at the nanometer scale occurring in cement paste. The changes of this fluid concrete paste to a load-bearing material are due to cohesion properties of the cement part and to the microstructure evolution over time. A false belief is to trust that cement is simply drying when it hardens; in fact it reacts with water (it hydrates) to form new amorphous and crystalline phases. The pore microstructure changes during the whole cement hydration. This evolution is a multi-scale development from the nanometer range up to tens of microns and is the key for the resulting material properties like workability, compressive and flexural strengths, durability, permeability, etc…

The microstructure and its modification with time result from the precipitation of different hydrates, coming from the reaction of the main anhydrous phases

(Ca_3SiO_5, Ca_2SiO_4, $Ca_3Al_2O_6$, $Ca_4Al_2Fe_2O_{10}$,[1] sulfate carriers, such as calcium sulfates) with water [Taylor 1997]. Indeed, nowadays all ordinary cement contain sulfate carriers. They are added after the clinker production during milling in order to regulate the set of cement and to avoid the flash set. In presence of sulfate ions, the anhydrous aluminate phases react to form Alumino-Ferrite tri-substituted hydrates (AFt) [Taylor, 1997; Goetz-Neunhoeffer, 2006; Moore, 1970]. Cement is a multiphase powder and therefore its hydration is complex and sensible to many parameters. During the hydration, anhydrous phases and water disappear and hydrates fill the space among grains. But the volume occupied by hydrates is generally lower than the corresponding stoichiometric volume of cement and water. Voids are consequently formed. Hardened cement is therefore a porous inhomogeneous material. The broad size range of hydrates leads consequently also to a very broad pore size distribution. Polymers and any organic or inorganic additives that concrete manufacturers may add in order to modify the fresh state as well as reaction kinetics (retarders, accelerators) will impact the crystallization of hydrated phases, the pore structure and therefore the microstructure development [Zinng 2008]. As the hydration is the key process, a summary about the most important reactions occurring during the cement hydration will be exposed. Special attention is paid to the correlation of hydrate formation with the onset of cohesion and the starting point of hardening.

The hydration of anhydrous silicate phases, leading to the precipitation of the Calcium Silicate Hydrates (C-S-H) is the major reaction in cement. A more detailed overview of C-S-H precipitation will be made hereafter. Generally most hydrates nucleate heterogeneously and a few homogeneously [Barret, 1974]. However, nucleation and growth kinetics are strongly dependent on ion concentrations in the pore solution. Furthermore, nucleation of hydrates can be easily disturbed by adding fine minerals. These effects facilitate modification of the microstructure during the nucleation phase and in the end improved concrete properties. Data generated in our laboratories will be presented to highlight the various possibilities offered to concrete manufacturers in order to accelerate cement hydration.

Motivated by the desire to enrich our understanding of the physical transformations in cement paste, scientists have developed over two decades, new characterization techniques allowing investigations of structural modifications over a broad size range. In parallel, due to the need of forecasting concrete properties for long service lifetimes, interesting models have been proposed to simulate numerically the cement hydration and the expected microstructure. An overview of established as well as emerging analytical and modeling technologies will be given.

2.2 Correlation between Hydrates, Microstructure and Cohesion Properties in Cement

The hydration of hydraulic materials is responsible for the transformation of concrete or mortar from the viscoelastic state to a load-bearing material. For the sake of simplicity only the main reactions which are responsible for the onset of cohesion up to the hardening are presented.

---

[1] In cement community, it is usual to use acronyms to designate cement phases. These phases are in cement notation: C_3S, C_2S, C_3A, C_4AF.

First, immediately after the mixing of cement with water, gravitational forces make grains settle, they slowly interconnect and then coagulate [Jiang, 1996]. Finally they form an infinite network of grains running in the whole paste volume [Nachbaur, 2001]. The duration is depending on the water to cement ratio, but occurs always within the first 10 minutes [Winnefeld, 2002]. When no superplasticizer is used, the forces between grains are strongly attractive [Plassard, 2005] and a relative high energy is needed to break up the physical gel-like network formed. But this coagulation step is a reversible transformation. That means that the cement paste recovers its fluidity if mixed once again and will coagulate again immediately when mixing is stopped. This is well characterized by dynamic rheometry measurements where the storage modulus becomes higher than the viscous modulus during the coagulation [Blask, 2001].

After a while, anhydrous phases start reacting with water; dissolve and the first hydrate phases appear. The very first ones come from aluminate reaction with sulfate carriers. The principal one of this series is the calcium trisulfoaluminate hydrate or commonly called ettringite, which precipitates as follows:

$$Ca_3Al_2O_6 + 3\ CaSO_4{,}2\ H_2O + 26\ H_2O \rightarrow Ca_6Al_2(SO_4)_3(OH)_{12}{,}26\ H_2O$$

Ettringite precipitation increases the number of possible contact points in the paste and also leads to an increased solid volume fraction since it binds a lot of water

Fig. 4 Cryogenic SEM picture of a cement, the water-to-cement ratio is 0.5, containing 7% of aluminum sulfate quenched 6 minutes after the mix with water. All around the grains, small hexagonal crystals of ettringite have precipitated. The pores and therefore distance between grains are rapidly bridged by these crystallites. A rapid stiffening of the paste is then observed and sprayed concrete is set.

molecules. For instance, in modern sprayed concretes the set accelerator technology is based on aluminum salts, Ettringite precipitation is used for achieving setting time within the first 10 minutes (Fig. 4).

Nevertheless and even in the case of sprayed concrete, ettringite precipitation alone does not allow reaching high enough mechanical strength required by concrete structures. Another hydrate is required for achieving higher resistances. This is accomplished by "C-S-H", the calcium-silicate-hydrate coming from the reaction of the two main anhydrous silicate phases in ordinary Portland cement, the tricalcium silicate (Ca_3SiO_5) and the dicalcium silicate (Ca_2SiO_4) which represent about 60wt% and 20wt % of cement respectively. During their dissolution, the aqueous solution becomes more and more concentrated in calcium, hydroxide and silicate ions up to reaching a maximal supersaturation with respect to C-S-H of 1,5 up to 10. As a consequence C-S-H precipitates. The C-S-H precipitation consumes less calcium and hydroxide than the dissolution provides and the solution gets still richer in calcium hydroxide until supersaturation with respect to the second hydrate phase the Portlandite ($Ca(OH)_{2(s)}$) is reached which then also precipitates. The C-S-H stoichiometry can vary according to the calcium hydroxide concentration in solution [Greenberg, 1965]. This variation implies large differences in the crystalline structure, a review of different C-S-H structures is made in the following reference [Richardson, 2008]. In normal cement, the calcium to silicon ratio is close to 1.7. Consequently, tricalcium and dicalcium silicate hydrations can be summarized as follows:

$$Ca_3SiO_5 + (1{,}3 + x)\, H_2O \rightarrow (CaO)_{1{,}7}\text{-}(SiO_2)\text{-}(H_2O)_x + 1{,}3\, Ca(OH)_2$$
$$Ca_2SiO_4 + (0{,}3 + x)\, H_2O \rightarrow (CaO)_{1{,}7}\text{-}(SiO_2)\text{-}(H_2O)_x + 0{,}3\, Ca(OH)_2$$

x is not well-defined but recent studies [Allen, 2007] give a closer estimation of 1.8.

C-S-H is the elementary brick of a nano-structural masonry which glue together cement grains. These cohesive capabilities are due to two special characteristics. Firstly, C-S-H crystallites are very small (60x20x40 nm^3) [Gauffinet, 1998]. Hence, they develop a huge specific surface area [Allen, 2007; Jennings 2000] and therefore can interconnect elements at the smallest scale. Secondly, when C-S-H is immersed in calcium rich solution at basic pH, extreme highly attractive forces arise between C-S-H crystals [Plassard, 2005; Jönsson, 2004]. Moreover, these forces are robust against pore solution variations [Jönsson, 2005] and make the cohesion also robust for all types of cement. As soon as C-S-H has precipitated, the structure becomes more resistant, stiffer and harder and this transformation is irreversible.

2.3 Gypsum Hydration

Gypsum is used as construction material since ancient times. First proofs of usage as construction material are in Anatolia and Syria, 9000 years ago. Today stucco, screed, plasterboard, filler, and many more products for the construction industry are based on this binder. Gypsum also is also used as sulfate carrier in OPC with about 5wt%.

Gypsum, $CaSO_4 \cdot 2\ H_2O$, is typically named dihydrate referring to the water content per mol $CaSO_4$. In different calcination processes dihydrate is heated above 100 °C, thus removing some of the crystal water and hemihydrate $CaSO_4$ 0.5 H_2O, so-called plaster of Paris or stucco is formed. It may also be found as the rare mineral bassanite. The water content is limited to 0.5 per mol $CaSO_4$, but can be further reduced by drying to the water free hemihydrate. It is used as drying agent as it rapidly rehydrates to hemihydrate. Simple heating of dihydrate in a rotary kiln or kettle yields a microcrystalline product with fast setting time and high water demand, so called β-hemihydrate. More elaborate production processes use autoclaves and additives such as HNO_3 to yield well crystalline α-hemihydrate products of low water demand and smooth workability. By the calcination of gypsum at higher temperatures, all the crystal water is removed and anhydrate, $CaSO_4$, is formed. This $CaSO_4$ anhydrate shows slow hydration and low water demand. Huge deposits of gypsum and anhydrate occur naturally and are mined industrially. Other sources for gypsum are industrial processes such as flue gas desulfurization, the synthesis of phosphoric acid, fluoric acid and others.

2.4 The Hydration of Plaster

The hydration of gypsum is dominated by two processes. The calcined phase hemihydrate and anhydrate respectively dissolve in water and dihydrate precipitates and crystallize. Although the solubility of $CaSO_4$ in water in equilibrium is limited to 2.06 g/L, hemihydrate can create supersaturated solutions of 8 g/L [Singh Middendorf 2007; Freyer Voigt 2003]. The dissolution of the calcined phases is exothermic, while the re-crystallization of dihydrate is only slightly exothermic. Thus the hydration reaction can be followed by heat flow calorimetry, where mainly the dissolution of the hemihydrate or anhydrite is monitored. Fig. 5 shows the heat flow curve of a typical hemihydrate. We can divide the reaction into 4 parts. Part I is dominated by wetting of the surfaces, saturation of any water deficient hemihydrate and hemihydrate dissolution. A supersaturated solution is generated. Part II shows less heat flow, which is due to a relatively slow dissolution rate since the ion concentrations are closer and closer to theoretical solubility of the considered hemihydrate. During this phase dihydrate nuclei are formed from the supersaturated solution and onto impurity particles surfaces. Within Part III a sufficient nuclei amount has precipitated and dihydrate crystals further grow. In order to feed the dehydrate crystallization, further Ca^{2+} and SO_4^{2-} ions are needed and therefore more hemihydrate is dissolved. The dissolution increase is measured with the higher heat flow. Part IV finally shows the final consumption of hemihydrate which has been fully converted in dihydrate. The conductivity achieved is then proportional to the solubility of dihydrate (2.06 g/L.).

The reaction can be observed by a variety of analytical methods. Raman spectroscopy and X-ray diffraction can follow the phase development of hemihydrate and dihydrate. By measuring the conductivity, the pore solution concentrations can be determined and can reveal the supersaturation state with respect to dihydrate reached in part I and II.

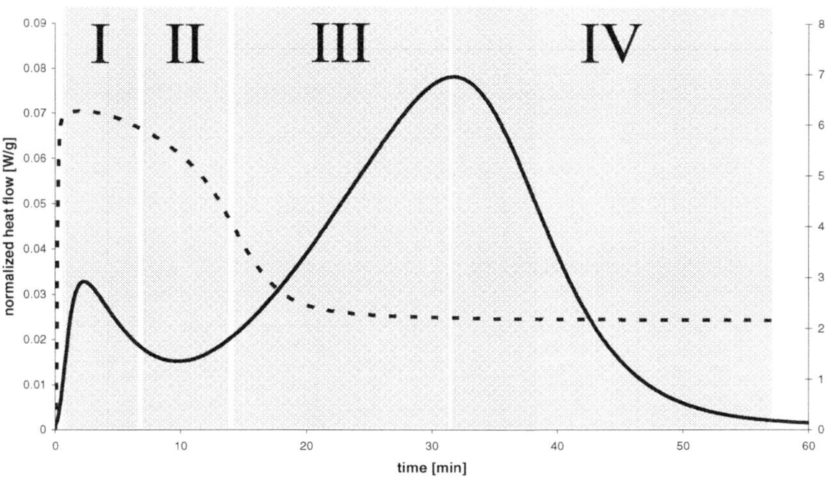

Fig. 5 Heat flow calorimetry diagram (line) of a typical hemihydrate, mixed in situ; conductivity dotted line. Due to the inherent delay in the response of the calorimeter, the time axes are not fully comparable.

3 Analytical Tools and Models

3.1 Experimental Techniques to Characterize the Microstructure Development

Investigations linking the micro-structural organization within binder materials in general and all mechanical/physical properties need methods accessing structural details from the nano- to the micrometer scale. We aim to give here a rapid overview of the most useful techniques with a special emphasis on the emerging ones or on new developments. The advantages and drawbacks of each technique are summarized in table 1. This reveals that a full characterization of the cement structure require the use of various methods and techniques.

Electron microscopy is now a standard technique for investigating the cement microstructure [Diamond, 1993; Stutzman, 2004; Scrivener, 2004; Galluci, 2007]. Despite its easy-to-use practicability it has also a couple of drawbacks limiting its use in certain cases. The three main disadvantages are the artifacts due to the sample preparation, the local representativeness of pictures and finally the problems inherent to a 2D-imagery for investigating 3D-characteristics. Nevertheless, this century is the witness of further improvements enabling better analysis of cement microstructure. As an example, a lot was done in environmental scanning electron microscopy [Möser, 2002 & 2003] and cryogenic microscopy [Fylak, 2006; Zinng, 2008] to avoid as much as possible artifacts due to the sample preparation. Additionally computing techniques were developed to interpret and transpose

Table 1 Overview of different techniques related to the observation of cement microstructure.

Characterization Technique	Minimal object size observable	Information provided	Main Advantage	Main Drawback
S.E.M.	10 nm	Morphology, topography, crystallography	Easy-to-use	2D Vizualisation + degeneresence of the microstructure
Cryogenic S.E.M.	50nm	Morphology, topography, crystallography	Possibility to observe determine layer thickness on grains, Low impact on microstructure due to sample preparation	2D Vizualisation
Micro-Tomography X	500 nm	Topography, Connectivity, Morphology	Non Invasive	Required Synchroton
Cryo-FIB-nanotomography	20 nm	Topography, Connectivity, Morphology	High resolution combined to the pssibility to study 50³um³	Long time processing
Nano-Indentation	100 nm	Elasticity, Density, Porosity	Only technique assessing the mechanical properties at the nanoscale	Low statistical reproducibility
S.A.X.S.	1 nm	Surface Area, size and shape of particles	Non Invasive	Low penetration depth
S.A.N.S.	0,5 nm	Surface Area, Degree of hydration, size and shape of particles	Non Invasive, higher depth of penetration than SAXS	Only available at neutron facilities, flux very low
Dynamic N.M.R. methods (relaxometry & Diffusometry)	1 nm	Surface Area, Porosity	Non Invasive	Does not work without paramagnetic species

2D-data to 3D in order to catch the finest microstructure details. Actually electron microscopy is a well suited method when the expected information is correlated to structures with a size above 100nm.

Recent papers showed that nanoindentation [Constantinidines, 2004] coupled or not with other techniques [Jennings, 2007] are a powerful probe to explore how C-S-H particles are stacked and packed. Besides the existence of two types of C-S-H heaps in cement was confirmed, one having low and one with high density. This kind of characterization is also particularly valuable to follow the nanoporosity evolution in connection with the C-S-H precipitation. This nanoporosity is then connected to the micro-mechanical properties and the pore diffusivity.

Improved by fresh developments in tomography, 3D-imagery makes outstanding progresses in cement interpretation [Bentz, 2000; Burlion, 2006]. With high-energy X-Ray sources, it is today possible for researchers to reconstruct tomograms with resolution about one micron [Galluci, 2007]. Despite its relative low resolution compared to the traditional microscopy, this technique facilitates non-invasive sample analysis and presents the invaluable opportunity to visualize the connectivity within pastes and to follow it in-situ without any artifact related to the sample preparation. For highest resolution, it is necessary to look for other techniques. Currently, the highest resolution (15 nm) (20 nm in table 1) obtained on cementitious samples was achieved by focused ion beam nanotomography (FIB-nanotomography). Moreover, large sample blocks of 50x50x50nm³ can be analyzed contrary to other high-resolution technique (TEM-Nanotomography), offering a new horizon in sample topology as a more precise definition of interfaces at the sub-micrometer level is possible [Holzer, 2006].

X-Ray and neutron scattering are well-established techniques for in-situ studies of cement pore structure [Allen, 1991; Kriechbaum, 1989]. Compared to the classical mercury porosimetry, these techniques possess the advantage of being non-invasive. With neutron source facilities, analysis of elastic and quasielastic spectra allows consequently the monitoring of the silicate phase hydration [Häussler, 2002; Faraone, 2004; Peterson, 2005]. Water binding and surface area evolution during hydrate precipitation can be simultaneously followed in a same sample. Recently, SAXS benefits of a popularization of laboratory equipments making this technique more and more accessible and attractive. Nevertheless, the highest resolutions based on X-Ray scattering as well as neutron scattering experiments require rare and expensive external facilities which make these characterization methods still infrequently used in cement analysis.

NMR has been an ever-present and key technique in the characterization of binders, from the post-war period with Pakes's article [Pake, 1948] showing proton NMR spectra of gypsum, until today where dynamic NMR methods deliver new outstanding results on cement paste porosity. The recent progresses made in cement field were reviewed in the following articles [Nestle, 2007 & Skibsted, 2008]. Two-dimensional NMR relaxometry provides unique information on exchange of water between the silicate surface and the capillary pores [McDonald 2005 & 2007]. Proton relaxation at the surface of hydrate is a probe (and non-invasive) for the determination of specific surface area of these hydrates and its evolution during the hydration [Zajac, 2007]. It also enables to study the pore size distribution over time. For instance, the fractal character of cement paste was clearly demonstrated with the power-law followed by the pore size distribution [Plassais, 2005]. New NMR techniques are certainly a high-performing tool in order to provide access to hydration processes at nano scale also in presence of admixtures [Rottstegge 2006].

3.2 *Multi Scale Computer Modeling Bridging Nano- to Macroscale*

Nowadays, the long-term durability and performances of concrete structures, especially under aggressive conditions, are now critical aspects of architectural design especially considering the growing necessity to build sustainable structures based upon more sustainable resources. Therefore, modeling or numerical simulations becomes increasingly helpful and valuable tool in order to forecast microstructure development from the early age until its maturity after many months or years. For this reason, knowledge about cement hydration kinetics is primordial for forecasting its performances. Consequently, it is desirable to model the cement structure evolution only with the cement phase composition and the grain size distribution. This point is basically the key and certainly the most difficult to realize due to the complexity of cement chemical reactions. For the moment, to the best of our knowledge, no current model is able to obtain a cement structure evolution without any fitting parameters. The best present methods are getting closer and closer to elucidating the elementary physico-chemical processes. For more than a decade, some outstanding structural models, based firstly on understanding cement hydration, have been proposed and for the first time brought real steps forward. As a consequence noteworthy accomplishments have been made linking reactivity, paste structure, thermodynamics and kinetics.

The most advanced platform gathering the modeling of rheology behavior, hydration kinetics, microstructure development, elastic properties was created by collaborators of NIST under the name "The Virtual Cement and Concrete Testing Laboratory" or VCCTL[2] [Bentz 2006]. A key innovation was to build a hydrate microstructure from thermodynamics laws which govern their nucleation, growth and dissolution mechanisms and by also taking into account diffusion process. The code for the hydration part was originally called CEMHYD3D [Haecker, 2005] and a later version was named HydratiCa [Bullard, 2008]. A future version is targeting the coupling of HydratiCa (for early times of hydration) and THAMES (for later times) in order to ensure a complete and multiscale assessment of concrete life.

A vector modeling approach was developed at the EPFL by S. Bishnoi et al. called μic [Bishnoi, 2008]. The purpose of this model is to be a powerful toolbox to simulate and to test different mechanisms of hydrates nucleation and growth and the corresponding microstructure evolution. The particular advantage of μic is its flexibility and gives to users a large degree of freedom to build various structures. For instance, the alite hydration was studied and it was proven that Avrami equations, broadly used in many systems, are not suitable for alite or cement [Bishnoi, 2009]. Other different hydration models for alite were tested in this article and appear to be inconsistent with experimental observations and consequently the full explanation of alite hydration remains unclear.

4 Nano-engineering of Nucleation

4.1 Design of C-S-H Nucleation and Growth

The most exciting cement mineral and which has also led to the largest collection of work is undoubtedly C-S-H. Unfortunately this hydrate is difficult to characterize and to quantify during the cement hydration. These problems were and are still basically the greatest barrier to overcome in order to perfectly seize the link existing between the development of this hydrate and all the mechanical changes occurring in a cement paste. The lack of investigation techniques was at the origin of many controversies in the past about its nucleation and growth process. Nowadays with the emergence of new characterizations, scientists' opinions tend to be more and more unified.

Among the largest single body of work, supported by many experimental investigations, Nonat and co-workers' study describes with the best accuracy the calcium silicate hydrates nucleation and growth during the early age of cement. They proposed that C-S-H nucleates [Garrault-Gauffinet, 1999] heterogeneously onto anhydrous silicate grains and grows according to two anisotropic rates: one parallel to the

---
[2] `http://vcctl.cbt.nist.gov/`

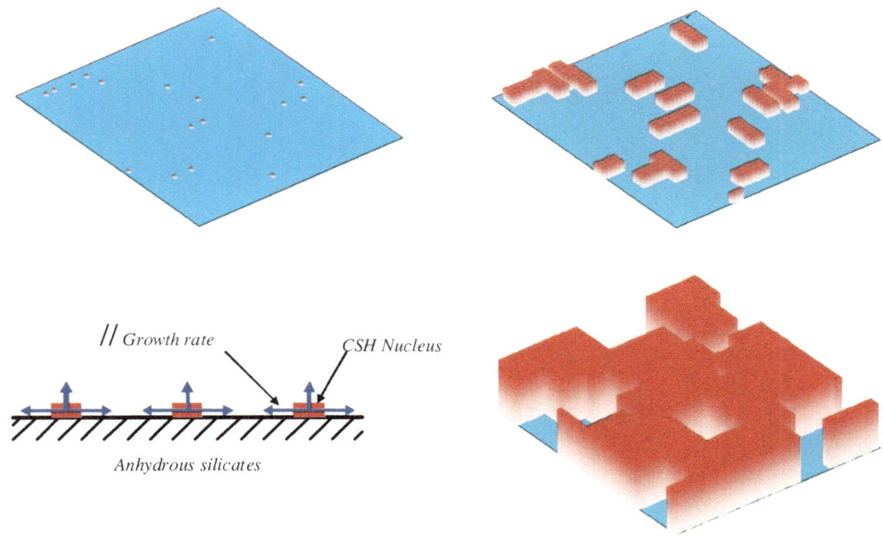

Fig. 6 Nucleation and growth of C-S-H clusters on a C_3S surface. They fit the degree of hydration of two grams of pure tricalcium silicate paste mixed with one gram of water, after one hour, 4.5 hours and 18 hours. These were created according to the model developed by Garrault, the 3D visualization made by Nicoleau.

grain and one perpendicular to the grain. Basically this is not a crystal growth, since crystals are very size-limited in normal conditions but a secondary nucleation of C-S-H particles around the first ones appeared during the primary nucleation [Garrault, 2001]. This means that the growth can be described by an iso-oriented aggregation of particles around the first ones, similar processes occur in many other systems [Niederberger, 2006; Nudelmann, 2007; Cölfen, 2008]. On Fig. 6, schematic representations illustrate the nucleation and then the growth of C-S-H on a C_3S surface.

The beauty of this model exists firstly in its simplicity since it utilizes only five parameters[3] that are directly connected with physical measures and secondly in its capacity to give precious information on C-S-H development in various systems [Garrault, 2001 & Nicoleau, 2004 & 2010]. Further aggregation around the first nuclei cause cluster growth on the grain surface which coalesce when they are big enough. This coalescence can be observed by NMR relaxation [Zajac, 2007] since the lateral surface of growing clusters disappears. After a certain time, C-S-H particles cover the whole surface of the cement grains and at this point kinetics become limited by a diffusion process of reactants through the newly formed C-S-H layer. Hydration is roughly summarized in Fig. 7 by the succession of three steps: the primary nucleation occurring during the induction period, the expansion of

[3] These parameters are: the surface area of growth, the number of nuclei, the parallel and perpendicular aggregation rates and finally the permeability of the C-S-H layer per unit of thickness.

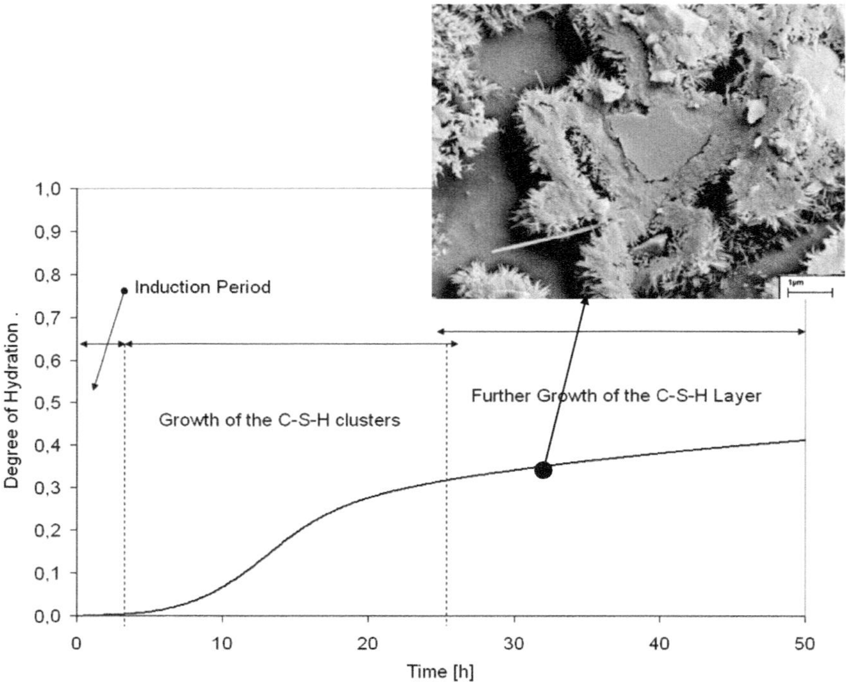

Fig. 7 Description of the tricalcium silicate hydration according to the degree of hydration. In the insert a cryogenic S.E.M. picture, in the middle of this one silicate grain covered by a C-S-H layer can be seen (Hydration of C_3S at a W/C=0.5). This silicate grain was cut into two parts during the sample preparation.

C-S-H clusters and finally the thickening of the C-S-H layer. The nucleation step is not only influencing the early hydration but astonishingly also the late hydration, since the diffusion rate of species through the layer depends strongly on its nano-porosity. One of the main factors increasing the nano-porosity is a stronger nucleation. Temperature [Zajac, 2007] or some additives play a role on the C-S-H nucleation and can affect the development of mechanical properties.

Precise acceleration of hydration is a difficult task and was always a question of dosage. Indeed, parameters positively influencing the hydration at the very early age can have a negative impact on the latter stages of hydration after many days or weeks. This means that an over-accelerated hydration in the first hours of setting, often very beneficial for getting rapid setting times, may have dramatic payback on 28 days compressive strength[4] and consequently on durability. Consequently it is a question of compromise for the concrete manufacturers whether to accelerate the setting or not, since advancing the early stage setting may lead to inferior later stage strength and durability. Naturally, modeling is a valuable toolbox in order to

[4] The strength at 28 days is a norm value indicating the mature strength of concrete.

understand the effect of accelerators and to design better ones [Nicoleau 2010]. An overview of some known accelerators is presented in the following.

Calcium chloride is the best known hardening accelerator used in not-reinforced concretes. Degree of hydration curves reported on Fig. 9 show the extreme acceleration brought by this additive in the first hours. The addition of calcium chloride has firstly an impact by increasing the number of primary nuclei and secondly the C-S-H growth mode may be also drastically changed. On Fig. 8, two cryogenic SEM pictures of two cement pastes hydrated after one hour reveal two grains exactly in the same configuration. The first one is a paste non-accelerated (a) and the second one is accelerated with 200 mM of calcium chloride (b). In the $CaCl_2$ case, the grain surface is already strongly covered by C-S-H (intermingled with ettringite) only one hour after the mix, while this is free of C-S-H on the reference.

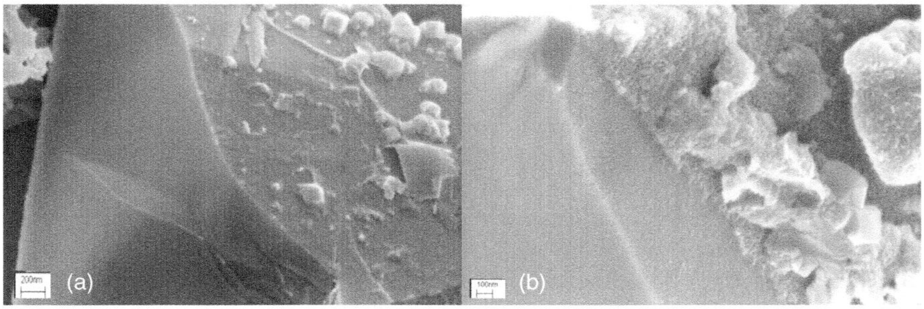

Fig. 8 (a) Cryo-SEM picture of a reference C_3S paste, and (b) a C_3S paste mixed with 200 mM $CaCl_2$ both taken after 60 min and at a water to cement ratio of 0.42. Both pictures are centered on a grain truncated during the sample preparation. Truncations are visible on the left part of pictures and grain surfaces on the right. C-S-H cannot be seen on (a) while an abundant precipitation occurred on (b).

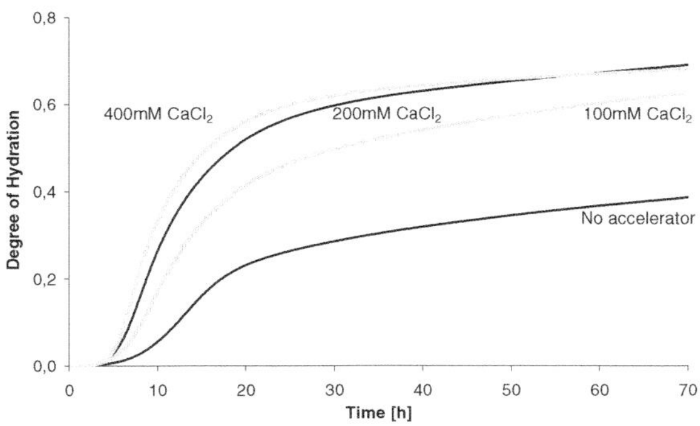

	Quantity of Nuclei (normed to the reference)	Agg. Rate //	Agg. Rate ⊥
Ref.	100%	6,1	17,2
+ 100mM $CaCl_2$	170%	6,8	13
+ 200mM $CaCl_2$	465%	4,9	25
+ 400mM $CaCl_2$	400%	6	40

Fig. 9 On the left, degree of hydration curves of C_3S paste (W/C=0.42) accelerated with different amounts of $CaCl_2$ and on the right parameters corresponding to the fit of these curves according to the Garrault's model presented in the following article [Garrault, 2010].

Many other ions play a role on the nucleation step, for instance by reducing or by increasing the number of C-S-H primary nuclei [Garrault, 2010 & Sauvaget, 2001] on anhydrous silicate phase grains. Concerning the known additives sodium metasilicate or water glass is one of the most efficient since it provides a massive quantity of silicate ions which will react immediately with the calcium ions provided by the dissolution of anhydrous phases. Sodium metasilicate was used a couple of decades ago as set accelerator, also for sprayed concrete. It was replaced later by other salts due to the problematic increase of alkali concentration in concrete and therefore the high risk related to the alkali-silica reaction, a concrete disease leading to rapid deterioration of concrete structure. The extra C-S-H nuclei gives an outstanding boost to the growth during the first hours but has a negative effect at later times, reflected generally in bad compressive strengths at 28 days.

Another way to promote the nucleation from the very beginning is to provide a very fine material with which C-S-H nuclei have a good affinity. These materials, sometimes nano-sized provide an immense specific surface area. Very low dosages are sufficient to double or triple the surface area of solid particles in the mix. The number of locations, where nucleation occurs, is often increased and their distribution in the paste changes. Therefore these nano-materials can be seen as nucleation promoters. Many examples can be found in literature on calcium carbonate [Soroka, 1976], titanium dioxide [Goertz, 2007], nano-silica [Belkowitz, 2009] or pozzolans [Korpa, 2008], or more innovative on carbon nanotubes [Kowald, 2007]. The number of extra nuclei per m^2 of material which can precipitate is depending first of all on the crystal affinity for this material [Boistelle, 1985]. As C-S-H nucleation follows the classical nucleation theory, this affinity can be estimated by comparing the necessary time to nucleate C-S-H (Eq. 1) from supersaturated solutions. This is directly related to the energy difference (factor F(β)) between homogeneous and heterogeneous nucleation (Eq. 3). A schematic representation on Fig. 10 explains the link between the wetting angle α formed by a C-S-H nucleus on a substrate surface and the energy decrease accompanying the heterogeneous nucleation.

$$\ln t_{ind} = \frac{\Delta G_{r^*}}{RT} - \ln k_n \quad (1)$$

with t_{ind} necessary to form a critical nucleus, ΔG the enthalpy formation of this nucleus and k_n a constant.

$$\text{with} \quad \Delta G_r = -RT \ln \beta \quad (2)$$

β the degree of supersaturation of the aqueous solution before the nucleation.

$$\Delta G_{r^*}^{hete.} = F(\alpha).\Delta G_{r*}^{\mathrm{homo}} \quad (3)$$

$$\text{with } F(\alpha) = 0{,}5 - 0{,}75\cos\alpha + 0{,}25\cos^3\alpha \quad (4)$$

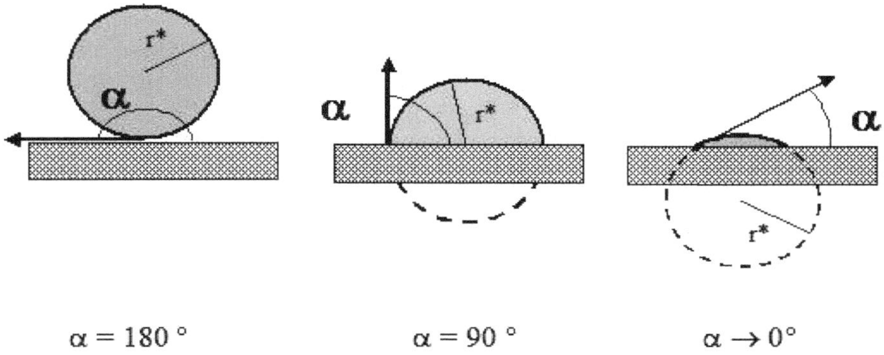

Fig. 10 Illustration of a nucleus formation of a critical radius r^* from a supersaturated solution. The higher the affinity for the substrate, the smaller the wetting angle is. The first case corresponds to the homogeneous nucleation. The heterogeneous nucleation is accompanied of a volume decrease of the critical nucleus.

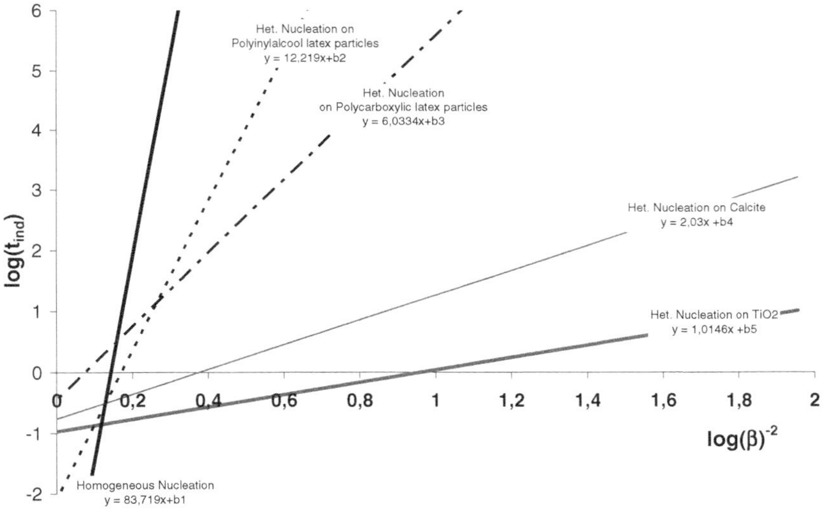

Fig. 11 Curves $log(t_{ind})=f(log\beta^2)$ corresponding to the homogeneous nucleation of C-S-H in supersaturated solutions and 4 different heterogeneous ones: on hairy latex particles with carboxylate functional groups, on hairy latex particles with hydroxyl functional groups [Nicoleau, 2004], on calcite and on titanium dioxide [Nicoleau, 2007].

It is actually possible to nucleate C-S-H onto various substrates, organic or inorganic. In order to illustrate this purpose, we gather some of our own results on nucleation studies in presence of different additives. Fig. 11 presents the nucleation curves obtained in the homogeneous case and in presence of either two different latex particles, calcite or finally titanium dioxide. It seems that the very general common point among all these additives is their negatively charged surface in alkaline solutions. The slope of the homogeneous curve can allow us to calculate the interfacial energy of the crystal-solution γ, it is found between 10 and 15 mJ/m² [Nicoleau, 2004; Garrault 1998]. By assuming that nuclei have a spherical cap (Fig. 10), a simple division between the slope of the homogeneous nucleation curve and the slopes of the different heterogeneous nucleation curves determined in presence of an additive, gives the diminution of nucleation enthalpies (F(α)) between both types of nucleation. Therefore the wetting angle of one C-S-H nucleus on this aforesaid additive (Fig. 12) can be calculated according to the equation 3.

In addition to our own results, we include in Fig. 12 the C-S-H nucleus affinity for dicalcium silicate (Belite) from reference [Garrault, 1998]. The authors also showed that this affinity should be even greater for tricalcium silicate (α<30°), the main anhydrous component in cement. This means that it is really difficult to find an additive having a higher affinity for C-S-H than the anhydrous silicate phases already present in all cements.

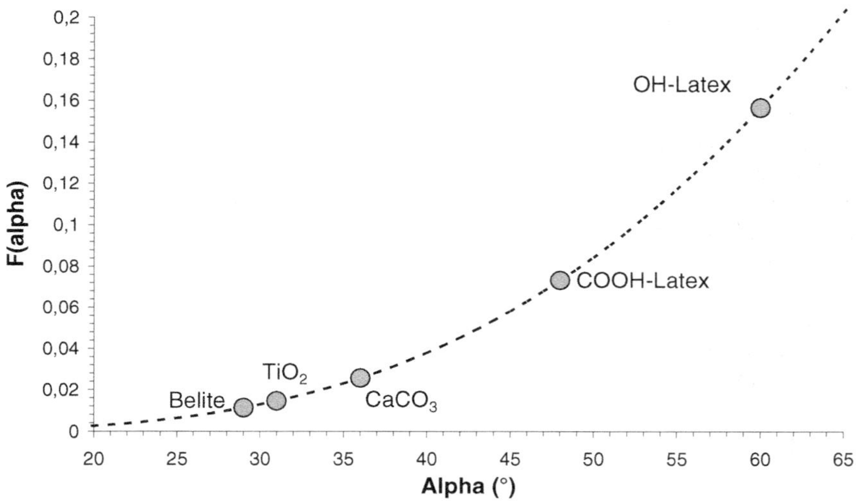

Fig. 12 Curve F(α) plotted according to the equation 4. The particular points represent affinities of a C-S-H nucleus for the various substrates invoked on the previous figure.

The last way to activate the nucleation is to provide synthetic nuclei or seeds. Accelerators based on seeding technology have a substantial advantage compared to other additives because seeds, if properly stabilized, favor strongly the homogeneous nucleation in the pore solution and not the heterogeneous one on cement grains. Therefore, synthetic nuclei do not accelerate the C-S-H layer formation and do not favor the limitation of growth in the diffusion-limited step. The seeding technology is not new. Many scientists have already attempted to enhance the mechanical properties by seeding concrete [Duriez, 1956]. Despite the fact that literature is rich in interesting articles related to this topic [Alizadeh, 2009; Thomas, 2009], not a single product based on a seeding material has come to the market.

Fig. 13 Cryogenic SEM pictures of *(a)* a cement paste quenched 11 hours after the mix and *(b)* a cement paste mixed with 0.3% by weight of C-S-H seeds (X-SEED®100) quenched only 6 hours after the mix. These times correspond to necessary hydration times in order to reach the same degree of hydration determined by isothermal calorimetry. The W/C ratio is 0.5. On picture (a), honeycomb-like C-S-H covers the grain surface whereas it precipitates far way the grain on picture (b). The use of stabilized C-S-H seeds favor heterogeneous growth.

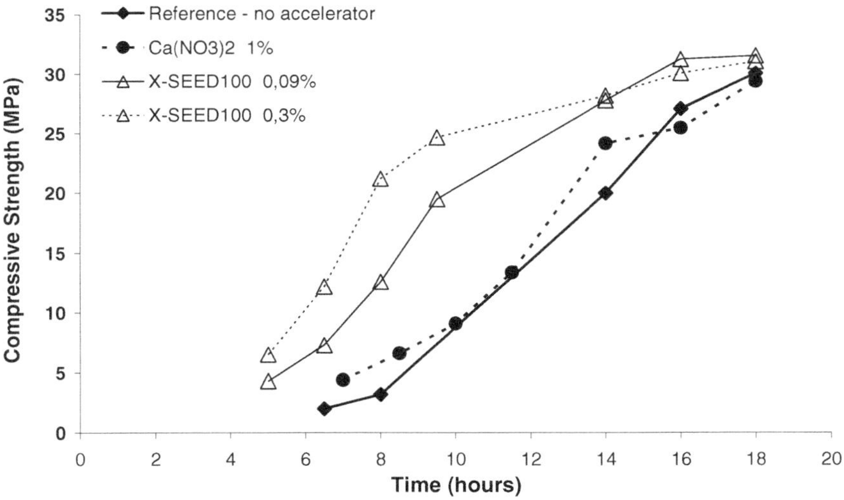

Fig. 14 Early compressive strengths measured on concrete cubes containing 400 kg/m³ of a low-alkali cement 52.5R mixed at constant water to cement ratio of 0.44 and cured at 10 °C in presence of two accelerators: calcium nitrate and stabilized C-S-H seeds. Dosages in accelerator are given in percentage of actives weight by weight of cement.

Very recently, a new product appeared, based on suspended synthetic nano-sized C-S-H nuclei. Such a suspension shows excellent accelerating properties [Nicoleau, 2010], without any drawback at later times, and promotes the nucleation and growth in the pore solution, as it can be observed on Fig. 13. The increase of compressive strength is huge and generally valuable compared to other state-of-art accelerators like calcium nitrate, especially at very low temperature where calcium salts can underperform (Fig. 14).

5 Nano-modification of Crystal Growth

5.1 Nano-modification of Gypsum Growth

To adjust the application performances of cementitious as well as gypsum based binders various additives are used. The most important properties to be controlled are the setting time by retarders as well as accelerators, as well as the strength and durability. All this is achieved by controlled crystallization of the hydrated phases accomplishable only at the nano level [Lewry Williamson 1994] and will be discussed for gypsum in this part.

A series of studies deal with the influence of additives on the crystal habit of dihydrate crystals. The influence on the crystallization at the nano level can be analyzed during the different stages of hardening. The starting point is the formation of nuclei, which can be suppressed or stimulated. The growth of the hydration product dihydrate is likely to be retarded by adsorbed molecules or accelerated by higher ionic strength in solution. The whole hydration also depends on the dissolution rate of hemihydrate or anhydrate. The dissolution of these materials is also moderated by additives [Mueller, Stark 2004]. Inorganic additives like alkali or ammonia salts of inorganic acids like sulfuric phosphoric or hydrochloric acid are typical accelerators for the crystallization process [Koslowski, Ludwig 1999]. The soluble salts increase the solubility of calcium sulfates in water [Yadav, 1995] and increase the concentrations of Ca^{2+} and SO_4^{2-} in solution. Due to different adsorption constants on dihydrate crystal faces, the crystal shape changes according to

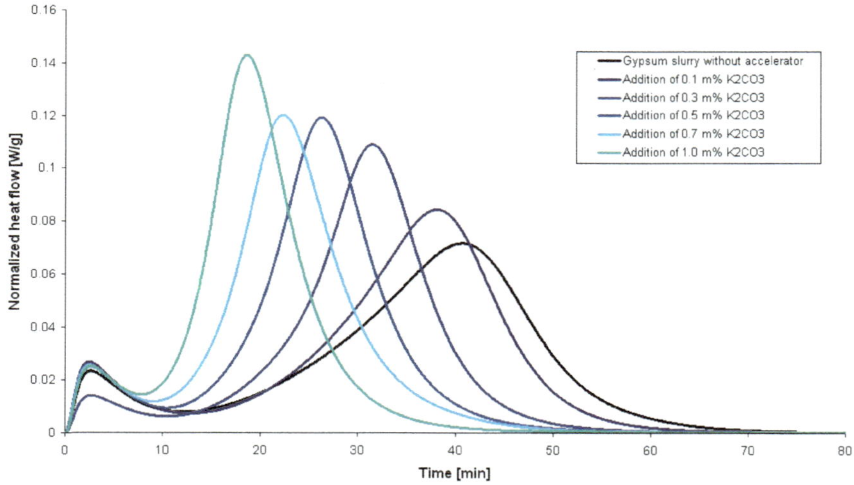

Fig. 15 Heat flow calorimetric study of the acceleration effect of K_2CO_3.

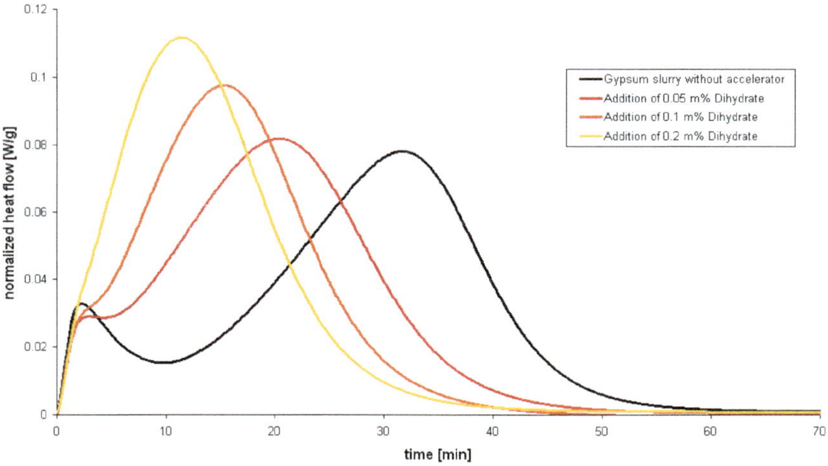

Fig. 16 Heat flow calorimetric study of the acceleration effect of dihydrate seeds.

ion type and concentration. A typical overall trend is to observe changes from a needle-like aspect ratio to a more isotropic one, which has negative consequences on mechanical properties by less gypsum crystals entanglement. Also temperature and pH play an important role and influence all stages of the plaster hydration [McCartey Alexander 1958; Bertoldi 1978 & Konak 1976].

On Fig. 15 and Fig. 16 two different acceleration mechanisms are demonstrated. The addition of K_2CO_3 increases the solubility of calcium sulfate in solution due to a raised pH value and elevated solution salinity. The heat flow, used as indicator for the progress of the hydration, shows a shorter and sharper hydration reaction with increasing K_2CO_3 concentration, which can be interpreted by faster crystallization and faster dissolution of hemihydrate. Nevertheless the initial period is still present and not shortened, indicating that nucleation step is not suppressed.

Addition of finely ground dihydrate in Fig. 16 shows a different acceleration mechanism in contrast to the previous example. Here, the time to fully hydrate the hemihydrate is not drastically shortened, but the gypsum growth starts immediately. Indeed, some small amounts of dihydrate material have been added and act as seeding material. Consequently, the usual period of nucleation is significantly reduced, until its complete suppression. The fine gypsum seeding material is more cost-effective for the gypsum producers than other inorganic compounds. Hence, this is the industrial method of choice.

Retarders are as important as accelerators in order to control the setting of gypsum based binders. Usage of Carboxylic acids or their alkali salts as retarders is based on the adsorption of the carboxylic groups on the surfaces of the forming dihydrate crystals and also on the hemihydrate [Singh, Middendorf 2005]. The adsorbed organic molecules not only retard the formation of new crystalline layers on the dihydrate crystals, but also prevent the hemihydrate from dissolving. As the molecules adsorb in different amount on the different faces of crystals the morphology will change [Badens Boiselle 1999]. Modeling of the adsorption processes is possible but only relevant for special systems and can not be generalized [Hill Plank 2004]. Water soluble polymers such as PAA or PMAA are known as retarders [Smith Alexander 1970; Amjad Hooley 1986]. The retardation effect is also used in the field of scale inhibition for example with amines or phosphonic acids [Liu Nancollas 1975].

However the hydration kinetic also influences the mechanical properties of standard set plaster. By grinding hemihydrate or introducing gypsum seeds an acceleration of about 30 % is easily achievable, resulting in an increased hardness of 30 % to 50 % [Amathieu Boistelle 1986]. The accelerated crystallization process results in the formation of smaller crystals. When keeping total porosity to a constant level the size of the resulting nano/micro pores is also smaller. This results in higher strength levels of the hardened material. To analyze the porosity of hardened gypsum binder, direct methods are preferable. Fig. 17 shows a broken sample of hardened gypsum. With Cryo-SEM methods it is possible to observe the fine structure of pores and micro pores. However the preparation is still difficult as gypsum has low hardness, cutting or polishing is not practicable and each image can only show a small section. XRD-tomography is a possibility to get information of the whole sample but the resolution allows only minor insight into the micro pore structure smaller than 2 μm.

Dynamic NMR gives the opportunity to analyze the finest pore structure below the micrometer scale [Nestle 2009]. The use of superplasticizer allows the formulator to make workable gypsum pastes with much less water. It has been

Fig. 17 Cryo SEM picture of a broken sample of hydrated plaster.

demonstrated that this water-to-gypsum ratio reduction decreases the porosity. Hence, the lower porosity increases then the mechanical properties.

5.2 From Hydration to Crystallization

Beside the evolution of the hydrated phases, habit and crystal size distribution may also influence the strength development in the final work material. In the case of gypsum the dense intergrowth of dihydrate crystals is responsible for the final properties of the material. The observation of the favorable crystal shape can be followed by SEM methods. Fig. 18 shows two typical crystal shapes of dihydrate. Both precipitated from solution with similar ion concentrations but under different growth rates controlled by the evaporation rate. Slow evaporation tends to favor plate-like crystals, while fast evaporation often results in needle-like structure.

Fig. 18 SEM pictures of typical gypsum crystals grown from solution.

5.3 From Microscopic to Macroscopic

Dry plaster is characterized by a linear elastic macroscopic behavior and it has been reported that its hardness and elastic modulus decrease with increasing porosity. Under constant loading, it undergoes permanent deformation that can be accentuated by moisture. The hardness, elasticity and porosity of a gypsum sample are the consequence beside the shape and size also of the intergrowth and interlocking of dihydrate crystals [Soroka Serada 1968]. Crack propagation is due to heterogeneous microstructure consisting of entanglement of elongated crystals and a bimodal distribution of pores with microscopic interconnected pores and large spherical air bubbles. It involves different mechanisms acting at different scales [Meille Fantozzi 2003]:

- Crack bridging by gypsum crystals, acting locally behind the crack tip
- Secondary cracking operating in a macroscopic process zone around the crack tip
- Substantial macroscopic crack branching that contribute to wake effects.

[Saâdaoui Caspar 2000]. The compressive strength of set plaster becomes significantly lower by wetting or even high relative humidity. Generally speaking, this can be explained with layers of water on the different surfaces acting as lubricant in between the dihydrate crystals, as proven by AFM methods [Finot Goudonnet 1997]. The fracture behavior of set plaster results mostly from the weakening of forces at the interface between gypsum crystals rather than a crystal break down. Macroscopic crack propagation occurs by linkage with micro cracks originating from weak regions [Saâdaoui Fantozzi 2005]. The sustainability of gypsum-based materials is determined not only by the strength but also largely by their deformability. Under permanent load, elastic and plastic deformations occur in gypsum, which primarily depend on raw materials quality, on placing and hardening conditions, on the hardened materials shape and on the applied load [Sattler 1974; Odler Roessler 1989].

Besides control at the nano level, adjustments may be performed at a bigger scale. In order to control the rheology of gypsum paste, additives like superplasticizers contribute to the flow of the paste. Foaming and defoaming agents control the macro pore structure.

6 Summary

In 2010 it is not new to affirm that nanotechnologies are one of the greatest vectors of innovation in cementitious systems. Never the less, there was a real breakthrough of these technologies into the real life of consumers only in the past few years. Despite its historical existence, we illustrated that with new developments in material analysis, the key parameters in cement hydration can be identified and hence new molecular architectures can help to tune the parameters that govern concrete properties.

It has been shown that two key starting points for nano-engineering and nanomodification of construction materials exist: the nucleation step and the crystallization period of the materials. Both steps substantially determine the transition of the material from a fluid suspension to a hardened structural material. Active components which change those steps are either supramolecular or colloidal (particle) nano-seeding-additives. But it is not easy to find suitable substances as they considerably change the subtle balance of dissolution, precipitation, surface energy, critical seed size, etc. Only the combination of mechanics, thermodynamics, polymer chemistry and material engineering areas results in more and more advanced, top-performing and highly sustainable solutions. And every step down in nano-size technology, nano-analytics and understanding of those mechanisms happening on the nano-scale will still open up the opportunity to improve those processes leading to more and more advanced, efficient and powerful construction materials.

References

Akkaya, Y., Shah, S.P., Ghandehari, M.: Influence of fiber dispersion on the performance of microfiber reinforced cement composites. ACI Special Publications 216: Innovations in Fiber-Reinforced Concrete for Value 216, 1–18 (2003)

Alizadeh, R., et al.: Hydration of tricalcium silicate in the presence of synthetic calcium-silicate-hydrate. J. of Mat. Chem. 19, 7937–7946 (2009)

Allen, J., et al.: Composition and density of nanoscale calcium-silicate-hydrate in cement, Nature Materials, Published Online: March 25 (2007); doi:10.1038/nmat1871

Allen, A.J.: Time-resolved phenomena in cements, clays and porous rocks. J. Appl. Cryst. 24, 624–634 (1991)

Amathieu, L., Boistelle, R.: Improvement of the mechanical properties fo set plasters by means of four organic additives inducing {101} faces. J. Cryst. Growth 79, 169–177 (1986)

Amjad, Z., Hooley, J.: Influence of polyelectrolytes on the crystal growth of calcium sulfate dihydrate. J. Coll. Int. Sci. 111(2), 496–503 (1986)

Badens, E., Veesler, S., Boiselle, R.: Crystallization of gypsum from hemihydrate in presence of additives. J. Cryst. Growth 198/199, 704–709 (1999)

Barret, P., et al.: Contribution to the study of the kinetic mechanism of aluminous cement setting. In: CCR, vol. 4, pp. 545–556 (1974)

Belkowitz, J., et al.: The investigation of nano silica in the cement hydration process, American Concrete Institute, SP-267(Nanotechnology of concrete: The next Big Thing Is Small), pp. 87-100 (2009)

Bentz, D.,et al.: Virtual testing of cement and concrete, ASTM Special Technical Publication, STP169D,1548–1555 (2006)

Bentz, D.P., et al.: Microstructure and transport properties of porous building materials. II: Three-dimensional X-ray tomographic studies. Materials and Structures 33, 147–153 (2000)

Bertoldi, G.A.: Modifying the shape and crystal size of precipitated calcium sulphate dihydrate and the conditions of dehydration. ZKG Int. 12, 626–629 (1978)

Bishnoi, S., Studying nucleation and growth kinetics of alite hydration using μic, CCR, Vol. 39, pp. 849-860 (2009)

Bishnoi, S.: Vector Modeling of Hydrating Cement Microstructure and Kinetics, Thesis of EPFL (2008)

Blask, O.: Connection between rheological and microstructural changes during hydration of cement pastes. In: Proceedings ICMA 2001 (2001)

Boistelle, R.: Concepts de la cristallisation en solution, Crosnier J & et al. Editions (1985)

Bullard, J.W.: A determination of hydration mechanisms for tricalcium silicate using a kinetic cellular automaton model. J. Am. Ceram. Soc. 91(7), 2088–2097 (2008)

Burlion, N., et al.: X Ray microtomography: Application to microstructure analysis of a cementitious material during leaching process. In: CCR, vol. 36, pp. 346–357 (2006)

Constantinides, G., et al.: The effect of two types of C-S-H on the elasticity of cement-based materials: results from the nanoindentation and micromechanical modeling. In: CCR, vol. 34, pp. 67–80 (2004)

Cölfen, H., Antonietti, M.: Mesocrystals and Nonclassical Crystallization. John Wiley & Sons, Chichester (2008)

Diamond, S., et al.: Microstructure of hardened cement paste – A new Interpretation. J. Am. Ceram. Soc. 76(12), 2993–2999 (1993)

Duriez, M., et al.: Possibilités nouvelles dans le durcissement rapide des ciments, mortiers et bétons, Ann. Inst. tech. du BTP, 9ième année, N°98 (1956)
Faraone, A., et al.: Quasielastic and inelastic neutron scattering on hydrated calcium silicate pastes. J. Chem. Phys. 121(7) (2004)
Fehling, E., Schmidt, M., Stürwald, S. (eds.): Ultra High Performance Concrete (UHPC), Second International Symposium on Ultra High Performance Concrete. University Press GmbH Kassel, Kassel (2008)
Finot, E., Lesniewska, E., Mutin, J.-C., Goudonnet, J.-P.: Reactivity of gypsum faces according to the relative humidity by scanning force microscopy. Surface Science 384, 201–217 (1997)
Freyer, D., Voigt, W.: Crystallization and Phase Stability of CaSO4 and CaSO4 – based salts. Monatshefte für Chemie 134, 693–719 (2003)
Fylak, M., et al.: Electron microscopic studies of hydration reactions of Portland cements by Quantomix-test substrate and cryotransfer preparation. In: Proc. GDCh-Monographie, Tagung Bauchemie, vol. 36, Gesellschaft Deutscher Chemiker (2006)
Gallucci, E., et al.: Crystallisation of calcium hydroxide in early age model and ordinary cementitious systems. In: CCR, vol. 37, pp. 492–501 (2007)
Galluci, E., et al.: 3D experimental investigation of the microstructure of cement pastes using synchrotron X-Ray microtomography (uCT). In: CCR, vol. 37, pp. 360–368 (2007)
Garrault, S., et al.: Tricalcium silicate hydration modeling and numerical simulations. In: Proceedings of CONMOD 2010, Lausanne (2010)
Garrault-Gauffinet, S., et al.: Experimental investigation of calcium silicate hydrate (C-S-H) nucleation. J. Cryst. Growth 200, 565–574 (1999)
Garrault, S., et al.: Hydrated layer formation on tricalcium and dicalcium silicate surfaces: Experimental study and numerical simulations. Langmuir 17, 8131–8138 (2001)
Garrault, S., et al.: Formation of the C-S-H layer during early hydration of tricalcium silicate grains with different sizes. J. Phys. Chem. B 110(1), 270–275 (2008)
Gauffinet, S., et al.: Observation direct de la croissance d'hydrosilicate de calcium sur des surfaces d'alite et de silice par microscopie à force atomique. C.R. Académie des Sciences de paris 327(4), 213–236 (1998)
Goertz, B., et al.: Investigation of the increasing strength effect of a nanoscale TiO2 additive during hardening of concrete as well as determination of the reactivity of TiO2 in the hydration process. In: Proceeding of Tagung Bauchemie, GDCh-Monographie, vol. 37, pp. 329–335 (2007)
Goetz-Neunhoeffer, F., et al.: Refined ettringite (Ca6Al2(SO4)3(OH)12.26H2O) structure for quantitative X-ray diffraction analysis. Powder Diff. 31(1), 4–11 (2006)
Greenberg, S.A.: Investigation of the colloidal hydrated calcium silicates. II Solubility relationships in the calcium oxide-silica-water system at 25°C. J. Phys. C. 69(1), 182–188 (1965)
Haecker, C.-J., et al.: Modeling the linear elastic properties of Portland cement paste. In: CCR, vol. 35, pp. 1948–1960 (2005)
Hill, J.-R., Plank, J.: Retardation of setting of plaster of paris by organic acids: understanding the mechanism through molecular modeling. J. Comp. Chem. 25(12), 1438–1448 (2004)
Holzer, L., et al.: FIB-Nanotomography of Particulate Systems – Part I: Particle Shape and Topology of Interfaces. J.A.C.S. 89(8), 2577–2585 (2006)
Häussler, F., et al.: SANS and calorimetry – useful combinable nethods for appled basic research on hydrating C3S and cement pastes. LACER 7 (2002)

Jennings, H.M., et al.: A multi-technique investigation of the nanoporosity of cement paste. In: CCR, vol. 37, pp. 329–336 (2007)
Jennings, H.M.: A model fort he microstructure of calcium silicate hydrate in cement paste. In: CCR, vol. 30, pp. 101–116 (2000)
Jiang, S.P., et al.: Studies on mechanism and physico-chemical parameters at the origin of the cement setting. II. Physico-chemical parameters determining the coagulation process. In: CCR, vol. 26(3), pp. 491–500 (1996)
Jönsson, B., et al.: Onset of cohesion in cement paste. Langmuir 20, 6702–6709 (2004)
Jönsson, B., et al.: Controlling the cohesion of cement paste. Langmuir 21, 9211–9221 (2005)
Konak, A.R.: Effects of Additives on the Crystallization of Gypsum. Kristall und Technik 11(1), 13–15 (1976)
Korpa, A., et al.: Pozzolanic reactivity of nanoscale pyrogene oxides and their strength contribution in cement-based systems. Adv. Cem. Res. 20(1), 35–46 (2008)
Koslowski, T., Ludwig, U.: The effect of admixtures in the production and application of building plasters. ZKG int. 5, 274–286 (1999)
Kowald, T., et al.: Active influence of hydration and the micro-/nanostructure with use of carbon based nanotubes. In: Proceeding of Tagung Bauchemie,GDCh-Monographie, vol. 37, pp. 37–45 (2007)
Kriechbaum, M.: Fractal structure of Portland cement paste during age hardening analyzed by small-angle x-ray scattering. Prog. Colloid. Polym. Sci. 79, 101–105 (1989)
Lewry, A.J., Williamson, J.: The setting of gypsum plaster. Part I: The hydration of calcium sulfate hemihydrate; Part II: The development of microstructure and strength; Part III: The effect of additives and impurities. J. Mater. Sci. 29, 6085–6090 (1994)
Liu, S.T., Nancollas, G.H.: A kinetic and morphological study of the seeded growth of calcium sulfate dihydrate in the presence of additives. J. Coll. Int. Sci. 52(3), 593–601 (1975)
McCartey, E.R., Alexander, A.E.: The effect of additives upon the process of crystallization. J. Colloid Sci. 13, 383–396 (1958)
McDonald, P.J., et al.: 2D correlation relaxometry studies of cement pastes performed using a new one-sided NMR magnet. In: CCR, vol. 37, pp. 303–309 (2007)
McDonald, P.J., et al.: Surface relaxation and chemical exchange in hydrating cement pastes: A 2D NMR relaxation study. Phys. Rew. E 72, 011409-1–011409-9 (2005)
McGowan, A.: Relationship between Moisture Content and Mechanical Properties of Gypsum Sheathing. In: Proceedings of 11th Canadian Conference on Building Science and Technology, Banff, Alberta (2007)
Meille, S., Saadaoui, M., Reynaud, P., Fantozzi, G.: Mechanisms of crack propagation in dry plaster. J. Eur. Ceram. Soc. 23, 3105–3112 (2003)
Moore, A.E., et al.: Crystal structure of Ettringite. Acta Cryst. 26, 386–393 (1970)
Möser, B., et al.: A new model of ordinary Portland cement hydration derived by means of ESEM-FEG. Mat. Sci. of Concrete, Spec. Cement and concrete trends and challenges, 89–107 (2002)
Möser, B., et al.: High resolution imaging of wet building material samples in their natural state using ESEM. In:Proceedings of 11th ICCC, Durban, pp. 289–302 (2003)
Mueller, M., Fischer, H.-B., Hummel, H.-U., Stark, J.: Gypsum Crystals and their Morphology. Chemine Technologija 3(33), 43–49 (2004)
Nachbaur, L., et al.: Dynamic mode rheology of cement and tricalcium silicate pastes from mixing to setting. In: CCR, vol. 31, pp. 183–192 (2001)

Nestle, N., Kutschera, M., Nicoleau, L., Leitl, M., Bräu, M.: Mineralische Bindemittelsysteme als Hochleistungsmaterialien. Phys. Unserer Zeit 4(40), 203–209 (2009)
Nestle, N., et al.: Liquid-phase self-diffusion in hydrating cement pastes – results from NMR studies and perspectives for further research. In: CCR, vol. 37, pp. 398–413 (2007)
Nicoleau, L., et al.: Accelerated calcium silicate hydrates growth experiments and modeling. In: Proceedings of CONMOD 2010, Lausanne (2010)
Nicoleau, L.: Interactions physico-chmiques entre le latex et les phases miérales constituant le ciment au cours de son hydratation, Thesis of Université de Bourgogne (2004)
Nicoleau, L.: New calcium-silicate-hydrates network. In: Proceedings of the 1st Int. Conf. in North America on Nanotechnology in Cement and Concrete, Irvine, USA, Ca (2010)
Nicoleau, L.: Unpublished laboratory results measured by BASF (2007)
Niederberger, M., et al.: Oriented attachment and mesocrystals: non-classical crystallization mechanisms based on nanoparticles assembly. Physical Chemistry Chemical Physics 8, 3271–3287 (2006)
Nudelman, F., et al.: Lessons from biomineralization; comparing the growth strategies of mollusk shell prismatic and nacreous layers in Atrina rigida, Faraday Discussions, Spiers Memorial Lecture (2007)
Odler, I., Rößler, M.: Relationships between pore structure and strength of set plaster. ZKG int. 8, 419–424 (1989)
Pake, G.E.: Nuclear resonance absorption in hydrated crystals: fine structure of the proton line. J. Chem. Phys. 16, 327–336 (1948)
Peterson, V.K.: Hydration of tricalcium and dicalcium silicate mixtures studied using quasielastic neutron scattering. J. Phys. Chem. B 109, 14449–14453 (2005)
Plassais, A., et al.: Microstructure evolution of hydrated cement pastes. Phys. Rew. E 72, 041401-1–041401-8 (2005)
Plassard, C.: Nanoscale experimental investigation of particle interactions at the origin of the cohesion of cement. Langmuir 21, 7263–7270 (2005)
Yadav, S., et al.: Predicting solubility of gypsum in aqueous salt solutions. J. Ind. Soc. Soil Sci. 43(2), 191–196 (1995)
Richardson, I.: The calcium silicate hydrates. In: CCR, vol. 38, pp. 137–158 (2008)
Rottstegge, J., et al.: Solid state NMR investigations on the role of organic admixtures on the hydration of cement pastes. Cem. & Conc. Comp. 28, 417–426 (2006)
Sattler, H.: Elastic and plastic deformations of plaster units under uniaxial compressive stress. Materiaux et Construction 7(39), 159–168 (1974)
Sauvaget, C.: Influence de phases sulfatées des clinkers sur leur réactivité, Thesis of Université de Bourgogne (2001)
Saâdaoui, M., Meille, S., Reynoud, P., Fantozzi: Internal friction study of the influence of humidiy on the set plaster. J. Eur. Cer. Soc. 25, 3281–3285 (2005)
Saâdaoui, M., Reynaud, P., Fantozzi, G., Peronnet, F., Caspar, J.P.: Slow crack growth study of plaster using the double torsion method. Ceramics Int. 26, 435–439 (2000)
Scrivener, K.L.: Backscaterred electron imaging of cementitious microstructures: understanding and quantification. Cem. & Conc. Composites 26, 935–945 (2004)
Shah, S.P., Konsta-Gdoutos, M.S., Metaxa, Z.S., Mondal, P.: Nanoscale modification of cementitious materials. In: Bittnar, Z., Bartos, P.J.M., Nemecek, J., Smilauer, V., Zeman, J. (eds.) Proceedings of the Third International Symposium on Nanotechnology in Construction, pp. 125–130. Springer, Heidelberg (2009)

Singh, N.B., Middendorf, B.: Calcium sulphate hemihydrate hydration leading to gypsum crystallization. Progress in Crystal Growth and Characterization of Materials 53, 57–77 (2007)

Singh, N.B., Vellmer, C., Middendorf, B.: Effect of carboxylic acids on the morphology, physical characteristics and hydration of α–hemihydrate plaster. Ind. J. Eng. Mat. Sci. 12, 337–344 (2005)

Skibsted, J., et al.: Characterization of cement minerals, cements and their reaction products at the atomic and nano scale. In: CCR, vol. 38, pp. 205–225 (2008)

Smith, B.R., Alexander, A.E.: The effect of Additives on the process of crystallization. J. Colloid Sci. 34, 81–90 (1970)

Sobolev, K., Flores, I., Hermosillo, R., Torres-Martínez, L.M.: Nanomaterials and nanotechnology for high-performance cement composites. In: Proceedings of ACI Session on Nanotechnology of Concrete: Recent Developments and Future Perspectives, Denver, USA, November 7 (2006)

Soroka, I., et al.: Calcareous fillers and compressive strength of Portland cement. In: CCR, vol. 6(3), pp. 367–376 (1976)

Soroka, I., Serada, P.J.: Interrelation of Hardness, Modulus of Elasticity, and Porosity in Various Gypsum Systems. J. Am. Ceram. Soc. 51(6), 337–340 (1968)

Stutzman, P.: Scanning electron microscopy imaging of hydraulic cement microstructure. Cem. & Conc. Composites 26, 957–966 (2004)

Taylor, H.F.W.: Cement Chemistry, 2nd edn. Thomas Telford Publishing (1997)

Tesárek, P., Drchalová, J., Kolísko, J., Rovnaníková, P., Černý, R.: Mechanical, Hygric and Thermal Properties of Flue Gas Desulfurization Gypsum. Acta Polytechnica 44(5-6), 83–88 (2004)

Thomas, J.J., et al.: Influence of nucleation seeding on the hydration mechanisms of tricalcium silicate and cement. J. Phys. Chem. C 113(11), 4327–4334 (2009)

Trettin, R., Kowald, T.: Nanotubes for highperformance concretes. Concrete Plant and Precast Technology 71(2), 20–21 (2005)

United States Geological Survey: Cement Statistics and Information. U.S. Department of the Interior | U.S. Geological Survey, USGS, National Center, Reston, VA 20192, `http://minerals.usgs.gov/minerals/pubs/commodity/cement/`

Vitruvius, M.P.: de Architectura, Book I..X, around 25 B.C

Winnefeld, F.: Rheological behavior of Portland cement pastes during early hydration. In: Proceedings of 24th International Conference on Cement Microscopy, San Diego, pp. 18–31 (2002)

Zajac, M.: Etude des relations entre vitesse d'hydratation, texturation des hydrates et résistance mécanique finale des pâtes et micro-mortiers de ciment Portland, Thesis de l'Université de Bourgogne (2007)

Zingg, A., et al.: The microstructure of dispersed and non-dispersed fresh cement pastes – new insight by cryo-microscopy. In: CCR, vol. 38(4), pp. 522–529 (2008)

Zinng, A.: Cement-superplasticizer interaction: link between macroscopic phenomena and mircostructural data of the early cement hydration, Thesis of the ETH Zurich (2008)

Next-Generation Nano-based Concrete Construction Products: A Review

Anal K. Mukhopadhyay

Abstract. Nanotechnology in the construction industry ranked eighth out of ten most significant areas of applications. Although nano-products are costly, the research in nano-technology is gaining momentum with the motivation of immediate profitable return from high value products. This chapter documents potential nano-products that can improve the performance and durability of concrete with high service life. Nano-products have the ability to manipulate structure at the nanometer scale, which leads to the generation of tailored, multifunctional, cementitious composites with superior mechanical performance and durability having a range of novel properties such as: low electrical resistivity, self-sensing capabilities, self-cleaning, self-healing, high ductility, and self-control of cracks. This chapter covers the primary areas of nano-engineering and nano-modifications of cementitious systems such as (i) incorporation of nano-scale spherical materials (e.g., nano-SiO_2, TiO_2, Al_2O_3, Fe_2O_3 etc.) and nano-tubes or fibers [(carbon nanotube (CNT) and carbon nano-fibers (CNF)] and nano-clay into cementitious materials during mixing, (ii) application of nano-porous thin film on aggregate surfaces before concrete mixing in order to improve interfacial transition zone (ITZ) in concrete, (iii) nano-engineering of concrete pore solution and controlled release of chemical admixtures, (iv) improve the workability in self-consolidated concrete (SCC), (v) further improvement of reactive powder concrete (RPC), (vi) applications in building materials, (vii) nanotechnology based devices etc. The discussion for each area of application covers (i) research findings by different researchers / agencies, (ii) possible mechanisms of improvement, (iii) potential areas of applications. Finally, challenges and future direction of nano-based products in construction field is briefly described.

1 Introduction

Nanotechnology is the creation of materials and devices by controlling of matter at the levels of atoms, molecules, and supramolecular (nano-scale) structures [Roco

Anal K. Mukhopadhyay
CE/TTI 508D, Materials and Pavement, Texas Transportation Institute,
Texas A&M University, College Station Texas 77845
e-mail: anol@tamu.edu

K. Gopalakrishnan et al. (Eds.): Nanotechnology in Civil Infrastructure, pp. 207–223.
springerlink.com

et al. 1999]. In other words, it is the use of very small particles of materials to create new large scale materials [Mann 2006].

Concrete is the one of the most common and widely used construction materials. It properties have been well studied at macro or structural level without fully understanding the properties of the cementitious materials at the micro level. The rapid development of the characterization techniques makes it possible to characterize cementitious materials at micro/nano-scale level. The better understanding of the structure and behavior of concrete at micro/nano-scale could help to improve concrete properties and make concrete more durable. It is expected that the strengths and durability of a concrete could be enhanced if the overall porosity and capillary pore sizes in cement paste are reduced or the diffusion of pore solution and aggressive chemicals is hindered by the introduction of some additives with a similar range of capillary pore sizes.

Successfully mimicking nature's bottom-up construction processes is one of the most promising directions. Nano-engineering encompasses the techniques of manipulation of the structure at the nanometer scale to develop a new generation of tailored, multifunctional, cementitious composites with superior mechanical performance and durability potentially having a range of novel properties such as: low electrical resistivity, self-sensing capabilities, self-cleaning, self-healing, high ductility, and self-control of cracks. The primary areas of nano-engineering and nano-modifications of cementitious systems include (i) incorporation of nano-scale spherical materials (e.g., nano-SiO_2, TiO_2, Al_2O_3, Fe_2O_3 etc.) and nano-tubes or fibers [(carbon nano-tube (CNT) and carbon nano-fibers (CNF)] and nano-clay into cementitious materials during mixing to improve mechanical performance as well as add novel properties, (ii) application of nano-porous thin film on aggregate surfaces before concrete mixing in order to improve interfacial transition zone (ITZ) in concrete, (iii) nano-engineering of concrete pore solution and controlled release of chemical admixtures, (iv) improve the workability in self-consolidated concrete (SCC), (v) further improvement of reactive powder concrete (RPC), (vi) applications in building materials, (vii) nanotechnology based devices etc.

Construction industry will be one of the major potential consumers of nanostructured materials [Bartos 2005], although, nanomodified highway materials are mostly at the laboratory stage. Application of nano-based products has the potential to overcome some of the common problems (if not all) in concrete. Table 1 listed the common problems and related solution strategies. Nanotechnology has the potential to enhance the desirable properties of concrete while helping to address some of the challenges facing the construction industry.

This chapter documents applications of nano-based products in construction areas based on the data and information collected from current literatures. The information would be beneficial to both construction engineering education and research and promote greater use of the nano-based products in large scale field applications.

Table 1 Solution Strategies for Common Concrete Problems through Nanotechnology Applications.

Problem	Solution strategies
Cracking due to brittleness	Improve ductility – (i) nano-reinforcement, nano-bridging of organic and cementitious materials, (ii) transformation of concrete form a strain-softening material to a strain hardening material through the use of new generation nanofibers
Crack formation due to dimensional instability from thermal (plastic shrinkage) and moisture fluctuation (drying shrinkage)	Reduce shrinkage by incorporating suitable nano-products
Insufficient resistance to chemical attacks such as ASR/DEF, sulfate attack, corrosion, freeze-thaw etc. primarily due to high permeable nature of concrete	Reduce permeability by proper utilization of nanotechnology – self repair / sealing, admixtures, self-healing, use of nano-products, interface management, nano-engineering of concrete pore solution
Estimating life-expectancy	Self monitoring through the use of nano-products

2 Incorporation of Nano-scale and Nanostructured Materials

Incorporation of nano-materials into the matrix to improve concrete mechanical properties emerged as a promising research field of nano-composite. Nano-scale particles are characterized by a high surface area to volume ratio and some of them are highly reactive. Most of the research work till date was conducted with nano-silica (nano-SiO2) [Bjornstrom et al. 2004, Ji 2005, Jo B 2007, Li 2004, 2006, 2007, Qing 2007, Lin KL 2008, Lin DF 2008, Sobolev 2009, Qing 2008, Kuo 2006] and nano-titanium oxide (nano-TiO2) [Li 2006, 2007]. A few studies on incorporation of nano-iron (nano-Fe2O3) [Li 2004], nano-alumina (nano-Al2O3) [Li Z 2006], and nano-clay particles [Chang 2007, Kuo, 2006] are also reported. Manufacture of nano-sized cement particles and the development of nano-binders [Lee 2005] is another area where limited numbers of investigations were carried out. Nano-engineering of concrete can take place in one or more of the three locations such as (a) in the solid phases, (b) in the liquid phases, (c) at the interfaces between liquid-solid and solid-solid [Garboczi 2009].

2.1 Incorporation of Nano-SiO_2 during Mixing

Research showed that the compressive and flexural strengths of cement mortars containing SiO_2 and Fe_2O_3 nano-particles were both higher than those of plain cement mortar [Li et al. 2004, Hui Li 2004]. The experimental results show that the compressive strengths of mortars with nano-silica (NS) were all higher than those of mortars containing silica fume at 7 and 28 days. An addition of 10% nano-SiO_2 with dispersing agents resulted 26% increase of 28 days compressive strength whereas the increase of strength was 10% with 15% silica fume [Li H et al. 2004]. Another research showed that the addition of small amounts of NS (i.e., 0.25%) caused 10% increase of compressive strength and 25% increase of flexural strength at 28 days [sobolev et al. 2009].

The SEM microstructural study of the mortar specimens with and without nano-particles revealed the mechanisms for improved performance with nano-SiO_2. When a small quantity of the nano-particles is uniformly dispersed in the cement paste, the hydrated products of cement deposit on the nano-particles due to their great surface energy, i.e., act as nucleation sites. Nucleation of hydration products on nano-particles further promotes and accelerates cement hydration [Bjornstrom et al. 2004, Lin 2008]. The addition of colloidal silica resulted acceleration of C_3S dissolution and rapid formation of C-S-H phase in cement paste (Bjornstrom et al. 2004). The other mechanisms of improved performance are (i) nano-particles fill the nano-size pores of the cement paste, and (ii) nano-SiO_2 reacts with $Ca(OH)_2$ (i.e., pozzolanic reaction) and generates additional C-S-H [Jo 2007]. Both processes are influenced by the particle size and the proper dispersion of the nano particles within the cement paste, with the colloidal dispersions being more effective than the powder [Gaitero et al. 2010]. A reduction in $Ca(OH)_2$ content and increase in C-S-H content in cement mortar as a result of nano-SiO_2 addition was noticed through DTA and XRD testing (Tang et al. 2003). With the addition of 3 wt% nano-SiO_2, significant improvement of early age ITZ structure with respect to reduction in content, crystal orientation degree, and crystal size of portlandite crystals was reported by Qing et al. [2003]. Increase of chemically combined water content and heat of hydration and decrease of CH content in presence of nanometer SiO_2 powder was noticed by Lu et al. [2006]. The microstructural studies by NMR, BET, and MIP indicated that Portland cement composite with nanosilica produce more solid, dense, and stable bonding framework [shih et al. 2006]. In other study [Dolado et al. 2005], it is reported that the improvement in strength due to nanosilica addition was not related to pozzolanic reaction but due to the formation of denser microstructures through growth of silica chains in C-S-H.

A combined effect of the above mechanisms produces an uniform dense microstructure with improvement not only in the cement paste but also in the ITZ. Effective dispersion of nano-particles is key to achieve the full benefits of adding nano-particles in cementitious system. Self-aggregation especially at high dosage of nano-particles is a common concern [Ozyildirim and Zegetosky 2010], which sometimes leads to in-homogeneous microstructure development and poor performance. The application of superplasticizer and high speed dispergation were found to be effective in proper dispersion of nano-silica particles [Flores et al. 2010]. High-intensity, high-shear mixing with the use of a proper dispersant would be helpful in thorough mixing with minimal clumping [Ozyildirim and Zegetosky 2010],

It is expected that permeability (with respect to gas, liquid, ionic movement) of concrete with nano-SiO_2 should be low enough to increase its durability and service life. Incorporation (i.e. 1.5 wt%) of nanosilica (i.e., amorphous silicic acid with avg. particle size of 15 nm) additives has caused a decrease in water penetration depth, gas permeability, and diffusion depth (Wagner et al. 1994). The water permeability test showed that the nano-SiO2 concrete has better water permeability resistant behavior than the normal concrete (Tao Ji 2005). The nanoindentation study showed that the volume fraction of the high-stiffness C-S-H gel increased significantly with addition of nano-silica [Mandal et al. 2010]. Volume fractions of high-stiffness C-S-H were 38% and 50% for samples with 6% and 18% nanosilica, respectively. This has significance on durability of concrete. Gaitero et al. [2008] reported that high stiffness C-S-H is more resistant to calcium leaching. ^{29}Si MAS-NMR spectra of the cement paste with nano-silica showed that nano-silica increases the average chain length of C-S-H gel. Therefore, addition of nano-silica has a positive impact on durability.

2.1.1 Nano-SiO_2 in High Volume Fly Ash Concrete

Nano-SiO_2 could significantly increase the early age compressive strength of high volume fly ash concrete [Li 2004]. Significant increase (i.e., 81%) of 3 days compressive strength was observed in nano-SiO_2 added high volume fly ash concrete (HVFC) in comparison with HVFC without any nano-addition. The addition of fly ash alone leads to higher porosity at early ages, while nano-SiO_2 actually lowers the concrete porosity through pore size refinement at early ages. The enhancement of pozzolanic activity of the fly ashes due to the presence of nano-SiO_2 was observed form heat of hydration test data. The maximum temperature due to heat of hydration was 61°C for the concrete with 50% FA incorporating 4% nano-SiO2 whereas it was 65°C for plain portland cement concrete (PCC) and 51°C for HVFC. The benefits of using HVFC in terms of better durability and long term mechanical properties had already been established but low early age strength of HVFC is a drawback. The addition of nano-SiO_2 has a great potential to overcome this drawback of HVFC. The composite addition of nano-SiO_2, fly ash, and silica fume was found to be very effective way to achieve good performance and an economic way to use nano-SiO_2 [Feng et al. 2004].

2.1.2 Incorporation of Nano-SiO_2 in Hardened Cement Paste

Reactive nanoparticles can be electrokinetically transported to reduce the permeability of hardened cement paste [Cardenas et al. 2006] through some kind of chemical reactions. Nanosilica (i.e., 20 nm size) and nanoalumina (2 nm size) particles dispersed in simulated pore fluids were used to make colloidal nanoparticles. It was observed that 5 minutes treatment using 5 V of potential applied over a span of 0.15 m is sufficient to drive nanoparticles into the pore system. The coefficients of permeability for each paste were reduced by 1-3 orders of magnitude.

2.2 Incorporation of Nano-TiO_2

TiO_2 (especially anatase polymorph) has identified as a potential nonomaterial with wide applications due to its strong oxidizing capability under near UV-radiation, chemical stability (both acidic and basic environments), chemical inertness in absence of UV light, absence of toxicity.

A few studies have shown that nano-TiO_2 can accelerate the early-age hydration of Portland cement [Jayapalan et al. 2009], improve compressive and flexural strengths [Li H et al. 2007]. The abrasion resistance of concrete containing nano-particles (both nano-TiO_2 and nano-SiO_2) for pavement is experimentally studied [Hui Li Zhang et al. 2006]. The abrasion resistance of concrete containing nano-TiO_2 is better than that containing the same amount of nano-SiO_2. The effectiveness of nano-TiO_2 (NT) in enhancing abrasion resistance increases in the order: 5% NT < 3% NT < 1% NT (with the decrease on NT content).

Concrete containing nano-TiO_2 has proven to be very effective for the self-cleaning of concrete as well as de-polluting the environments. Nano-TiO_2 triggers a photocatalytic degradation of pollutants (e.g., NOx, carbon monoxide, VOCs, chlorophenols, and aldehydes from vehicle and industrial emissions [Vallee et al. 2004, Murata et al. 1999, Chen 2009]. A clean TiO_2 surface in the presence of sunlight enables the removal of harmful NO_x gases from the atmosphere by oxidation to nitrates. The photocatalytic concrete pavement blocks were found to be very effective to remove NO_x (source of serious air pollution) through photocatalytic reaction of TiO_2 [Kamitani et al. 1998, Murata et al. 2002]. The surface reactions in detailed are explored using X-ray photoelectron and Raman spectroscopy [Dalton et al. 2002]. In Europe and Japan, nano-TiO_2 based "self-cleaning" concrete products are commercially available for use in the building facades and in concrete paving materials. The performance is confirmed under an UV intensity similar to outdoor level, however, long term performance under outdoor exposure condition is yet to be established.

The flexural fatigue performance of concretes containing nano-particles is significantly improved and the sensitivity of their fatigue lives to the change of stress is also increased. The concrete containing nano-TiO_2 (1 wt%) showed the best flexural fatigue performance (better than concrete containing polypropylene fibers), which has been extensively used in paving concrete [Li Hui et al. 2007].

2.3 Incorporation of Nano-Al_2O_3

Nano-Al_2O_3 was found to be very effective to increase the modulus of elasticity of cement mortar. With 5% nano- Al_2O_3 (~ 150 nm particle size), the elastic modulus increased by 143% at 28 days whereas the increase of compressive strength was not very obvious [Zhenhua Li et al. 2006]. A proper mixing procedure was selected in order to ensure adherence of nano- Al_2O_3 particles on the sand surfaces. It is believed that during cement hydration, these nano-alumina particles were available to fill the pores at the sand-paste interfaces and created a dense ITZ with less porosity. Effective densification of ITZ by nano- Al_2O_3 particles was mainly responsible for significant increase of elastic modulus of mortars. With the increase of nano-alumina contents, the elastic modulus of mortars increases when nano-alumina content is less than 5%. At higher replacement level (e.g., > 5%), agglomeration of nano-particles caused ineffective densification of ITZ and as a result, the elastic modulus of mortars decreases.

2.4 Incorporation of Nano-ZrO2

The effect of synthetic nano-ZrO_2 powder addition in cement on the strength development of portland cement paste was studied by Fan et al. 2004. Reduction in porosity and permeability, enhancement in compressive strength, and improvement in microstructure of cement paste were observed due to the addition of Nano-ZrO_2 powder in cement. Both pore filling and bridging action were identified as possible mechanisms for improvement.

2.5 Calcium Carbonate Nano Particle Addition

Conduction calorimeter based test results [Sato and Diallo 2010] indicated that the addition of nano-$CaCO_3$ significantly accelerated the rate of heat development and shortened the induction period of C_3S hydration. It was proposed that nano-$CaCO_3$ either broke down the protective layer of the C_3S hydration to shorten the induction period or accelerated the C-S-H nucleation (i.e., seeding effect) in the paste solution of high ionic concentration. Calcium carbonate particles with surface area $\geq 10\ m^2/g$ were used in the production of interior and exterior construction materials such as plaster, cement render, mortar and concrete [Cervellati et al. 2006]. The improvement in hardened properties such as good mechanical resistance, improvement in impact resistance, improvement in flexural strength, high permeability to water vapor but low permeability to liquid water, high porosity, low specific weight, good sound absorption etc. were obtained.

2.6 Early Age Strength Increase of Belite Cement

Belite cement is a environment friendly (reduce CO_2 addition) and energy efficient cement and offers superior durability property. Although, long tern strength gain of belite cement is either comparable or even better than ordinary Portland cement, low early strength due to slow hydration rate is a limitation for its widespread use. Addition of nano-particles to accelerate belite hydration at early ages was studied by different researchers [Dolado et al. 2007, Campillo et al. 2007]. Different nano-particles (e.g., nanotechnol) were added to belite cement and both the early age mechanical properties and microstructure modification were studied. The results showed that the addition of nanoparticles can overcome the drawback of this type of eco-friendly cements, which will enable them to be competitive to OPC.

2.7 Reinforcements of Nanotubes/Nanofibers

Carbon nanotubes/nanofibers (CNTs/CNFs) are potential candidates for use as nano-reinforcements in cement-based materials. CNTs/CNFs exhibit extraordinary strength with moduli of elasticity in the order of TPa and tensile strength in the range of GPa, and they have unique electronic and chemical properties [Ajayan 1999, Salvetat et al. 1999, Srivastava et al. 2003]. Cementitious materials (e.g., concrete) are typically behave as brittle materials with low tensile strength and prone to crack. Incorporation of fibers into cementitious materials is a common practice to increase tensile strength, ductility and improve durability. The interfacial interactions between carbon nanotubes and cement hydrates produce high bond strength. Carbon nanotubes act as bridges across cracks and voids, which ensures load-transfer in tension [li Geng et al. 2005].

Research has shown that flexural strength and stiffness of cementitious materials can be significantly increased by adding low concentration (e.g., 0.025% by weight of cement) of homogeneously dispersed multiwall carbon nanotubes (MWCNTs). It is reported that adding small amount of carbon nanotube (1%) by weight could increase both compressive and flexural strength (Mann, 2006).The addition of treated (treaded with a mixed solution of H_2SO_4 and HNO_3) or untreated CNTs to cement paste results considerable decrease in volume electrical resistivity and a distinct enhancement in compressive sensitivity. The cement paste with treated CNT reinforcement showed higher mechanical strength, higher compressive sensitivity and lower electrical conductivity than those with untreated CNT [Li Geng et al. 2007].

Research has revealed that incorporation of macrofibers and microfibers in cementitious system can control cracking through bridging and load transfer across cracks and pores. Although, microfibers delay the propagation of microcracks, they do not stop their initiation. CNFs are able to bridge nanocracks and pores and achieve good bonding with the cement hydration products. Incorporation of an optimal amount of CNFs (close to 0.048 wt%) was shown to improve flexural strength of the cementitious matrix significantly [Metaxa et al. 2010]. To develop high-performance nanofiber-cement nanocomposites, a homogeneous distribution of the nanofibers in cementitious matrices must be achieved. Clogging of CNF in cement paste due to improper distribution of CNF fibers is a common concern.

2.8 Nano Clay Composite

Nanoclay particles have shown promise in enhancing the mechanical performance, the resistance to chloride penetration, and the self-compacting properties of concrete and in reducing permeability and shrinkage [Chang et al. 2007, Kuo et al. 2006, Morsy et al. 2009, He and Shi 2008]. Organo-modified montmorillonites (OMMT) which have been widely used in polymer/clay nano-composites (PCN) are employed as fillers and reinforcements in cement mortars [Kuo et al. 2006]. The hydrophilic montmorillonite (MMT) micro-particles cannot be directly used as reinforcements in cement and concrete because (i) water absorbed in the interlayer regions between silicate sheets will cause detrimental expansion and (ii) the interlayer alkali cations of MMT micro-particles are harmful to the durability of cement mortar and concrete. The OMMT micro-particles modified by a cationic-exchange reaction become hydrophobic and thus can be utilized to improve the strength and permeability of cement mortar and concrete. The compressive and flexural strengths of cement mortars can be increased up to 40% and 10%, respectively and the coefficients of permeability of cement mortars could be 100 times lower if an optimal dosage (less than 1%) of OMMT micro-particles is added. The OMMT micro-particles around capillary pores can obstruct the diffusion of pore solution and aggressive chemicals and thus reduce the permeability of cement mortar and concrete. MIP results showed that the accessible pore volume is significantly reduced due to the obstruction of OMMT micro-particles around capillary pores. The optimal dosage of OMMT micro-particles approximately increases with the water/cement ratio in a mix design. Clusters of OMMT micro-particles were observed from SEM micrographs when the dosage of OMMT micro-particles is larger than 1%. Addition of Nano-montmorillonite composite (liquid form with planar diameter of about 100 nm) in cement paste (0.4 and 0.6 wt%) causes increase of compressive strength (~ 13.24%) and decrease of permeability coefficient (~ 49.95%) with more dense solid materials and stable bonding framework in the microstructure [Chang et al. 2007]. Additionally, non-modified, nano-sized smectite clays were observed to act as nucleation agents for C–S–H and to modify the structure of C–S–H [Lindgreen et al. 2008, Kroyer et al. 2003].

Micro- and nano-clays are used to design slipform paving concrete to be both flowable and shape stable at very small dosages (1.0%) [Tregger et al. 2010]. Nano-clay was found to be effective to increase shape stability with only a minimal loss in flowability. One nanoclay (i.e. C1) showed the greatest increase in green strength and compressive yield stress of the paving concrete as well as the straightest edges during the minipaver test.

3 Self-healing Polymer to Control Microcracking

Healing polymers, which include a microencapsulated healing agent and a catalytic chemical trigger (Kuennen 2004) could be especially applicable to fix the microcracking in bridge piers and columns. When the microcapsules are broken by a crack, the healing agent is released into the crack and contact with the catalyst. The polymerization happens and bond the crack faces. But it requires costly epoxy injection.

4 Self-sensing of Concrete Stress

Concrete with nano-Fe2O3 can have self-diagnostic ability of stress as well as improvement of compressive and flexural strengths [Li et al. 2004, Xiao and Ou 2004]. It was observed that the volume electric resistance of cement mortar changes with the applied load in presence of nano-Fe_2O_3 (30nm particle size). On the other hand the plain cement mortar is poor in monitoring its stress. The resistance linearly decreased with the increase of the compressive loading for mortars with Nano-Fe_2O_3 more sharply with 5% Nano-Fe_2O_3. Based on this observation, it is logical to postulate that concrete with nano- Fe_2O_3 should be capable to sense its own compressive stress. This property can be used for structural health monitoring in real time without the use of any embedded or attached sensors, which can be considered as a potential application in constructing smart structures.

Han et al. [2004] studied change of specific resistance under compression for cement paste containing two types of nano-TiO_2 particles (i.e., anatase and rutile based) and nano-carbon fiber. They observed that the cement paste containing anatase (i.e., polymorph of TiO_2) based nano-TiO_2 shows pressure-sensitivity property whereas paste containing rutile based nano-TiO_2 doesn't show that property. Cement paste containing carbon fiber shows the best pressure-sensing property with lowest specific resistance. The rate of reduction of specific resistance for paste with nano-TiO_2 (anatase) was 7-10% whereas it was 17-35% for paste with carbon fiber.

5 Self Consolidating Concrete (SCC)

The dispersion / slurry of amorphous nanosilica is used to improve segregation resistance for self-compacting concrete [Bigley and Greenwood 2003]. Addition of nanoparticles such as silica in concrete has been realized an effective way to develop high-performance and self-compacting concrete with improved workability and strength.

6 Reactive Powder Concrete (RPC)

A process for synthesis of cement with nano-scale particle sizes has recently been developed. It is demonstrated that nanocements with nano-scale particle sizes and varying phase compositions can be tailored. Addition of nano-cement in reactive powder concrete (RPC) systems has the potential to further enhance the properties of an already optimized system. The replacement of a small fraction of the conventional cement with these nano-cements causes faster cement hydration reaction and reduces the induction period [Dham et al. 2010]. The additional benefit of addition of nanocement is creating denser microstructure which causes higher compressive and tensile strength. Reduction in permeability through the improvement of ITZ is expected to be a potential merit of nanocement addition in RPC. The improvement in strength was not accompanied by an increase of elastic modulus. Further research is needed to analyze the effect of nanocement addition on the long term durability of the RPC systems.

7 Nanoporous Thin Film Technology to Improve Concrete Performance

As discussed above, most nanotechnology research has focused on characterizing concrete when nanosilica particles are dispersed in the cement paste. Nanoparticles added during cement mixing affect only the microstructure of the paste without making any significant improvement in the strength of the interfacial transition zones (ITZ). The addition of nanoparticles as nanoporous thin films (NPTF) on aggregate surface before concrete mixing was found to be an effective way to improve the ITZ and thereby the performance of concrete (Munoz and Meininger 2010). Water suspension of nanoparticles (i.e., colloidal suspension) are used to coat aggregates through dip- or spray-coating methods. The technology necessary to apply NPTF on aggregates is already available in the market. This is a cost effective method as small quantify of nanoparticulate additives is needed to obtain significant results as opposed to conventional powder addition method. The improvements in compressive, tensile, and flexural strengths and reduction in drying shrinkage was observed through the incorporation of NPTFs in mortar and concrete. The overall modulus of elasticity increased in mortars with nano-silica coated aggregates is believed to be due to the improvements in the ITZ. The control image shows presence of high concentration of pores (~ 18-30% relative porosity) at the interface which makes a clear demarcation between ITZ and bulk paste. On the other hand, concentration of pores in the ITZs of the other three images with nano-particles shows a significant reduction (i.e., 8-22% relative porosity) of porosity. This improvement in performance can ameliorate longitudinal and transverse cracking, corner breaks, punchouts, and D-cracking in concrete pavement. The research on NPTF addition in concrete is in an early stage. Further work is needed to understand the mechanisms and the full scale impact of this technology.

8 Nano-engineering of Concrete Pore Solution

While applications of nanotechnology to concrete have previously focused almost exclusively on additions of solid nanoparticles (to the solid matrix), this study indicates that nano-engineering of the pore solution can also offer substantial benefits. Nano-sized molecules were highly successful at both increasing solution viscosity and decreasing electrical conduction (Bentz et al. 2008). By adding appropriate nano-sized viscosity modifiers to the pore solution, conductive and diffusive transport can be reduced. It is envisioned that a doubling of the service life of concrete can be achieved by increasing the pore solution viscosity by a factor of two as service life is directly proportional to the diffusion coefficient of an ingressing ionic species such as chloride or sulfate ions.

9 Controlled Release of Admixtures

Controlling the time of the availability of an admixture in cement systems is essential for its optimal performance. Adding admixtures during mixing may not be optimum for the desrired chemical effects and may sometimes require to delay release of compounds such as superplasticizers, and other additives. Development of new materials for program delivery and control of admixtures in concrete and other matrials presents a significant technilogical advance. Incorpopration of nanoscale compositite materials in order to control the timing of the release of chenical admixtures is reported [Raki et al. 2004, Raki and Beaudoin 2007]. A sulphonated napthalene formaldehyde-based superplasticizer (Disal™) was used to produce the controlled release formulation (CaDisal). Based on mini-slump measurements [Kantro 1980], it is observed that CaDisal provided a longer time for the superplasticizer to keep cement workability at a reasonable lebel after mixing. Slow release of the superplasticizer through the addition of this new composite additive not only maintained the workability of the fresh mix but also improved the strength development of mortar samples. Further studies on long-term durability of concrete samples containing the controlled release composite are ongoing.

10 Nanotechnology in Building

If nano-particles are integrated with traditional building materials, the new materials might possess outstanding or smart properties for the construction of super high-rise, long-span or intelligent civil infrastructure systems.

It is proposed that white cement containg TiO_s nano-particles can have photo-catalytic properties, which allow the maintenance of the aesthetic characteristics of architectural and decorative concretes over time [Cassar et al. 2003] along with removal of pollutants as an additional benefit. A white cement containing photo-catalytic self-cleaning nanoparticles has been used in the construction of the modernistic "Dives in Misdericordia" church in Rome. TiO_2 coatings (15 nm thick) have been applied to glass windows to photocatalytically decompose dirt and smut particles that adhere to the glass [skarendahl et al. 2003]. Other applications of nanotechnology in building sector include thin film solar cells, coatings with embedded nanoparticles to provide infrared and ultraviolet screening on windows, tailored molecules to improve cement, colloidal silica (10 nm particle size) for improving polymer-based floor coatings, improved insulation materials with a porosity of the order of 10-100 nm, nanofilters to purify water, improved paint coatings, and asphalt blends with better oxidation resistance.

11 Nanotechnology Based Devices

Nano Electro Mechanical Systems (NEMS) can be used for better quality control of concrete, which indirectly helps to improve its durability. NEMS are wireless nanomachines designed to measure (i) density and viscosity of green concrete during mixing and pumping, (ii) strength development, (iii) shrinkage stress, (iv) parameters affecting concrete durability such as temperature, moisture, chloride, pH and carbon dioxide [MaCoy et al. 2005].

Microelectromechanical systems (MEMS) has been used to measure temperature and internal relative humidity (RH) of concrete by utilizing microcantilever beams and moisture-sensitive thin polymer. MEMS were found to be (i) effective and sensitive in measuring concrete temperature and moisture and (ii) durable under corrosive environment and internal / external stresses. Monitoring temperature and moisture is very important in order to (i) monitor setting and hardening characteristics of concrete and (ii) predict the possibility of occurrence of concrete chemical distresses such as corrosion of steel reinforcement, freeze–thaw distress, carbonation and alkali-aggregate reaction. The areas that need further investigation are (i) long-term behavior and repeatability of MEMS embedded into concrete, (ii) wireless interrogation such as signal processing, powering, communication, location, orientation, data storage and computation capabilities.

12 Future Challenges and Direction

Some of the potenial areas of applications of nano-based products as well as future direction are listed below

1. Engineeried materials using nanotechnology, which will allow maximum use of locally availlable materials and avoid unnecessary transport. Design ductile, flexible, breathable, permeable or impermeable concrete properties on demand
2. Design concrete mix which is resistance to freeze-thaw, corrsion, sulfate, ASR and other environmental attacks
3. Develop speciality products such as products with blast resistent, conductive properties as well as temperature-, moisture- and stress-sensing abilities.

While nanotechlogy based construction products provide many advantages to the design and construction process, the following can be considered as future challenges

1. Nanotechnology becomes a double-edge sword to the construction industry. Production of nano materials requires high energy. The engineered materials should be sustainable as well as cost-and-energy effective. Use of nanotechnology creates an environmental challenge to the construction industry. For example, nanotube might cause a lung problem to construction workers.
2. Developing a better procedure to ensure proper dispersion of nano-materials in large scale field applications.
3. With more research and practice efforts, smart design and planning, construction projects can be made sustainable and avoid damages to environment.
4. Lack of adequate R&D funding
5. Slow adoption rates of new technologies
6. Low level of collaboration for multidisciplinary problems
7. Inadequacy of QC technologies
8. The image of cement based materials need further upliftment

References

Ajayan, P.M.: Nanotubes from carbon. Chem. Rev. 99, 1787–1799 (1999)

Bartos, P.J.M.: Nanotechnology in construction: A roadmap for development. In: Proceedings of the 2nd International Symposium on Nanotechnology in Construction, Bilbao, Spain, November 13-16, pp. 27–36 (2005)

Bentz, D.P., Snyder, K.A., Cass, L.C., Peltz, M.A.: Doubling the Service Life of Concrete. I: Reducing Ion Mobility Using Nanoscale Viscosity Modifiers (2008) (submitted to Cement and Concrete Composites)

Bjornstrom, J., Martinelli, A., Matic, A., Borjesson, L., Panas, I.: Accelerating effects of colloidal nanosilica for beneficial calcium-silicate-hydrate formation in cement. Chem. Phys. Lett. 392(1-3), 242–248 (2004)

Campillo, I., Guerrero, A., Dolado, J.S., Porro, A., Ibanez, J.A., Goni, S.: Improvement of initial mechanical strength by nanoalumina in belite cements. Materials Letters 61(8-9), 1889–1892 (2007)

Cardenas, H.E., Struble, L.J.: Electrokinetic Nanoparticle Treatment of Hardened Cement Paste for Reduction of Permeability. Journal of Materials in Civil Engineering 18(4), 554–560 (2006)

Cassar, L., Pepe, C., Tognon, G., Guerrini, G.L., Amadelli, R.: White cement for architectural concrete, possessing photocatalytic properties. In: International Congress on the Chemistry of Cement, Durban, South Africa, vol. 1 (2003); CD

Cassar, L.: Photocatalysis of cementitious materials: clean buildings and clean air. MRS Bulletin 29(5), 328–331 (2004)

Cervellati, G., Rosa, R.: Use of calcium carbonate particles with high surface area in production of plaster, cement, mortar and concrete. PCT Int. Appl. WO 2006134080, 40 (2006)

Chang, T.-P., Shih, J.-Y., Yang, K.-M., Hsiao, T.-C.: Material properties of portland cement paste with nano-montmorillonite. J. Mater. Sci. 42(17), 7478–7487 (2007)

Chang, T.-P., Shih, J.-Y., Yang, K.-M., Hsiao, T.-C.: Material properties of portland cement paste with nano-montmorillonite. Journal of Materials Science 42(17), 7478–7487 (2007)

Chen, J., Poon, C.-S.: Photocatalytic construction and building materials: From fundamentals to applications. Build Environ. 44(9), 1899–1906 (2009)

Dalton, J.S., Janes, P.A., Jones, N.G., Nicholson, J.A., Hallam, K.R., Allen, G.C.: Photocatalytic oxidation of NOx gases using TiO2: A surface spectroscopic approach. Environmental Pollution 120(2), 415–422 (2002)

Dham, M., Rushing, T.S., Helferich, R., Marth, T., Sengupta, S., Revur, R., Weiss, C.A., Cummins, K.: Enhancement of reactive powder concrete via nanocement integration. Journal of Transportation Research Record 2(2142), 18–24 (2010)

Dolado, J.S., Campillo, I., Erkizia, E., de Miguel, Y.R., de Ibarra, Y.S., Porro, A., Ayuela, A.: Effects of Nanosilica Additions on Cement Pastes. Applications of Nanotechnology in Concrete Design. In: Proceedings of the International Conference Applications of Nanotechnology in Concrete Design, pp. 87–96 (2005)

Dolado, J.S., Campillo, I., Porro, A., Sanchez, J.A., Goni, S., Guerrero, A.M., Moragues, A.: An study on the acceleration of belite cement hydration. Cemento-Hormigon 78(897), 24–30 (2007)

Fan, J.-J., Tang, J.-Y., Cong, L.-Q., Mcolm, I.J.: Influence of synthetic nano-ZrO2 powder on the strength property of portland cement. Jianzhu Cailiao Xuebao 7(4), 462–467 (2004) (in Chinese)

Feng, Q., Liang, C.-D., Liu, G.-M.: Experimental study on cement-based composites with nano-SiO2. Cailiao Kexue Yu Gongcheng Xuebao 22(2), 224–227 (2004) (in Chinese)

Flores, I., Sobolev, K., Torres-Martinez, L.M., Cuellar, E.L., Valdez, P.L., Zarazua, E.: Peformance of cement systems with Nano-SiO_2 particles produced by suign the sol-gel method. Journal of Transportation Research Record 1(2141), 10–14 (2010)

Garboczi, E.J.: Concrete nanoscience and nanotechnology: Definitions and applications. In: Bittnar, Z., Bartos, P.J.M., Nemecek, J., Smilauer, V., Zeman, J. (eds.) Proceedings of the NICOM3, Nanotechnology in construction, 3rd International Symposium on Nanotechnology in Construction, Prague, Czech Republic, pp. 81–88 (2009)

Han, B., Guan, X., Ou, J.: Specific resistance and pressure-sensitivity of cement paste admixing with nano-TiO2 and carbon fiber. Guisuanyan Xuebao 32(7), 884–887 (2004) (in Chinese)

He, X., Shi, X.: Chloride permeability and microstructure of Portland cement mortars incorporating nanomaterials. Transportation Research Board Record. Journal of the Transportation Research Board 2070, 13–21 (2008)

Hosseini, P., Booshehrian, A., Farshchi, S.: Influence of nano-SiO_2 addition on microstructure and mechanical properties of cement mortars for ferrocement. Journal of Transportation Research Record 1(2141), 15–20 (2010)

Li, H., Xiao, H.-g., Ou, J.-p.: A study on mechanical and pressure-sensitive properties of cement mortar with nanophase materials. Cement and Concrete Research 34, 435–438 (2004)

Jayapalan, A.R., Lee, B.Y., Fredrich, S.M., Kurtis, K.E.: Influence of additions of anatase TiO_2 nanoparticles on early age properties of cement-based materials. Journal of Transportation Research Record 1(2141), 41–46 (2010)

Ji, T.: Preliminary study on the water permeability and microstructure of concrete incorporating nano-SiO_2. Cement Concr. Res. 35(10), 1943–1947 (2005)

Jo, B.-W., Kim, C.-H., Tae, G.-h., Park, J.-B.: Characteristics of cement mortar with nano-SiO_2 particles. Construct Build. Mater. 21(6), 1351–1355 (2007)

Jo, B.-W., Kim, C.-H., Lim, J.-H.: Characteristics of cement mortar with nano-SiO2 particles. ACI Materials Journal 104(4), 404–407 (2007)

Gaitero, J.J., Campillo, I., Mondal, P., Shah, S.P.: Small Changes Can Make a Great Differences. Journal of Transportation Research Record 1(2141), 1–5 (2010)

Kamitani, K., Murata, Y., Tawara, H., Takeuchi, K.: Air purifying pavement: development of photocatalytic concrete blocks. In: Wu, Z. (ed.) International Symposium on Cement and Concrete, pp. 751–755. International Academic Publishers, Beijing (1998)

Kantro, D.L.: Influence of water-reducing admixtures on properties of cement pastes-a miniature slump test. Cem. Concr. and Aggre. 2(2), 95–102 (1980)

Kuennen, K.: Small Science Will Bring Big Changes To Roads. Better Roads (2004)

Kuo, W.-Y., Huang, J.-S., Lin, C.-H.: Effects of organo-modified montmorillonite on strengths and permeability of cement mortars. Cement Concr. Res. 36(5), 886–895 (2006)

Lee, S.J., Kriven, W.M.: Synthesis and hydration study of Portland cement components prepared by the organic steric entrapment method. Mater Struct. 38(1) (2005)

Li, G.: Properties of high-volume fly ash concrete incorporating nano-SiO2. Cement and Concrete Research 34, 1043–1049 (2004)

Li, G.Y., Wang, P.M., Zhao, X.: Mechanical behavior and microstructure of cement composites incorporating surface-treated multi-walled carbon nanotubes. Carbon. 43(6), 1239–1245 (2005)

Li, G.Y., Wang, P.M., Zhao, X.: Pressure-sensitive properties and microstructure of carbon nanotube reinforced cement composites. Cement & Concrete Composites 29(5), 377–382 (2007)

Li, H., Zhang, M.-h., Ou, J.-p.: Abrasion resistance of concrete containing nano-particles for pavement. Wear 260, 1262–1266 (2006)

Li, H., Zhangh, M.-H., Ou, J.-P.: Flexural fatigue performance of concrete containing nano-particles for pavement. International Journal of Fatigue 29(7), 1292–1301 (2007)

Li, H., Xiao, H.-g., Yuan, J., Ou, J.: Microstructure of cement mortar with nano-particles. Compos. B Eng. 35(2), 185–189 (2004)

Li, H., Xiao, H.-g., Ou, J.-p.: A study on mechanical and pressure-sensitive properties of cement mortar with nanophase materials. Cement. Concr. Res. 34(3), 435–438 (2004)

Li, H., Zhang, M.-p., Ou, J.-p.: Abrasion resistance of concrete containing nano-particles for pavement. Wear 260(11-12), 1262–1266 (2006)

Li, H., Zhang, M.H., Ou, J.P.: Flexural fatigue performance of concrete containing nanoparticles for pavement. Int. J. Fatig. 29(7), 1292–1301 (2007)

Lin, K.L., Chang, W.C., Lin, D.F., Luo, H.L., Tsai, M.C.: Effects of nano-SiO2 and different ash particle sizes on sludge ash-cement mortar. Journal of Environmental Management 88(4), 708–714 (2008)

Lin, D.F., Lin, K.L., Chang, W.C., Luo, H.L., Cai, M.Q.: Improvements of nano-SiO2 on sludge/fly ash mortar. Waste Manag. 28(6), 1081–1087 (2008)

Li, Z., Wang, H., He, S., Lu, Y., Wang, M.: Investigations on the preparation and mechanical properties of the nano-alumina reinforced cement composite. Mater Lett. 60(3), 356–359 (2006)

Li, Z., Wang, H., He, S., Lu, Y., Wang, M.: Investigations on the preparation and mechanical properties of the nano-alumina reinforced cement composite. Materials Letters 60, 356–359 (2006)

Lu, Z.-y., Xu, X.: Effect of nanometer SiO2 on hydration character of portland cement. Jianzhu Cailiao Xuebao 9(5), 581–585 (2006) (in Chinese)

Mann, S.: Nanotechnology and Construction. Nanoforum Report (2006), http://www.nanoforum.org (May 30, 2008)

McCoy, M., Betts, J., Norris, A., Saafi, M.: Nanotechonology in construction: Nano materials and devices offer macro improvements in concrete materials. In: RILEM Proceedings, PRO 45 2nd International Symposium on Nanotechnology in Construction (NICOM2), pp. 223–231 (2005)

Metaxa, Z.S., Konsta-Gdoutos, M.S., Shah, S.P.: Carbon nanofiber-reinforced cement-based materials. Journal of Transportation Research Record 2(2142), 114–118 (2010)

Mondal, P., Shah, S.P., Marks, L.D., Gaitero, J.J.: Comparative study of the effects of microsilica and nanosilica in concrete. Journal of Transportation Research Record 1(2141), 6–9 (2010)

Morsy, M.S., Aglan, H.A., Abd El Razek, M.M.: Nanostructured zonolite-cementitious surface compounds for thermal insulation. Construct Build. Mater. 23(1), 515–521 (2009)

Munoz, J.F., Meininger, R.C., Youtcheff, J.: New possibilities and future pathways of nanoporous thin film technology to improve concrete performance. Journal of Transportation Research Record 2(2142), 34–41 (2010)

Murata, Y., Obara, T., Takeuchi, K.: Air purifying pavement: development of photocatalytic concrete blocks. J. Adv. Oxidat. Technol. 4(2), 227–230 (1999)

Murata, Y., Tobinai, K.: Influence of various factors on NOx removal performance of permeability interlocking block based on photocatalysis. Journal of Structural and Construction Engineering 555, 9–15 (2002) (in Japanese)

Norris, A., Saafi, M., Romine, P.: Temperature and moisture monitoring in concrete structures using embedded nanotechnology/microeletronic mechanical systems (MEMS) sensors. Construction and Building Materials 22(2), 111–120 (2008)

Celik, O., Caroline, Z.: Exploratory investigation of nanomaterials to improve strength and permeability of concrete. Journal of Transportation Research Record 2(2142), 1–8 (2010)

Qing, Y., Zenan, Z., Deyu, K., Rongshen, C.: Influence of nano-SiO2 addition on properties of hardened cement paste as compared with silica fume. Construct Build Mater 21(3), 539–545 (2007)

Raki, L., Beaudoin, J.J., Mitchell, L.D.: Layered double hyroxides-like materials: nanocomposites for use in concrete. Cem. Concr. Res. 34, 1717–1724 (2004)

Raki, L., Beaudoin, J.J.: Controlled release of chemical admixtures. Canadian Patent # CA 2554347, US patent Application US 2007/0022916 A1 (2007)

Roco, M.C., Williams, R.S., Alivisatos, P.: Nanotechnology Research Directions: IWGN Research Report. Committee on Technology, Interagency Working Group on Nanoscience, Engineering and Technology (IWGN), National Science and Technology Council (1999)

Salvetat, J.-P., Bonard, J.-M., Thomson, N.H., Kulik, A.J., Forro, L., Benoit, W., Zuppiroli, L.: Mechanical properties of carbon nanotubes. Appl. Phys. Mater. Sci. Process 69, 255–260 (1999)

Sato, T., Diallo, F.: Seeding effect of nano-$CaCO_3$ on the hydration of tricalcium silicate. Journal of Transportation Research Record 1(2141), 61–67 (2010)

Shih, J.-Y., Chang, T.-P., Hsiao, T.-C.: Effect of nanosilica on characterization of Portland cement composite. Materials Science and Engineering, A: Structural Materials: Properties, Microstructure and Processing A424 (1-2), 266–274 (2006)

Skarendahl, A.: Nanotechnology developments for building and protection. Vbyggaren (Swedish Society of Civil and Structural Engineers), vol. 6, pp. 8–11 (2003)

Sobolev, K., Flores, I., Torres-Martinez, L.M., Valdez, P.L., Zarazua, E., Cuellar, E.L.: Engineering of SiO2 nanoparticles for optimal performance in nano cement-based materials. In: Bittnar, Z., Bartos, P.J.M., Nemecek, J., Smilauer, V., Zeman, J. (eds.) Nanotechnology in construction: Proceedings of the NICOM3, 3rd International Symposium on Nanotechnology in Construction, Prague, Czech Republic, pp. 139–148 (2009)

Srivastava, D., Wei, C., Cho, K.: Nanomechanics of carbon nanotubes and composites. Appl. Mech. Rev. 56, 215–230 (2003)

Tang, M., Ba, H., Li, Y.: Study on compound effect of silica fume and nano-SiOx for cementing composite materials. Guisuanyan Xuebao 31(5), 523–527 (2003) (in Chinese)

Tregger, N., Pakula, M., Shah, S.P.: Influence of micro- and nano-clays on fresh state of concrete. Journal of Transportation Research Record 1(2141), 68–74 (2010)

Vallee, F., Ruot, B., Bonafous, L., Guillot, L., Pimpinelli, N., Cassar, L., Strini, A., Mapelli, E., Schiavi, L., Gobin, C., Andre, H., Moussiopoulos, N., Papadopoulos, A., Bartzis, J., Maggos, T., McIntyre, R., Lehaut-Burnouf, C., Henrichsen, A., Laugesen, P., Amadelli, R., Kotzias, D., Pichat, P.: Cementitious materials for self-cleaning and depolluting facade surfaces. In: RILEM Proceedings (2005), PRO 41 RILEM International Symposium on Environment-Conscious Materials and Systems for Sustainable Development, pp. 337–346 (2004)

Wagner, J.-P., Hauck, H.G.: Nanosilica - an additive for High-strength Concrete. Wissenschaftliche Zeitschrift - Hochschule fuer Architektur und Bauwesen Weimar - Universitaet 40(5/6/7), 183–187 (1994) (in German)

Ye, Q., Zhang, Z., Chen, R., Ma, C.: Interaction of nano-SiO2 with portlandite at interface between hardened cement paste and aggregate. Guisuanyan Xuebao 31(5), 517–522 (2003) (in Chinese)

Ge, Z., Gao, Z.: Applications of Nanotechnology and Nanomaterials in Construction. In: First International Conference on Construction In Developing Countries (ICCIDC–I), Advancing and Integrating Construction Education, Research & Practice, Karachi, Pakistan, August 4-5 (2008)

Optimization of Clay Addition for the Enhancement of Pozzolanic Reaction in Nano-modified Cement Paste

Bjorn Birgisson and Mahir Dham

Abstract. This chapter presents a process for obtaining nano-size silicate platelets for use in cement paste from exfoliating Cloisite Na+ clay particles. The resulting silicate platelets require all surface electrical charges to be neutralized. The results showed that this is achievable through the use of organic ammonium chloride surfactant. The exfoliation process was evaluated through the use of X-ray diffraction and the optimum stable concentration of organic ammonium chloride was determined with the help of electrokinetic studies and turbidity measurements which determine the stability of the clay surfactant solution. The ultimate effects on hydrated cement paste are evaluated using compressive strength measurements for the subsequent replicates of samples prepared.

1 Introduction

In terms of volume used, concrete is the most successful material in the world. It is made up of aggregates bound together with hydrated cement. In general it contains between 10 and 15 wt. % cement. Globally 2.6 billion ton of cement is produced annually, giving more than 3 m^3 of concrete per person. Due to the size of the cement production, it consumes large amount of energy and is one of the largest contributor to carbon dioxide (CO_2) release. Cement production accounts for some about 4-5% of man-made CO_2 emissions. Thus the amount of cement used in concrete production must be minimized. The challenge is to get more strength out of the concrete with less cement.

One concept recently developed is to use nanoparticles made from exfoliated clays, carbon tubes and/or precipitated silica particles (nano silica). Results from

Bjorn Birgisson
Professor and Division Head, Department of Civil and Architectural Engineering, Brinellvägen 34, KTH Royal Institute of Technology, 10044 Stockholm, Sweden
e-mail: bjornbir@kth.se

Mahir Dham
Development Chemist, Mapei Corporation, Deerfield Beach, Florida 33442
e-mail: mahirdham@gmail.com

K. Gopalakrishnan et al. (Eds.): Nanotechnology in Civil Infrastructure, pp. 225–236.
springerlink.com

experiments show that by modifying the cement paste structure, it is possible to modify both the hydration path and the structure of the binder. As an example, nano-SiO_2 additions with fly ash or with fly ash and silica fume combined have shown more pozzolanic activity than traditional silica fume alone [Collapardi, et al. 2004; Li 2004; Li et al. 2005]. This is due to the fact that their size allows them to react more readily with the calcium hydroxide, thereby increasing C-S-H conversion [Ji 2005].

In recent years, the addition of nano-particles into cement paste is gaining attention due to their exceptional properties due to their high surface area and therefore high reactivity. Recent experiments have shown that nano silica do not only change the structure of the C-S-H but it will also change the path of and mode of hydration [e.g. Lagerblad and Fjällberg 2008], which in turn will influence the structure of the paste. Nano-silica (nano-SiO2) is an industrial product that is more fine-grained and thus reacts more readily with calcium hydroxide (CH) [Jo et al. 2007].

Nano-modification of the cement inherently affects the hydration process, which in turn affects the structure and the physical properties. The addition of the nano-clay particles also simulates a process similar to the pozzolanic reaction where silicates and water react with calcium hydroxide to form C-S-H. It has been shown that the rate of this reaction is proportional to the amount of surface area available for the reaction [Jo et al. 2007]. Basic pozzolanic reactions produced from the composite cements have been known to effect the production of CH and C-S-H, the two main hydration products that have been attributed to the strength of the cement [Escalante-Garcia and Sharp 2004; Taylor 1997; Shayan and Xu 2006]. Generally speaking, the pozzolanic reaction works by reducing the amount of CH produced during the hydration either by consumption of CH itself of its hydration components [Saeki and Monteiro 2005, Taylor 1997].

More recently exfoliated nano-size exfoliated silica sheets are obtained from montmorillonite smectite clays [Dham 2007]. Research by Dham (2007) has led to a process where the researchers were able to exfoliate clays with weak platelet bonds, resulting in exfoliated nano-sized montmorillonite clay (Cloisite Na+). Importantly, the cost of the nanosilicates is pound for pound less than for silica fumes, since the basic material is derived from natural ("as dug") clays. The process has also been used on phosphate waste clays in Florida with good results, which are currently being land-filled [Boyd et al. 2007].

In this chapter, the exfoliation process is described, along with the methods used to verify the exfoliation of Cloisite Na+ into nano-size silicate platelets that are suitable for use in concrete. An important component of this system is the neutralization of the surface charges of the resulting silicate platelets through the introduction of a suitable surfactant.

2 The Exfoliation Process

Montmorillonite clay itself is a sheet-like structure (Fig. 1) composed of tetrahedra SiO_4 and octrahedra AlO_6 at a 2:1 ratio. Each of these sheets sustains a charge on their surface and edges. Clays are characterized by their ion exchange properties

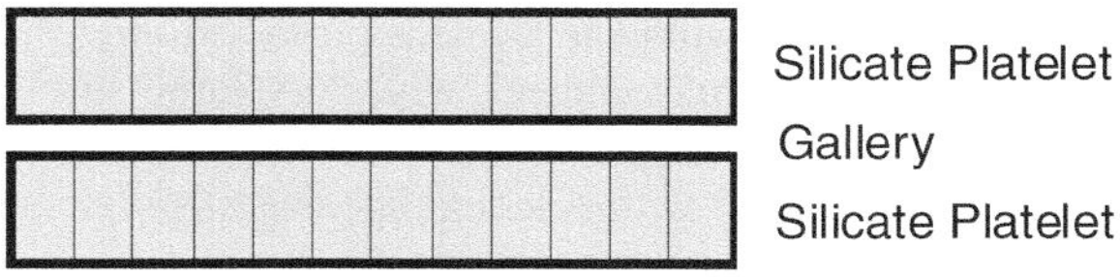

Fig. 1 Idealized layered silicate structure

that result in a high affinity to water, which makes the naturally incompatible for use in concrete in their unmodified form.

The advantage of the sheet-like configuration over other small particles is that the multi-layer silicate structure can be penetrated between layers by small molecules (see Fig. 2), forcing the silicate platelets apart. This process is called intercalation.

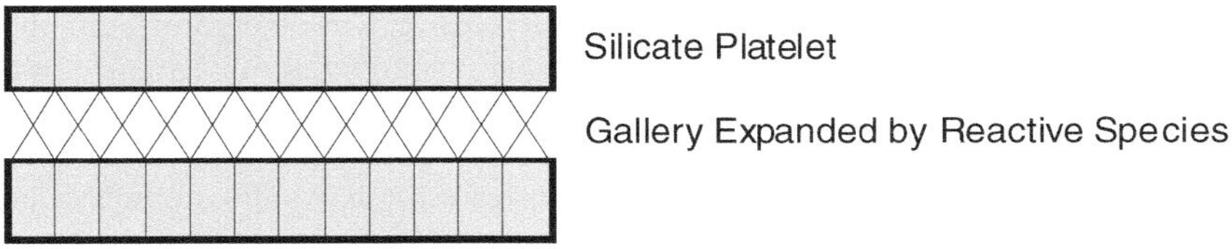

Fig. 2 Partial separation of silicate platelets (intercalation)

If the penetrating molecules reactive enough, this process may result in complete separation of the silicate layers (i.e. exfoliation), as shown schematically in Fig. 3.

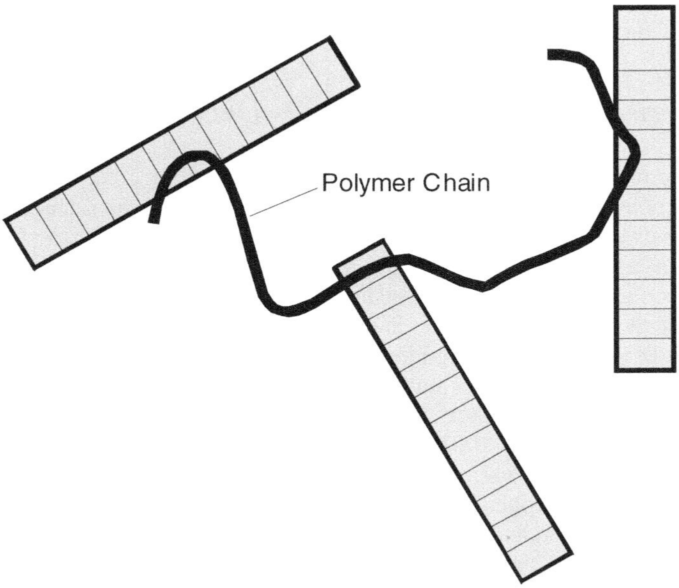

Fig. 3 Total separation of silicate platelets (exfoliation)

As a result of exfoliation, a small mass of montmorillonite clay can result in numerous small, thin (e.g. 20-nm) platelets, with very large surface area, that are fully separated. These nano-size silicate platelets may e.g. be used directly as pozzolans in concrete or bonded with selected species of polymers which tend to bond to all of these surfaces, creating a linkage effect among the silicate platelets. This bonding can also be considered as a flocculated material since one polymer chain can link several clay particles together. The resulting composite can exhibit vastly improved properties.

3 Materials and Methods

In this study, Cloisite Na^+ clay, which is also known as sodium montmorillonite, is added to the cement paste mixture along with a surfactant to ensure thorough dispersion of the clay. The surfactant used in this research was an organic ammonium chloride called 2-(Dimethylamino) ethyl acrylate, methyl chloride [Sigma Aldrich 2006; BMT Inc. 2008; Cementec Inc. 2008; Southern Clay Inc. 2008].

For complete exfoliation of clay, it is advantageous if synthesized charged organic/inorganic nanoparticles are used instead of conventional intercalating agents. For this reason, organic ammonium chloride was perceived to be most suited.

The first obstacle to overcome was determining the exact amount of surfactant necessary to produce a stable clay/surfactant solution. Initially, specific proportions of surfactant, with respect to clay, were investigated. Evaluation of the stability of the system became a necessity and thus electrokinetic phenomena (zeta potential) and system turbidity techniques were used to ascertain the optimal amount of surfactant required.

3.1 Electrokinetic Studies for Zeta Potential Measurements

The equipment used for evaluating the procedure included a Brookhaven Zeta Plus zeta potential meter. Zeta potential is the potential of the surface at the plane of shear between the particle and the surrounding media as the particle and media move with respect to each other. In the presence of an applied electric field, the charged surface (and the attached material) tends to move in the appropriate direction, while the counter ions in the mobile part of the double layer would have a net migration in the opposite direction. On the other hand, an electric field would be created if the charged surface and the diffuse part of the double layer were made to move relative to each other. The zeta potential facilitates easy quantification of the surface charge [Singh 2002].

For our replicates of various samples it was found that when the absolute value of zeta potential is around 50 mV the dispersions are very stable due to mutual electrostatic repulsion and when the zeta potential is below this value, coagulation (formation of larger assemblies of particles) occurs very quickly, causing rapid sedimentation. Even when the surface charge density is very high but the zeta potential is low, the colloids are unstable. Also, the velocity of heterocoagulation

(coagulation of different particles) depends on the zeta potentials of each type of particle. Therefore, zeta potential is an important parameter characterizing colloidal dispersion. The instrument used to measure the zeta potential is called a zeta potential meter.

3.2 Turbidity Measurements

Turbidity in water can be caused by suspended and colloidal matter such as clay, silt, finely divided organic and inorganic matter, and plankton and other microscopic organisms. Turbidity is an expression of the optical property that causes light to be scattered and absorbed rather than transmitted with no change in direction or flux level through the sample. The Turbidity Meter can measure this turbidity, in units of nephelometric turbidity units or NTU. This term originated because turbidimeters with scattered-light detectors located at 90° to the incident beam are called nephelometers and so the units ascribed to their Turbidity measurements is NTU [EPA 1999].

The principle of this method is based on the intensity of light scattered by the sample under defined conditions. The higher the intensity of scattered light, the higher is the turbidity. For the replicate clay samples mixed with OAC, it was possible to determine the optimum amount of the surfactant required by the system to ensure a stable turbidity measurement.

The turbidimeter used in this research was a Hach Model 2100AN Laboratory turbidimeter capable of measuring Turbidity values from 0 to 10000 NTU. The optical system was comprised of a tungsten filament lamp, lenses and apertures to focus the light, a 90° detector to monitor scattered light, a forward scatter light detector, a transmitted light detector, and a back scatter light detector [Lab Manual 2006].

3.3 X-Ray Diffraction: Exfoliation of Clay Platelets

The X-Ray Diffraction (XRD) technique is used to measure atomic spacing between lattice layers in a crystal. This can be calculated from the X-Ray Diffraction test results using the Braggs Law equation, is as follows: $n\lambda = 2d \sin\theta$, where, λ = wavelength of the incidental X-Ray beam; d = the atomic spacing between the layers; θ = angle of incidence; n is taken as unity.

Three samples were examined with the X-Ray diffraction technique, using a Philips X'Pert system [MAIC 2009]. These samples were:

- Sample 1 consisted of Cloisite Na^+ alone.
- Sample 2 contained 9.5% by weight of Cloisite Na^+ and 9.5% by weight OAC, mixed in water and then dried.
- Sample 3 contained 5.6% by weight of Cloisite Na^+, 5.6 % by weight OAC, and 5.6% by weight of poly-vinal alcohol (PVA) polymer, mixed in water and then dried.

4 Results

Initial laboratory observations indicated that very low dosages of OAC surfactant were required to induce stability. An experimental matrix was formulated to evaluate a range of surfactant contents to be used per sample. This matrix is tabulated in Table 1.

Since the amount of surfactant to be used had a very low weight to be measured, a buffer solution of the surfactant was used where 1 g of surfactant was added to 1000 g of water.

Table 1 Experimental matrix

Sample Number	Clay (g)	Water (g)	Surfactant (mg/ g of clay)	Buffer (g)
1	1	98	1	1
2	1	97	2	2
3	1	96	3	3
4	1	95	4	4
5	1	94	5	5
6	1	93	6	6
7	1	92	7	7
8	1	91	8	8

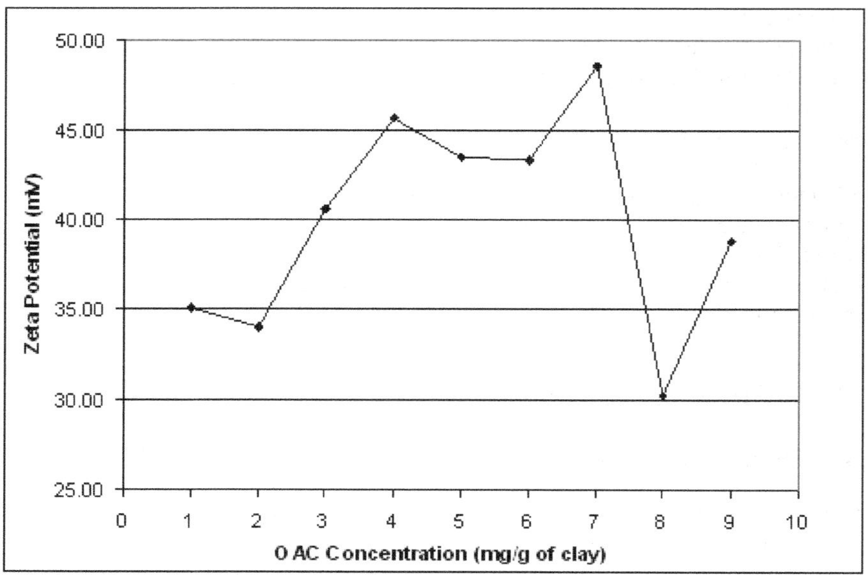

Fig. 4 The effect of OAC concentration on Zeta Potential

4.1 Electrokinetic Study Results

As stated before, the dispersion is at maximum stability if the surface charge is complete and the zeta potential of the solution is well above zero. This was how the experimental matrix was terminated. Once the optimal zeta potential was reached, it began to drop as the OAC concentration continued to increase. The results from the zeta potential analysis are depicted graphically in Fig. 4.

As inferred from the graph, OAC concentrations of 4 to 7 mg/g produced the highest zeta potential, close to 50 mV. Above 7 mg/g the zeta potential dropped off, indicating that the stable state had been exceeded. It can therefore be concluded from this test that the optimal OAC addition rate is between 4 and 7 mg of surfactant per g of Cloisite Na^+.

4.2 Turbidity Analysis Results

The turbidity meter results support the Zeta Potential data, as the turbidity for the replicates of samples between 4 and 7 mg of surfactant per g of Cloisite Na^+ exhibited the lowest values in NTU. This shows that the sample is most stable in that region. Fig. 5 shows the turbidity results.

Variation of turbidity over time can be a significant factor in the period immediately after combination of the materials. To examine this effect, turbidity measurements were regularly taken over an extended time frame. It was discovered that

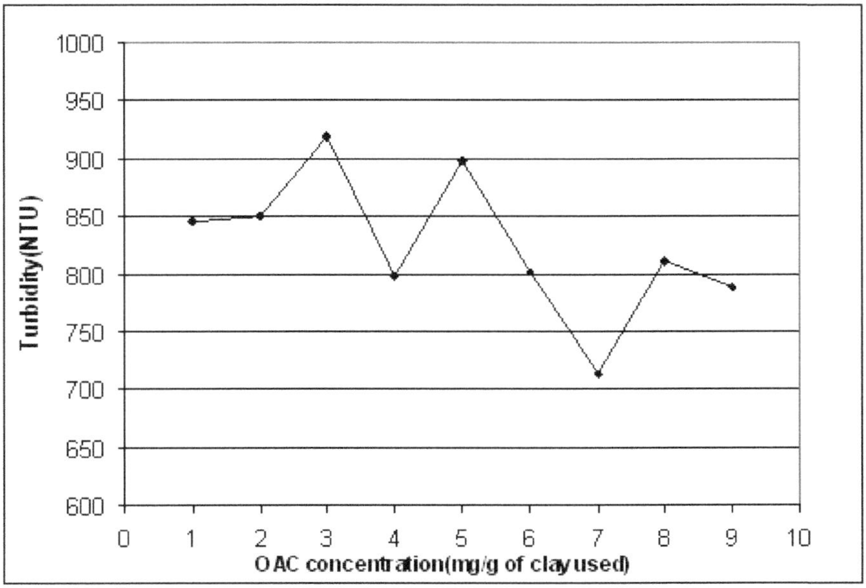

Fig. 5 The effect of OAC concentration on turbidity

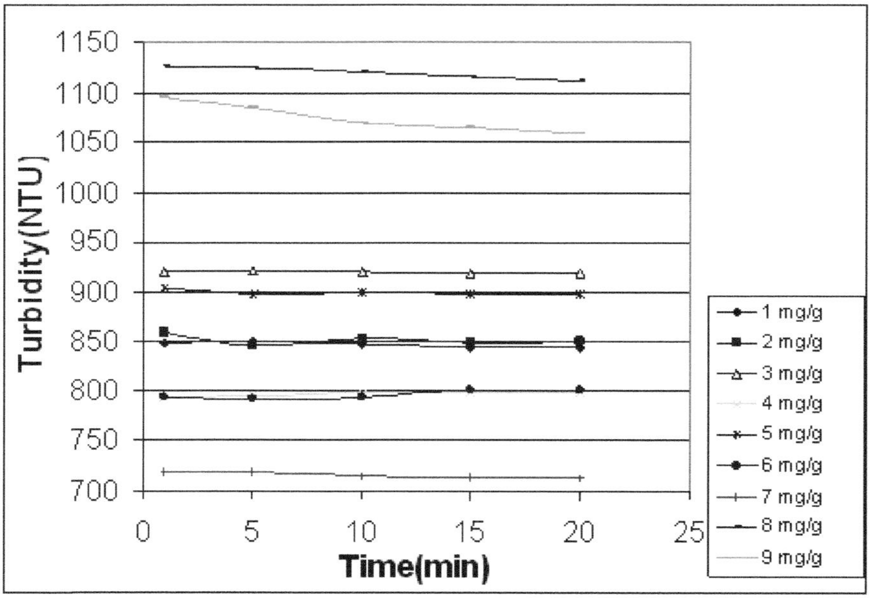

Fig. 6 Turbidity as a function of time for various samples

the turbidity levels stabilize at around 15 minutes. Fig. 6 represents plots of turbidity as a function of time for different dosages.

As inferred from the graph, OAC concentrations of 4 to 7 mg/g produced the highest zeta potential, close to 50 mV. Above 7 mg/g the zeta potential dropped off, indicating that the stable state had been exceeded. It can therefore be concluded from this test that the optimal OAC addition rate is between 4 and 7 mg of surfactant per g of Cloisite Na^+.

4.3 X-Ray Diffraction Results

Fig. 7 depicts the x-ray diffraction results for the three samples discussed previously, while the calculated d-spacings from various replicates are tabulated in Table 2. The significantly higher high d-spacing for the second sample (clay + OAC) proves that the surfactant is causing exfoliation of the clay, though addition of the polymer does not. This can be inferred from the d-spacing, which has increased with the shifting of the peaks of each curve. Exfoliation should lead to a large spreading of the clay galleries, which is readily determined with the x-ray diffraction technique.

Table 2 D-spacing for x-ray diffraction samples

	Clay	Clay + OAC	Clay + OAC + polymer
d-spacing	11.84	22.06	14.71

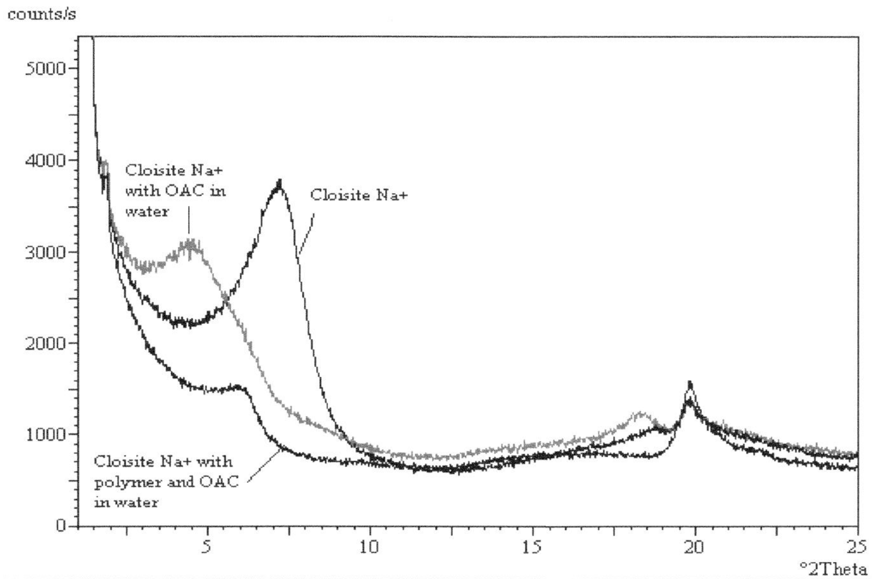

Fig. 7 Intensity as a function of 2θ angles for various samples

4.4 Mechanical Study: Stress-Strain Analysis

To evaluate the effects of the introduction of nano-size silicate platelets obtained from exfoliated Cloisite Na+ clay, various replicates of 7.5 cm x 15 cm (3in x 6in) cement paste cylinders were cast using Type I cement with w/c of 0.6 and tested under axial compression after 28 days of curing in lime solution. Importantly, cylinder specimens were selected over cube testing in order to allow for on-specimen strain measurements using 3 LVDT's spaced at 120 degrees on the specimens. From the results of the compression testing, stress verses strain graphs were plotted. All replicates had three samples and were subjected to an axial displacement rate of 6 in/min during testing.

The results of compressive strength testing of cement paste cylinders from various replicates containing varying proportions of exfoliated Cloisite Na^+ are depicted in Fig. 8 and summarized in Table 3. As seen in the figure, the addition of 2%, 5%, and 9% of exfoliated Cloisite Na^+ results in an increase in the compressive strength of the replicates of samples when compared to the control mix. Though the 2% and 5% replacement rates exhibit successively large increases in ultimate load, the 9% replacement sample, though still stronger than the control, actually begins to decrease in strength.

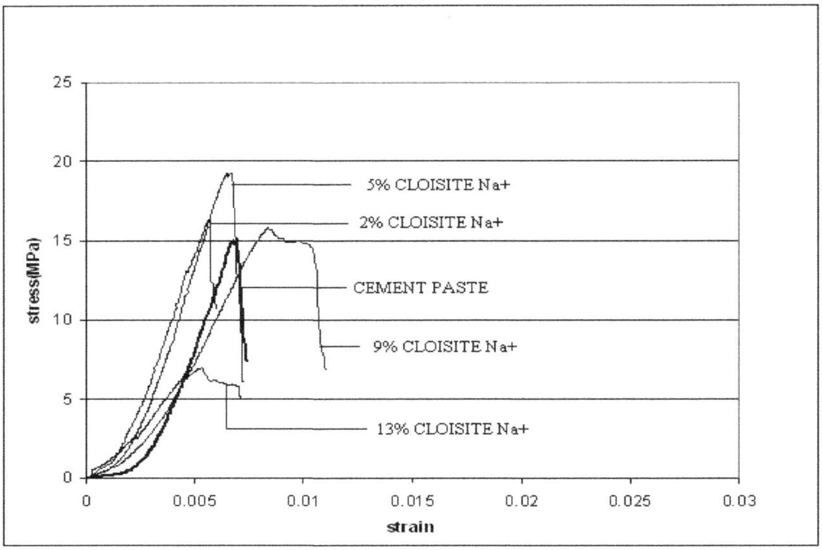

Fig. 8 Compressive stress versus measured axial strain as a function of strain for 7.5 cm x 15 cm (3 in x 6 in) cement paste cylindrical specimens with varying amounts of silicate platelets obtained from exfoliated Cloisite Na+ clay

The 13% replacement rate of the exfoliated Closite Na+ reduces the compressive strength in comparison to the control mix, thus indicating the overdosing with the exfoliated clay platelets can be detrimental to mechanical properties. This phenomenon could be attributed to:

1) Poor dispersion caused by an excessive amount of exfoliated clay platelets resulting in agglomeration of the platelets, producing weak points in the matrix that contribute to premature failure.
2) Insufficient shear rates during mixing of the higher exfoliated clay content pastes. This could lead to inefficient mixing and lack the ability to induce full exfoliation of the exfoliated clay when the amount of clay is too large.
3) The optimal point may have passed, meaning there may be more exfoliated clay present than that needed to react all of the calcium hydroxide within the cement paste.

Table 3 Maximum stress and strain values with respect to clay addition

Clay %	Maximum Stress (MPa)	Post Peak Strain	Failure Strain
0	14.81	0.0057	0.0052
2	16.34	0.0060	0.0056
5	19.22	0.0070	0.0061
9	15.82	0.0100	0.0074
13	7.00	0.0080	0.0053

5 Summary and Conclusions

A procedure for exfoliating Closite Na+ clay particles was presented. The resulting nano-size silicate platelets were used in cement paste, resulting in increased compressive strengths after 28 days of curing. The following conclusions could be generalized for all clay and surfactant systems:

- The surfactant must be chosen to complement the application and the type of material that it is to be mixed with in order to achieve optimal results.
- The optimal amount of surfactant required by the clay can be determined using zeta potential and turbidity measurements.
- Addition of surfactant modified clay to cement has the potential to increase compressive strength of the cement paste.
- Excessive amount of surfactant may cause the clay to agglomerate as the atoms could gain a charge and become attracted to each other.

Acknowledgements

The first author is grateful to Dr. Kasthurirangan (Rangan) Gopalakrishnan, Research Assistant Professor, Iowa State University for his contributions towards this chapter.

References

BMT Inc. Information from Boral Material Technologies Inc., 45 Northeast Loop 410, Suite 700 San Antonio, Texas 78216, USA (2008), Tel: 800-964-0951

Boyd, A.J., Birgisson, B., Beatty, C., Zaman, A.: Utilization of phosphatic clay waste in concrete. FIPR Project Number: 03-02-164, Florida Institute of Phosphate Research, Bartow, Florida (2007)

Cementec Inc. Information from Cementec Industries Limited, 288, 200 Rivercrest Drive SE, Calgary, Alberta, Canada T2C 2X5 (2008), Tel: (403) 720-6699

Collepardi, M., Collepardi, S., Skarp, U., Troli, R.: Optimization of Silica Fume, Fly Ash and Amorphous Nano-Silica in Superplasticized High-Performance Concrete, American Concrete Institute, May 1, pp. 495-506 (2004)

Dham, M.: Nanomodification of Cement Matrix for Enhanced Ductility and Fracture Properties in Concrete. Doctoral dissertation, University of Florida, 315 p. (2007)

EPA. The US Environmental Protection Agency Guidance Manual, Turbidity Provisions (April 1999)

Escalante-Garcia, J.I., Sharp, J.H.: The chemical composition and microstructure of hydration products in blended cements. Cem. Concr. Compos. 26(8), 967–976 (2004)

Ji, T.: Preliminary study on the water permeability and microstructure of concrete incorporating nano-SiO_2. Cement and Concrete Research 35(10), 1943–1947 (2005)

Jo, B.W., Kim, C.H., Tae, G.H., Park, J.B.: Characteristics of Cmenet Mortar with Nano-SiO2 Particles. Construction and Building Materials 21, 1351–1355 (2007)

Lab Manual, Laboratory Manual, Hach Model 2100AN Laboratory Turbidimeter (2006)

Lagerblad, B., Fjällberg, L.T.: Hydratation-Styrande mekanismer och model (Early hydration-mechanism and modell), CBI rapport 2 (2008) (in Swedish with english summary)

Li, G.: Properties of high-volume fly ash concrete incorporating nano-SiO_2. Cement and Concrete Research 34(6), 1043–1049 (2004)

Li, X., Gao, H., Scrivens, W.A., Fei, D., Thakur, V., Sutton, M.A., Reynolds, A.P., Myrick, M.L.: Structural and mechanical characterization of nanoclay-reinforced agarose nanocomposites. Nanotechnology 16, 2020–2029 (2005)

MAIC, Major Analytical Instrumentation Center, 107 MAEC Building, University of Florida, Gainesville (2009)

Mindess, S., Young, J.F., Darwin, D.: Concrete, 2nd edn., pp. 517–545. Pearson Education Inc., New Jersey (1996)

Papadakis, V.G., Tsimas, S.: Supplementary cementing materials in concrete: Part I: efficiency and design. Cement and Concrete Research 32(10), 1525–1532 (2002)

Rodríguez-Camacho, R.E., Uribe-Afif, R.: Importance of using the natural pozzolans on concrete durability. Cement and Concrete Research 32(12), 1851–1858 (2002)

Saeki, T., Monteiro, P.J.M.: A model to predict the amount of calcium hydroxide in concrete containing mineral admixture. Cement and Concrete Research 35, 1914–1921 (2005)

Shayan, A., Xu, A.: Performance of glass powder as a pozzolanic material in concrete: A field trial on concrete slabs. Cement and Concrete Research 36(3), 457–468 (2006)

Sigma Aldrich. Surfactant information, product number 496146, Sigma Aldrich product catalog (2006)

Singh, P.K.: Dispersion of nanoparticulate suspensions using self assembled surfactant aggregates. Ph.D. Dissertation, University of Florida, Gainesville (2002)

Southern Clay Inc., Information from Southern Clay, 1212 Church Street Gonzales, Texas 78629, USA (2008), Tel: (830) 672-2891

Taylor, H.: Cement Chemistry. Tomas Telford, London (1997)

Thomas, J.J., Jenings, H.M., Chen, J.J.: Influence of Nucleation Seeding on the Hydration Mechanisms of Tricalcium Silicate and Cement. Journal of Physical Chemistry 113, 4327–4334

Characterization of Asphalt Materials for Moisture Damage Using Atomic Force Microscopy and Nanoindentation

Rafiqul A. Tarefder and Arif Zaman

Abstract. Asphalt materials have traditionally been characterized for moisture damage at component (microscale) and laboratory specimen (macro) scales. Though characterizations at such scales are commendable and often necessary and well understood by the pavement engineering community, there are problems which cannot be solved using micro and macroscale characterizations. For example, moisture damage in asphalt pavement is caused by moisture interaction with asphalt-aggregate bonds, which occurs at the atomic or nanoscale level. Macro and microscales scales are inadequate for developing an understanding of the bond damage phenomena. As a result the moisture damage still remains as one of the most common but complex problems of asphalt concrete. This chapter describes the use of Atomic Force Microscopy (AFM) and nanoindentation techniques to gain accurate insight into moisture damage performance of asphalt materials at nanoscale. In particular, moisture damage is quantified in performance grade (PG) asphalt binders using AFM measured adhesion values. It is shown that a PG 76-28 binder is more resistant to moisture damage than a PG 70-22 binder. Using nanoindentation, it is shown that hardness increases due to wet conditioning of aggregate, whereas hardness decreases due to wet conditioning of mastic, which is a mixture of fines and asphalt binder. Modulus value measured by nanoindenter did not show any tend due to wet and dry conditioning. It is hoped that pavement materials engineers and researchers benefit from the nanoscale characterization presented in this chapter.

Rafiqul A. Tarefder
Assistant Professor, Department of Civil Engineering, University of New Mexico
Albuquerque, New Mexico 87113
e-mail: tarefder@unm.edu

Arif Zaman
PhD Candidate, Department of Civil Engineering, University of New Mexico
Albuquerque, New Mexico 87113
e-mail: arif@unm.edu

K. Gopalakrishnan et al. (Eds.): Nanotechnology in Civil Infrastructure, pp. 237–256.
springerlink.com

1 Introduction

Atomic force microscopy and nanoindentation are relatively new techniques for materials characterization. An AFM measures the surface roughness and adhesion force, whereas the nanoindentation measures the modulus and hardness of a material. Although application of these two techniques are not much yet in civil engineering, they are being widely used in sister disciplines such biomedical, aerospace, manufacturing, materials, physics, chemistry, and electrical engineering. In other disciplines, both techniques have been used mainly for characterization of materials a nanoscale (10^{-9} m or nanometer). In civil engineering as well as in pavement engineering, mostly macroscale techniques are used for characterization of materials. However, there are problems in asphalt pavement area that have not been understood and solved yet using macroscale techniques. For example, moisture-induced damage in asphalt concrete has been studied for over 80 years, it still remains an unsolved problem. Moisture damage in asphalt can occur through the loss of adhesion and/or the loss of cohesion. Loss of adhesion is caused by breaking of the adhesive bonds between the aggregate surface and the asphalt binder primarily due to the action of water. Loss of cohesion is caused by the softening or breaking of cohesive bonds within the asphalt binder due to the action of water or water diffusion. The breaking of bonds between two asphalt or aggregate molecules occurs at a nanoscale. As such, macroscale characterization becomes inadequate for understanding moisture interaction with asphalt-aggregate bonds. This chapter focuses on nanoscale characterization of asphalt materials for moisture damage using atomic force microscopy (AFM) and nanoindentation techniques.

In an AFM test, a sample surface is probed with a sharp AFM tip and the attractive or repulsive force between the tip and the sample surface is measured. In electrical engineering and applied physics disciplines, AFM is mainly used to measure the surface roughness of laboratory grown nano devices fabricated for the microelectronic technology (Hill et al. 2001). Another application of AFM is to measure adhesion/cohesion force, which is widely used by the chemical engineering and applied chemistry group. Recently, Lieber and coworkers from Harvard University did the pioneer work to introduce the functionalized AFM and this application of chemical functionalized AFM has been termed as Chemical Force Microscopy (CFM) (Frisbie et al., 1994, Noy et al. 1997). Beach et al. (2001) studies pull off force between hexadecanethiol monolayers, self-assembled on gold-coated silicon nitride cantilever tip and silicon wafer, using AFM. Okabe et al. (2000) used hydrophobic -CH3 and –COOH functional modified AFM probes to map the adhesion forces and image the samples. Masson et al. (2007) studied the asphalt stiffness with AFM. This study was included to non-functionalized AFM tips. Du et al. (2001) explored the elastic modulus and yield strength of polymer thin films with AFM. In the asphalt area, Pauli et al. (2003) has done AFM testing on Strategic Highway Research Program (SHRP) asphalt binder and calculated the surface energy. Masson et al. (2006) conducted phase image testing of asphalt using AFM and correlate them with chemically analyzed contents of asphalt such as saturates, naphthene aromatics, polar aromatics, asphaltenes. Tarefder et al. (2010) studied the effect of polymer modification of adhesion force using AFM.

In a nanoindentation test, an indenter indents a sample surface and the movement of the indenter is measured with an increasing load. Load versus indentation curve is analyzed for finding stiffness and hardness of materials. So far, nanoindentation technique has been mostly used for characterization of nanostructured materials. A nanostructured material is a material which has at least one constituent at a characteristic length-scale of the order of tens of nanometers or less. Engineering applications of a nanostructure material often call for its stiffness and hardness properties for performance evaluation. Yang and Zhang (2001) did study of nanoindentation creep on polymeric materials. Gu et al. (2007) studied the depth profiling behavior of polymeric coating with nanoindentation. Jäger and Lackner (2009) studied the elastic, viscous and plastic material behaviour with nanoindentation of the polymers. Mondal et al. (2007) conducted nanoindentation study on cement hardened paste to determine the mechanical properties. To date, nanoindentation tests on asphalt has been conducted by only a very few researchers in Europe (Ossa and Collop 2007, Ossa et al. 2005, Pichler et al. 2005). Jäger et al. (2007) identified the viscoelastic properties of asphalt with nanoindentation and considering the real tip geometry. Recently, Tarefder and Zaman (2010) studied the feasibility of Berkovich and spherical tips for indentation on asphalt binder and asphalt concrete. They described that the main feature of this device is that characterization of asphalt components such as mastic, and aggregate can be done without separating them from asphalt concrete or composite. Results from this experiment can be very useful to characterize, design and model asphalt performances in new way complementing the traditional way.

In this chapter, both AFM and nanoindentation techniques will be used for nanoscale characterization of asphalt binder and asphalt concrete samples for moisture damage.

2 AFM for Moisture Damage Study

In an AFM test, a small and sharp tip placed at the free end of a cantilever is brought to an asphalt film. As the tips approaches the film surface, the cantilever

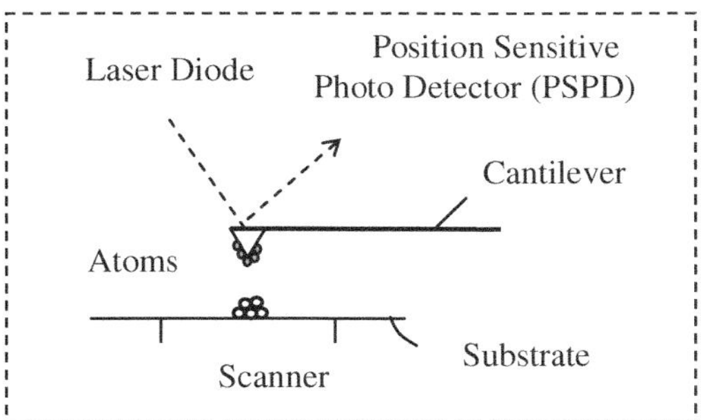

Fig. 1 Schematic of an AFM test

deflects due to the interaction between the AFM (atoms) and the asphalt surface (atoms). The magnitude of the deflection directly depends on the attraction or repulsion force between the molecules of the tips and the molecules of sample surface molecule. This deflection is measured by an optical lever consists of a laser diode and position sensitive photo detector shown in Figure 1. The measured deflection is used to calculate the force, F acting on the AFM tips using, $F = k\ \delta$; where δ is the deflection and k is the stiffness of the cantilever. Adhesion is defined as the force between atoms of an AFM tips and atoms of asphalt binder.

AFM testing of soft sample is complex. In soft materials, AFM tip can sticking in the sample. To avoid this problem, noncontact mode AFM tests were conducted on asphalt samples. However, the success of AFM testing on soft asphalt sample depends on the testing parameter setup. The values of those parameters determined through trials in this study and they are listed in Table 1. In the AFM, the gain value controls the error signal to generate a feedback signal. The gain parameter depends on a number of factors including the scan rate, scan size, and test sample topography. This study finds that a gain value of 0.1 is sensitive enough to change in cantilever deflection for asphalt testing. Cantilever deflection is smaller in asphalt/soft sample than that in metal/hard sample. The set point controls the amount of cantilever bending, during sample scanning. The minus value of set point, as shown in Table 1, ensures that the tip is not in contact with sample surface.

Table 1 AFM Testing Parameters

Sample	Gain	Set Point (P)	Scan Area (μm)	Scan Rate (Hz)	Drive (%)
PG 70-22 (dry)	0.1	-0.120	4.9 x 4.9	3	25
PG 70-22 (wet)	0.1	-0.051	0.5 x 0.5	2	25
PG 76-28 (dry)	0.1	-0.039	0.8 x 0.8	2.3	25
PG 76-28 (wet)	0.1	-0.426	4.7 x 4.7	1	45

Scan rate is the frequency of the back and forth movements of the sample beneath the AFM probe. A scan rate of 4 to 5 results in an image that appears smeared. This is because the feedback loop may not have enough time to respond to change in film roughness. Slow scan rate produces good resolution of the image as the feedback system finds enough time to respond, while fast scan rate can be time efficient. In this study, a scan rate between 1 and 3 Hz was shown to produce high quality image. The drive amplitude is the amplitude of the AC signal of the sine wave generator that drives the cantilever to vibrate. The drive amplitude was selected through trials in the range of 25 to 45 for asphalt imaging. When the drive amplitude was too large, the cantilever made an intermittent contact with the surface.

2.1 Tip Functionalization and Calibration

Silicon nitride tips are purchased from VEECO Instruments, Inc (Veeco Instrument Inc., 2007). This tip is a beam bounce cantilever (called RFESPA-CP MPP211) with a length of 125 μm, natural frequency of 90 kHz, and spring constant of 3 N/m. AFM tips were functionalized using carboxyl (-COOH), methyl (-CH3), ammin (-NH3) and hydroxyl (-OH) groups from the help of Novascan Technologies, Ames, IA. These functionals are known to be the major part of asphalt chemistry (Testa, 1995). The functionalized AFM test requires calibration of tips. As discussed previously the force in AFM is calculated for a specific tip, the tip constant is determined through calibration. Special module to calibrate the AFM tips supplied by Veeco was used in this study. In calibration procedure, the tip distances are measured on a sample on a sample of known force between the tip and the sample molecules. Tip constant is calculated from known force and displacement ($k = F/\delta$). This constant is used for converting displacement data of other samples to force; thereby producing force distance (F-D) curve.

2.2 Asphalt Sample Surface Imaging

This paper describes the results of AFM testing on two types of PG grade asphalt binders under wet and dry conditions. Noncontact mode AFM with Proscan software was used to conduct the AFM testing and to analyze the images. In AFM testing, sample image is produced as a first step. Figure 2 shows the 3D images of dry and wet PG 70-22 binder samples. It can be seen that wet sample has more spikes than the dry samples. These images were analyzed using WSXM software (Horacos et al. 2007). In image analysis, asphalt film roughness is measured using quantities such as average, maximum and root mean square (RMS) values of surface roughness. These quantities are listed in Table 2. It can be seen that PG 76-28 (Dry) sample shows the lowest RMS value (0.56 nm) and the all other RMS values are fairly close. In general, the RMS roughness values of the wet samples are higher than those of dry samples.

Table 2 Height and Roughness values of all samples

Sample	Maximum height (nm)	Average height (nm)	RMS Roughness (nm)
PG 70-22 (dry)	27.96	9.40	2.97
PG 70-22 (wet)	4.59	2.51	0.56
PG 76-28 (dry)	17.73	13.47	0.94
PG 76-28 (wet)	23.21	8.51	2.43

Figure 2 shows that wet samples have more spikes than dry samples. This is due to the fact that wet sample was damaged by water action, which has probably made the asphalt film surface more rough. The acceptable value of roughness for AFM sample is about 25 nm. As shown in Table 2, as all of the 4 samples have RMS roughness value less than the limiting 25 nm value.

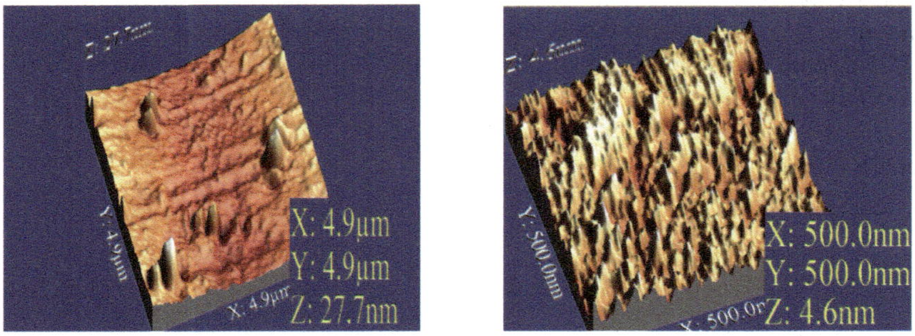

(a) PG 70-22 (dry) (b) PG 70-22 (wet)

Fig. 2 Three dimensional (3D) images of asphalt surface using AFM

2.3 Characteristics of F-D Curve of Asphalt

F-D curve is defined by the force between tip and sample plotted as a function of distance between the tip and sample surface. When the AFM tip approaches and retracts from the sample, it is deflected by the interaction with the sample. A force curve is such a plot of the force applied to the AFM tip as a function of the tip-sample displacement of the cantilever holder relative to the surface. Figure 3 shows the force distance curves comparison of dry and wet PG 76-28 samples. The black colored graph represents the dry sample and the orange colored graph represents the wet sample. The cycle in the force measurement starts at a tip-surface separation. At a large distance, no force acts between the tip and surface, but as the tip approaches, the distance decreases and attractive forces pull the cantilever tip towards the sample. At a certain point, the tip jumps or bends towards the sample surface. This jump or bending occurs when the gradient of attractive forces exceeds the spring constant plus the gradient of repulsive forces. Moving the tip further causes a deflection of the cantilever equal in amount to that produced if the sample were pushed. Next, the tip is withdrawn back to its starting position. During retraction, at one point, tip finally snaps outward. The lowest point of the retracting F-D curve shown in Figure 3 is the maximum force in the retracting curve, at which point the adhesion is broken and the cantilever breaks free from the surface. Adhesion between the tip and the sample is mainly due to van der Waals interactions and electrostatic forces (acid-base). Asphalt surface is

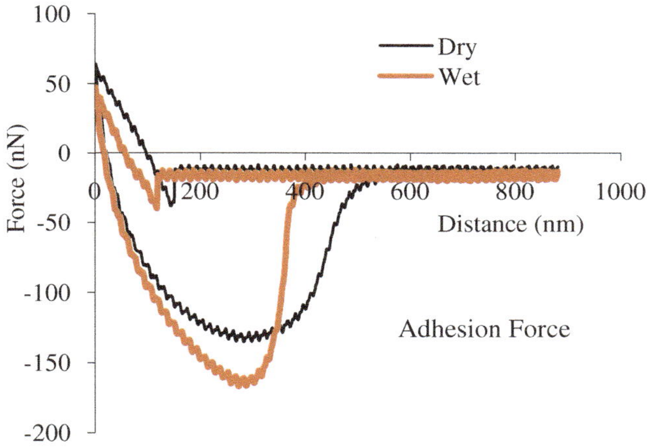

Fig. 3 Force distance graph of PG 76-28 sample with –COOH tip

charge neutral; adhesion force is mainly due to van der Waals interaction. From Figure 3, it can be seen that adhesion force in wet asphalt sample is higher than that in dry sample.

2.4 Dry vs. Wet Adhesion Forces

Each asphalt sample was tested at nine different locations. Table 3 shows the test results on dry and wet PG 70-22 samples adhesion with all the functionalized tips. All the ten (10) point's tests values as well as the average result are tabularized here to show the wet and dry adhesion forces. Due to space limitation, adhesion data of PG 76-28 samples are not shown. Form this table data, it is possible to show that test data are quite consistent and repeatable, which is not discussed further detail in this chapter.

The average adhesion values of dry and wet samples of PG 70-22 and PG 76-28 are plotted in the Figures 4 and 5. For PG 70-22, the adhesion values of all the wet samples are smaller than those of dry samples as shown in Figure 4. The adhesion value obtained by –COOH functionalized tip is higher that the adhesion value obtained by $–CH_3$ functionalized tips. Probably, tip type, –COOH being hydrophilic and $–CH_3$ being hydrophobic, has affected the results. For PG 76-28 binder, the adhesion values of all the wet samples are higher than those of dry samples as shown in Figure 5. Instead of –COOH, the –OH functional which is also a hydrophilic tip, has shown the highest adhesion values. The hydrophobic –CH_3 has shown a slight increase in adhesion force due to sample wetting indicating PG 76-22 is less susceptible to moisture damage compared to the PG 70-22.

Table 3 Dry and wet adhesion forces of PG 70-22 sample

Tip	Cond	Adhesion Forces (nN) (first 5 points)				
COOH	Dry	147	181	239	231	291
	Wet	377	512	361	488	378
CH3	Dry	60	73	89	38	57
	Wet	105	90	94	89	102
NH3	Dry	99	126	95	82	109
	Wet	201	120	147	172	203
OH	Dry	234	273	288	291	291
	Wet	483	362	310	409	272

Tip	Condition	Adhesion Forces (nN) (second 5 points)					Average Force (nN)
COOH	Dry	147	92	126	191	308	191
	Wet	377	367	422	499	457	435
CH3	Dry	60	65	34	43	56	57
	Wet	105	129	119	89	92	102
NH3	Dry	99	158	95	110	139	114
	Wet	201	183	205	227	204	181
OH	Dry	234	282	184	209	212	260
	Wet	483	414	376	477	419	386

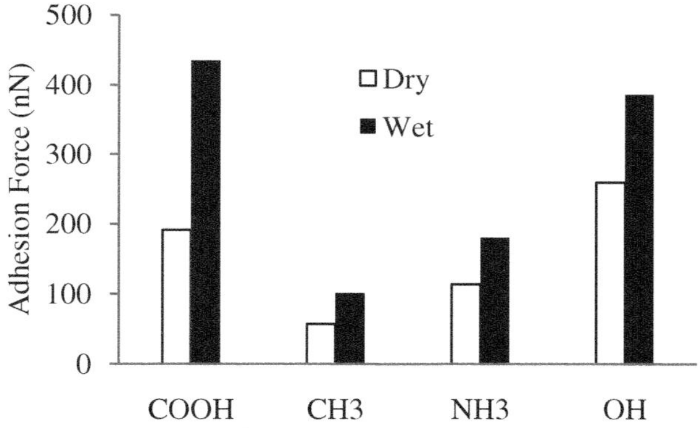

Fig. 4 Average dry vs. wet adhesion forces in PG 70-22 binder

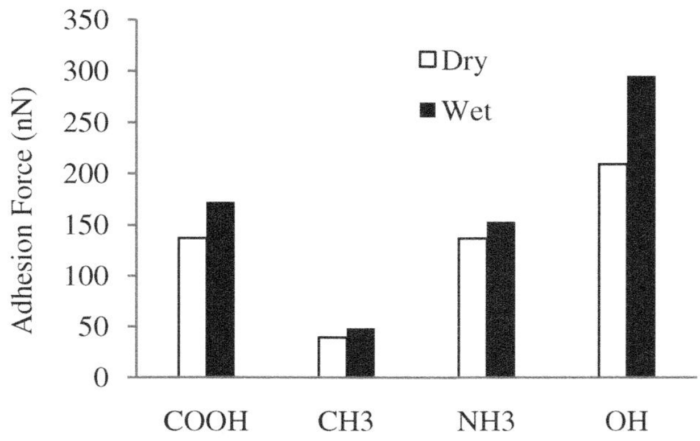

Fig. 5 Average dry vs. wet adhesion forces in PG 76-28 binder

2.5 Remarks Based on AFM Results

Based on the AFM testing of wet and dry asphalt sample, the following remarks can be made:

- The chemistry of the tip, functional groups in asphalt, affects its adhesion behavior.
- Wet sample shows higher adhesion force values compared to the dry sample. This indicates that damage occurs in wet asphalt film due to water action.
- The ratio of wet and dry adhesion force in PG 70-22 binder is higher than that in PG 76-28. This indicates that the PG 76-28 binders are more resistant to moisture damage than the PG 70-22 binders.

3 Nanoindentation for Moisture Damage Study

In a nanoindentation test, the movement of a diamond probe in contact with a sample surface is measured by applying an increasing load. Time, load, and displacement are recorded during the test. The elastic and plastic properties of the sample are determined from the load-displacement data. Stiffness and hardness are the most common mechanical properties measured from indentation data. Stiffness is the ratio of stress to strain. Hardness is the resistance to permanent deformation. It is defined as the maximum applied load divided by the projected plastic area of contact. Nanoindentation tests were conducted using a nanoindenter supplied by MicroMaterials Ltd. Wrexham, UK. Nanoindentation test was performed using spherical tips and the direction of indentation was horizontal. This nanoindenter is a pendulum-based electro-magnetical depth-sensing system, in which samples are mounted vertically as shown in Figure 5. This kind of horizontal indenter has advantages over the vertical indenter as the horizontal indenter does not use self weight on the sample during indentation. As shown in the Figure 5, a pendulum using a coil and magnet is rotated about its frictionless pivot so that the diamond probe penetrates the sample surface placed in a sample holder. Indenter displacement is measured with a parallel plate capacitor achieving sub-nanometer resolution. The balance weight is adjusted to balance the pendulum when different tip geometries are used with the indenter. The damping plates are used to damp out any oscillations that may occur during the indentation process.

In the horizontal indenter, a high resolution microscope and an AFM are available to take images of sample during indentation test. The nanoindentation system is enclosed in a thermally controlled cabinet. This cabinet provided protection from air disturbances, allowed the system to be thermally controlled and reduced error due to noise disturbance by providing sound insulation. Though asphalt is a temperature dependent material, all the indentation was conducted in this cabinet, however only at one temperature of 23.7 °C. This indenter has load resolution of 1 nN and depth resolution of 0.004 nm. Routine calibrations are done to keep the machine performing at its resolution level. Calibration includes: (a) load calibration which determines the forces that can be applied at the indenter tip during measurements, and (b) depth calibration which relates a known distance moved by the sample in contact with the indenter to a change in capacitance.

Fig. 6 Horizontal Nanoindenter at the University of New Mexico

3.1 Nanoindentation Theory

To determine the mechanical properties of the tested asphalt sample, the depth versus loading-unloading data was fitted using the following power law function (Oliver and Pharr 1992):

$$P = a(h - h_c)m \tag{1}$$

where h is any depth, h_c is the unrecoverable or plastic depth, a and m are constants. At $h = h_{max}$, $P = P_{max}$; therefore above equation can be expressed as:

$$h_c = h_{max} - \varepsilon(C.P_{max}) \tag{2}$$

where C is the contact compliance equal to the tangent at maximum load or the slope of the initial portion of unloading curve. The value of $\boldsymbol{\varepsilon}$ depends on the indenter geometry. The value of $\boldsymbol{\varepsilon}$ is 0.75 for a Berkovich indenter and 0.85 for spherical indenter. Asphalt sample hardness (H) can be determined from the peak load (P_{max}) and the projected area of contact, A:

$$H = P_{max} / A \tag{3}$$

The reduced modulus of the sample can be analyzed using the unloading portion of the depth-load curve as follows:

$$E_r = \frac{\pi^{0.5}}{2C\,A^{0.5}} \tag{4}$$

where C is the contact compliance and E_r is the reduced modulus. Reduced modulus is a measure of the elastic modulus of the sample. It can be thought of as an intermediate parameter accounting for the non-rigid effect of diamond indenters in contact with other materials. Finally, reduced modulus is converted to the Young's modulus using the following equation also known as Oliver and Pharr Equation (Oliver and Pharr 1992):

$$\frac{1}{E_r} = \frac{1 - \nu_s^2}{E_s} - \frac{1 - \nu_i^2}{E_i} \tag{5}$$

where

E_s = Young's modulus of sample
ϑ_s = Poisons ratio of sample
E_i = Young's modulus of indenter (1141 GPa)
ϑ_i = Poisons ratio of indenter (0.07)
E_r = Reduced modulus of sample obtained from Eq. (4)

Fortunately, a NanoTest software developed by the nanoindentation manufacturer uses the above method of Oliver and Pharr for automatic curve fitting and analysis of the raw indentation data. The hardness value is related to loading curve and the elastic modulus value is related to unloading load-displacement curve of the nanoindentation.

3.2 Nanoindentation on Asphalt Samples

Nanoindentation tests were conducted on three asphalt concrete samples that were designed using performance grade binder PG 76-28 and PG 70-22. Three mixes that uses the aforementioned PG binders were collected from the local plant. Of them, SP-III mix was made of PG 76-28 binder, and SP-B and SP-C mixes used PG 70-22 binder. Each of the mixes is compacted into 6 in. diameter cylinders by a Superpave gyratory compactor using a 600 kPa (87,02 psi) vertical pressure (AASHTO T 312). Using a water-cooled laboratory saw, one-inch thick disc is sliced from the center of each cylinder in an attempt to acquire samples with uniform air voids.

Smooth surface of the cube is very important for nanoindentation experiment. Because the contact area is measured indirectly from the depth of penetration, a rough surface may cause errors in the determination of the area of contact between the indenter and the specimen (Johnson 1985). Fine laboratory saws at Geology Department at the University of New Mexico were used to cut and prepare thin AC cubes (12 mm x 12 mm x 6 mm). The cube surfaces were polished by a grinding machine rotating at angular speed of 150 rpm with a sequence of SiC papers of decreasing abrasiveness (100, 200, 400, 800, 1000, 1200 and 1400-grit) under continuous water cooling. Each step was carried out for 150 second. Only one phase (surface) of the cube sample was polished. Finally, the specimens were washed in a water bath to remove any remaining dusts. Figure 7 shows polished asphalt concrete cube that was used for nanoindentation testing. The sample dimension was about 25 mm x 25 mm x 3 mm.

One set of samples were tested under dry, and the other set was tested under wet condition (AASHTO T 283). Freeze and thaw cycles are the main elements of this method to make the sample ready for moisture related damage testing evaluation. The nanoindentation test samples are kept in a vacuum jar to vacuum for 10 minutes under 25 kPa pressure. Then the samples are kept in fridge at -20°C temperature for 24 hours. After that they are placed in 60°C hot water bath for another 24 hours. From the hot water bath they are kept in room temperature bath for another two hours. The samples are kept in zip lock bag so that it may not drop the available moisture before final testing.

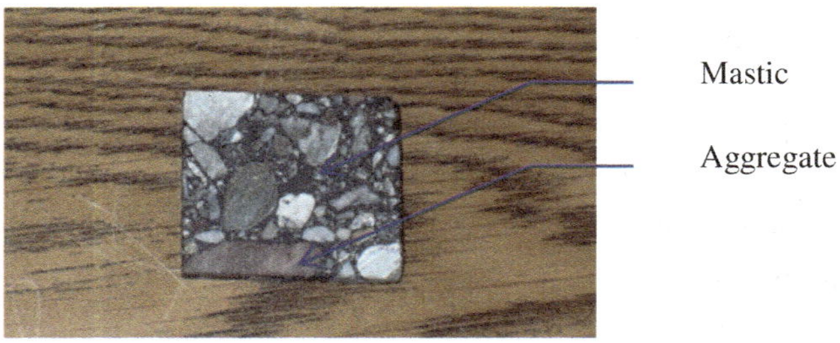

Fig. 7 Asphalt concrete cube for nanoindentation tests

3.3 Sample Image after Nanoindentation

The nanoindentation on both aggregate and mastic phases of asphalt concrete cube sample were performed using a nanopositioner attached with our nanoindentation equipment. The photographs of asphalt samples after nanoindentation are shown in Figures 8(a) and 8(b). It can be seen that a significant impression on mastic sample from Figure 8(a). Where as the impression on aggregate is recovered quickly as shown in Figure 8(b). Aggregate sample is much stronger than the mastic sample and hence a little deformation is observed here. The deformation was permanent on the mastic as the picture was taken almost 1 hour after the test.

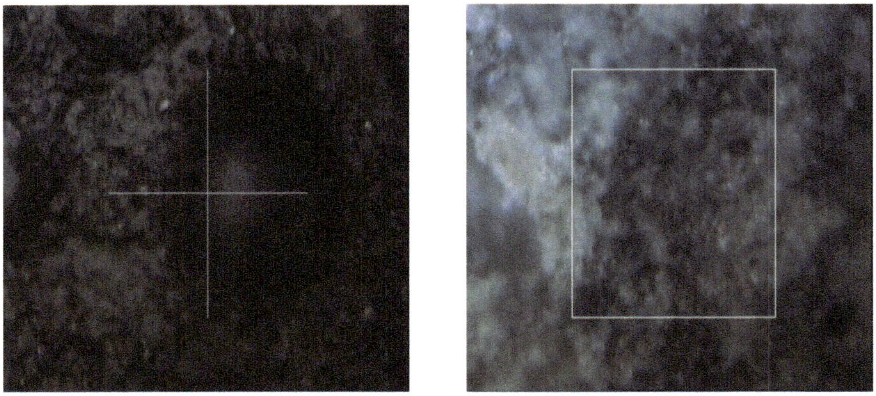

(a) Mastic (b) Aggregate

Fig. 8 Mastic and aggregate samples after nanoindentation

3.4 Nanoindentation Results and Discussions

In a standard indentation test, a range of load is set on a specified grid points, and only successful indentation results are used for analysis. Whether an indentation was successful, that means whether the good data obtained is good or bad, is determined by whether Oliver and Pharr method can be applied to the load vs indentation raw data. As a result, the amount of load may vary for successful indentation. Using Oliver and Pharr equation, hardness and Young's modulus of dry and wet samples at several grid points are calculated. Since the contact area of the nanoindenter is small, several grid points were selected on samples with a heterogeneous composition on later scale of about 40 μm. The indents are located at least 40 μm (micrometer) apart to avoid the influence of residual stresses from adjacent impressions. The load and displacement resolution of the indenter are 1 nN (NanoNewton) and 0.01 nm (nanometer) respectively. Both load and depth controlled tests are conducted on asphalt. On the asphalt concrete sample, the locations of aggregate and mastic are determined using the nano-positioner attached with our nanoindenter equipment.

The range of aggregate modulus and hardness values and their average are presented in Table 4. From the results, it can be concluded that no one level of modulus or hardness can be selected as per representative value of mastic and aggregate modulus and hardness. In addition, wet to dry ratio of hardness and modulus are shown in Table 4. From the results, it can be seen that wet to dry ratio is greater than one in all the samples except SP-III. This could not be explained from this study. It is possible that surface roughness of the aggregate has affected such results. Perhaps the surface became rougher after wet conditioning of the samples and resulted in high hardness and young's modulus values.

Table 4 Hardness and Young's modulus of wet and dry aggregate

Sample	Condition	Hardness Ranges (MPa)	Avg. Hardness (MPa)	Wet/Dry Ratio
SP-B	Dry	1249 to 1347	1271	17.5
	Wet	3253 to 56732	22332	
SP-C	Dry	7 to 9	8	4.3
	Wet	2 to 94	34	
SP-III	Dry	65 to 311	198	8.9
	Wet	302 to 4050	1776	
Sample	Condition	Young's Modulus Ranges (MPa)	Avg. Modulus (MPa)	Wet/Dry Ratio
SP-B	Dry	24243 to 51819	1271	10
	Wet	49192 to 810478	394989	
SP-C	Dry	133 to 260	197	32.8
	Wet	902 to 17439	6479	
SP-III	Dry	20341 to 25822	20481	0.3
	Wet	6898 to 3711	5721	

The range of mastic modulus and hardness values and their average are presented in Table 5. Overall, hardness and Young's modulus values of wet samples are smaller than those of the dry samples. It shows the weakness of asphalt concrete mastic against the moisture. Hardness of SP-B and SP-III mastics are much higher than the hardness of SP-C sample. However, wet to dry hardness of SP-C sample is smaller than that of other two mixes. This indicates that SP-C mastic is less susceptible to moisture damage than the SP-B and SP-III mastics. Though SP-C and SP-B mixes contains a PG 70-22 binder, the gradation of fine mastic particles (passing number 200 sieve materials) are not same. Probably, binder content and percentage passing number 200 sieve materials have affected the mastic's moisture damage behaviour. Mastic has been studied separately, not being in asphaltic composite, using dynamic shear tests (Kim and Little 2004). The previous study by Liu (2005) showed the Young's modulus of asphalt concrete is about 2800 MPa in tension. Whereas the Young's modulus of mastic is within the range of 200-1300, except for SP-B in dry condition. It can be noted that Liu (2005)'s tests were at macroscale on asphalt concrete samples through conventional laboratory testing. It is not possible to separate the load on mastic from the load on aggregate when applied load is on an overall asphalt concrete sample. As a result, modulus and hardness values from previous study were not available to compare with the values obtained in this study.

From Tables 4 and 5, no trend in the unloading stiffness or modulus value was observed in the dry or wet conditioned samples. Probably stiffness is not a damage related phenomenon.

Table 5 Hardness and Young's modulus (MPa) of wet and dry mastic

Sample	Condition	Hardness Ranges (MPa)	Avg. Hardness (MPa)	Std. dev. (MPa)
SP-B	Dry	24 to 162	80.5	72
	Wet	0.6 to 3.3	1.6	1.5
SP-C	Dry	1.2 to 2.6	2.1	0.77
	Wet	0.4 to 3.5	1.6	1.62
SP-III	Dry	0.5 to 7	3.9	2.47
	Wet	1.3 to 138	53.1	74.27
Sample	Condition	Young's Modulus (MPa)	Avg. Modulus (MPa)	Std. dev. (MPa)
SP-B	Dry	2550 to 8999	4851	3599
	Wet	73.92 to1708	663	907
SP-C	Dry	185 to 377	281	96.1
	Wet	38 to 376	267	63.8
SP-III	Dry	108 to 740	487	309.1
	Wet	319 to 2307	1364	997.6

3.4.1 Dry vs. Wet SP-B Aggregate and Mastic

Nanoindentation results for wet and dry SP-B samples are presented in Figures 9(a) and 9(b). The nanoindentation on aggregate and mastic were conducted separately using the nano-positioner of the nanoindenter. It can be noted here that the behaviour of the mastic or aggregate minerals at different grid locations found to vary. Ideally, one should report the range of wet and dry samples' modulus and hardness, which was done in Table 4 and 5 in this study. However, the authors attempted to present their indentation results using graphs and plots. Thus the comparison made in Figure 9 is only a general trend that was observed in this study. From Figure 9(a), it can be seen that the same amount of load (20 mN) caused about 800 nm of deformation on wet aggregate and almost 3000 nm deformation on dry aggregate. There is a gain in hardness value in wet sample due to moisture effect. Also a longer loading curve can be seen for dry sample compared to that of the wet sample. The unloading curve of the wet sample is stiffer that the unloading curve of dry aggregates. Figure 9(b) presents the behaviour of load versus depth curves for the mastic of dry and wet SP-B sample. The maximum load on dry samples was about 0.53 mN and that caused a damage depth of less than 2000 nm whereas the maximum load of 0.27 mN on wet sample caused about 3000 nm of depth. This indicates a weakness of mastic part of asphalt concrete against the negative consequences of moisture.

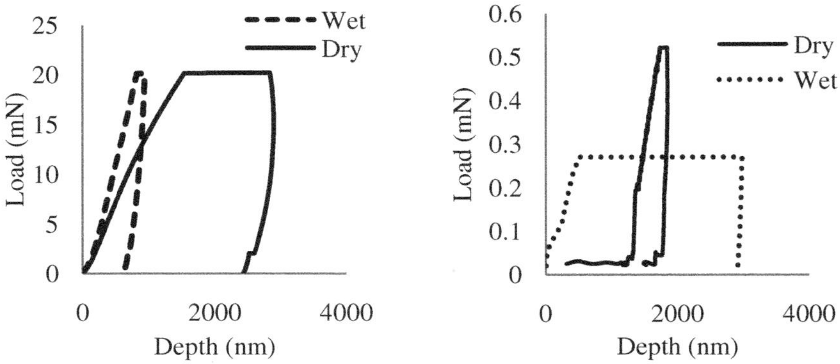

(a) aggregate (b) mastic

Fig. 9 Nanoindentation tests on SP-B sample

3.4.2 Dry vs. Wet SP-C Aggregate and Mastic

The load and penetration depth curves of SP-C samples are shown in Figure 10(a) and 10(b). For load value of 0.12 mN, Figure 10(a) shows that the SP-C wet aggregate shows smaller indentation depth than the dry aggregate. A load of (0.12 mN) caused about 375 nm of depth in wet sample whereas it was about 1000 nm in dry sample. The wet-dry behaviour of aggregate of SP-C is similar to wet-dry behaviour of SP-B aggregate. The load-depth curves of SP-C mastic are shown in Figure 10(b). Hardness value of SP-C mastic decreases due to sample wetting. The 0.12 mN load caused 400 nm depth on dry sample mastic and almost 500 nm depth on wet sample mastic. This is a clear indication of damage due to moisture in mastics of sample SP-C.

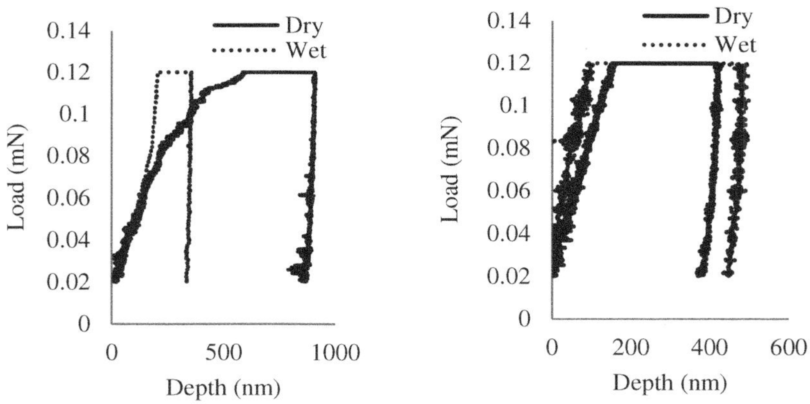

(a) aggregate (b) mastic

Fig. 10 Nanoindentation tests on SP-C sample

3.4.3 Dry and Wet SP-III Aggregate and Mastic

Figure 11(a) and 11(b) shows some successful indentation, however at two different magnitudes of load being used on mastic samples. From Figure 11(a), it can be seen that dry aggregate shows much smaller hardness value compared to the wet aggregate. As shown in Figure 11(a), the penetration depths on SP-III dry aggregates (due to 0.6 mN load) and wet aggregates (due to 0.26 mN load) are 40 mN

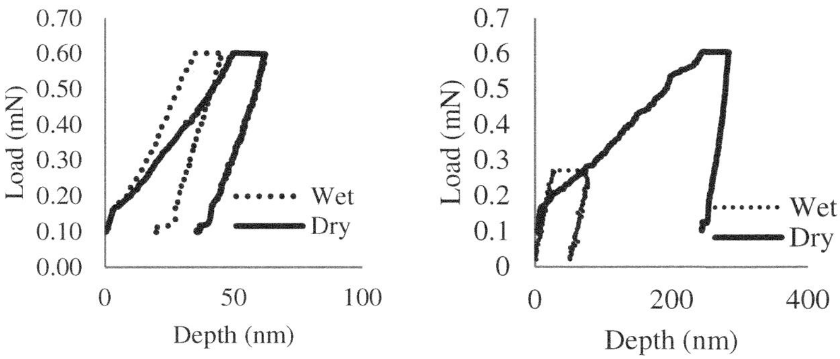

(a) Aggregate (b) Mastic

Fig. 11 Nanoindentation tests on SP-III sample

and 60 mN, respectively. Figure 11(b) compares load-indentation behaviour of dry and wet mastic samples of SP-III mix. It is observed that SP-III mastic is much harder in wet condition than in dry condition. A load of 0.26 mN caused a about 50 nm deformation on wet conditioned sample. Whereas dry SP-III mastics show 250 nm indentation depth under 0.6 mN load. The behaviour of SP-III aggregate is opposite to that of the behaviour of SP-B and SP-C aggregate. SP-III mastic used PG 70-28 binder, and the mastics of SP-B and SP-C mixes use a relatively softer binder PG 70-22.

3.5 Remarks Based on Nanoindentation Results

Based on the results presented in this study, the following conclusions can be made:

- In all three mixes, the hardness value of aggregate increases due to moisture conditioning. Probably, wet condition modifies aggregate surface mineral in a way that increases the hardness value in wet conditioned aggregate.
- The mastics of SP-B and SP-C have shown similar behaviour, that is hardness value decreases in mastic due to wet condition. Both of these mastics contain a PG 70-22 binder. However, the mastic of SP-III has show opposite trend. Its hardness increases due to moisture conditioning. This mastic contains a higher performing grade (PG 76-28) binder.
- The modulus or stiffness values obtained from nanoindentation did not show any relation with moisture damage.

4 Concluding Remarks

Asphalt is one of most complex materials in civil engineering. Asphalt's behavior varies with temperature, and moisture, and loading frequency and amplitude. There are problems such as moisture damage, aging are poorly understood phenomena in asphalt engineering. These complicated problems in a complex material such as asphalt have traditionally been studied using micro and macro scale laboratory testing and modeling techniques. With the advent of nanoscale testing equipment and device such as atomic force microscope and nanoindenter, it is now possible to characterize asphalt materials at a minute scale to better understand moisture damage, which has been demonstrated in this chapter.

References

Beach, E., Tormoen, G., Drelich, J.: Pull-off forces measured between hexadecanethiol self-assembled monolayers in air using an atomic force microscope: analysis of surface free energy. Journal of Adhesion Science and Technology 16(7), 845–868 (2002)

Du, B., Tsui, O., Zhang, Q., He, T.: Study of Elastic Modulus and Yield Strength of Polymer Thin Films Using Atomic Force Microscopy. Langmuir 17(11), 3286–3291 (2001)

Frisbie, C.D., Rozsynai, L.F., Noy, A., Wrighton, M.S., Lieber, C.M.: Functional Group Imaging by Chemical Force Microscopy. Science 265(5181), 2071–2074 (1994)

Frisbie, C.D., Rozsynai, L.F., Noy, A., Wrighton, M.S., Lieber, C.M.: Functional Group Imaging by Chemical Force Microscopy. Science 265(5181), 2071–2074 (1994)
Gu, X., Michaels, C., Drzal, P., Jasmin, J., Martin, D., Nguyen, T., And Matrin, J.: Probing photodegradation beneath the surface: a depth profiling study of UV-degraded polymeric coatings with microchemical imaging and nanoindentation. J. Coat. Technol. Res 4(4), 389–399 (2007)
Hill, D., Blasco, X., Porti, M., Nafría, M., Aymerich, X.: Characterising the surface roughness of AFM grown SiO_2 on Si. Microelectronics Reliability 41(7), 1077–1079 (2001)
Horacos, I., Fernandez, R., Gomez-Rodriguez, J., Colchero, J., Gomez-Herrero, J., Baro, A.: WSXM: A software for scanning probe microscopy and a tool for nanotechnology. Review of Scientific Instruments 78(1), 013705–013708 (2007)
Jäger, A., Lackner, R.: Finer-Scale Extraction of Viscoelastic Properties from Nanoindentation Characterised by Viscoelastic–Plastic Response. Strain 45(1), 45–54 (2009)
Jäger, A., Lackner, R., And Eberhardsteiner, J.: Identification of viscoelastic properties by means of nanoindentation taking the real tip geometry into account. Meccanica 42(3), 293–306 (2007)
Johnson, K.L.: Contact Mechanics. Cambridge University Press, Cambridge (1985)
Kim, Y.R., Little, D.N.: Linear Viscoelastic Analysis of Asphalt Mastics. J. Materials in Civil Engineering, 122–132 (2004)
Liu, C.: Van Der Waals Force and Asphalt Concrete Strength and Cracking. Journal of Engineering Mechanic, ASCE 131(2), 161–166 (2005)
Masson, J., Leblond, V., Margenson, J., Bundalo-Perc, S.: Low-temperature bitumen stiffness and viscous paraffinic nano and micro-domains by cryogenic AFM and PDM. Journal of Microscopy 227(3), 191–202 (2007)
Masson, J.-F., Leblond, V., Margeson, J.: Bitumen morphologies by phase-detection atomic force microscopy. NRCC-48157, Journal of Microscopy 221, part. 1, 17–29 (2006), doi:10.1111/j.1365-2818.2006.01540.x
Mondal, P., Shah, S., Marks, L.: A Reliable Technique to Determine the Local Mechanical Properties at the Nano-Scale for Cementitious Materials. Cement and Concrete Research 37(10), 1440–1444 (2007)
Noy, A., Veznov, D., Lieber, C.: Chemical Force Microscopy. Annu. Rev. Mater. 27, 381–421 (1997)
Okabe, Y., Akiba, U., Fujihira, M.: Chemical force microscopy of'-CH3 and -COOH terminal groups in mixed self-assembled monolayers by pulsed-force-mode atomic force microscopy. Applied Surface Science 157(4), 398–404 (2000)
Oliver, W., Pharr, G.: Improved technique for determining hardness and elastic modulus using load and displacement sensing indentation experiments. Journal of Material Research 7(6), 1564–1580 (1992)
Ossa, E.A., Collop, A.C.: Spherical Indentation Behavior of Asphalt Mixtures. ASCE Journal of Materials in Civil Engineering 19(9), 753–761 (2007)
Ossa, E.A., Deshpande, V.S., Cebon, D.: Spherical indentation behaviour of bitumen. Acta Materialia 53, 3103–3113 (2005)
Pauli, T., Grimes, W., Huang, S., Robertson, E.: Surface Energy Study of SHRP Asphalt Film by AFM. Petroleum Chemistry Division Preprints 478(1), 14 (2003)
Pichler, C., Jäger, A., Lackner, R., Eberhardsteiner, J.: Identification Of Material Properties From Nanoindentation: Application To Bitumen And Cement Paste. In: 22nd DANUBIA-ADRIA Symposium on Experimental Methods in Solid Mechanics, Monticelli Terme, Parma, Italy (2005)

Tarefder, R.A., Arifuzzaman, M.: Nanoscale Evaluation of Moisture Damage in Polymer Modified Asphalts. ASCE Journal of Materials in Civil Engineering 22(7), 714–725 (2010)

Tarefder, R.A., Zaman, A., Uddin, W.: Determining Hardness and Elastic Modulus of Asphalt by Nanoindentation. ASCE Int. J. Geomech. 10(3), 106–116 (2010)

Testa, S.: Chemical Aspect of Cold-Mix Asphalt Incorporating Contaminated Soil. Journal of Soil Contamination 4(2), 1–17 (1995)

Yang, S., Zhang, Y., Zeng, K.: Analysis of nanoindentation creep for polymeric materials. Journal of Applied Physics 95(7), 3655–3666 (2004)

Nanoclay-Modified Asphalt Binder Systems

Julian Mills-Beale and Zhanping You

Abstract. Nanomaterials use in asphalt and concrete pavement infrastructure is gaining ground among researchers and scientists. In this chapter, an attempt is made to provide insight on nanomodification of transportation infrastructure, with the primary focus on asphalt binder systems. It further reveals ongoing work being conducted on the use of nanoclay materials to enhance the mechanical properties of the asphalt binder cement system. From the preparation of the nanoclay-modified asphalt binder to the Superpave™ characterization process, it has been proven that nanoclay materials holds great potential in the design and construction of sustainable and durable asphalt pavement infrastructure.

1 Introduction

The advent of nanotechnology seeks to revolutionize different fields of engineering disciplines with positive results in the not too distant future. The science and technology of nanotechnology is receiving massive attention from diverse disciplines. In the field of civil transportation infrastructure, nanomaterials are becoming common in the development of new, enhanced and better performing cement systems – Portland cement and asphalt binder cement.

The reason why transportation research scientists and engineers are getting interested in nanomaterials to improve the mechanical and structural durability of concrete and asphalt pavement road infrastructure is not far-fetched. Nanotechnology allows for the creation of new cementitious concrete and asphalt pavement materials at the molecular level which can influence greatly interactions at the macroscopic level. Pavement scientists, researchers and engineers met in August 2006 to establish a "Roadmap for Research" with a focus on the utilization of Portland cement concrete

Julian Mills-Beale
Civil and Environmental Engineering, Michigan Technological University, 1400 Townsend Drive, Houghton, MI 49931
e-mail: jnmillsb@mtu.edu

Zhanping You
Civil and Environmental Engineering, Michigan Technological University, 1400 Townsend Drive, Houghton, MI 49931
e-mail: zyou@mtu.edu

K. Gopalakrishnan et al. (Eds.): Nanotechnology in Civil Infrastructure, pp. 257–270.
springerlink.com

and asphalt concrete. This workshop was organized under the auspices of the US National Science Foundation (NSF) with the theme "Nanomodification of Cementitious Materials".

This chapter deals with literature review on the nanomodification of civil transportation infrastructure with particular focus on asphalt pavements, and current projects involving the use of nanoclay particles for the preparation of a promising nanomaterial – nanoclay - for enhancing the field performance of asphalt pavements. It further highlights on some of the key challenges and considerations necessary as the industry plans to move towards the implementation of nanomaterials in asphalt pavements.

2 Nanomaterials in Transportation Infrastructure

The promising future that nanotechnology holds for the transportation infrastructure industry lies in the fact that the design, modification and material characterization at the nanolevel translates into better performance properties at the macroscale level in fields like energy and electronics [MacInnes 1981; Nishide 2004; Furukawa 1993]. This interesting technology is being adopted by the asphalt pavement infrastructure industry.

2.1 Nano-scale Enhancement of Concrete Pavement Systems

A number of research activities worth mentioning have been conducted in the last decade to examine the appropriateness of using nanomaterials in concrete pavement technology.

In modeling concrete pavements, some researchers have established that the hydrated cement past (HCP) component of concrete mixtures is a nanomaterial [Mehta 1986]. It is reported that the structure of calcium silicate hydrate also has striking similarity with clay particles [Taylor 1997]. Taylor further stated that the typical calcium silicate hydrate consists of ultra-thin layers of solids separated by water-filled gel pores. This critical nano-sized material structure influences the potential migration of water and its consequent effect on shrinkage and cracking of the concrete pavement.

Nano-sized TiO_2 and SiO_2 have been used in the design and construction of concrete pavements [Li et al. 2006]. The research proved hinted that the addition of nano-sized TiO_2 and SiO_2 increases significantly the flexural fatigue performance of concrete mixtures used in concrete pavements. Another interesting finding was the fact that the double-parameter Weibull distribution is applicable to characterize the fatigue lives behavior of concretes containing nano-particles. Other researchers worldwide have pursued notable projects which seeks to answer the pertinent questions of how and why nanomaterials should be used in concrete [Balaguru 2005; Fu et al. 1996; Li et al. 2007; Li et al. 2006; Li et al. 2004].

2.2 Nano-Scale Enhancement of Asphalt Pavement Systems

In order to better understand and interpret the physio-chemical interactions of asphalt nanomaterials, some researchers are developing tools to model the phenomenon [Pauli et al. 2006]. Significant among the work being conducted by Pauli et al. are: 1) to develop standard practices to characterize the asphaltene and maltene wax phases in the asphalt binder system using nano-thin-film chromatography techniques; 2) to develop nanoindentation of thin-film asphalt mastics and pavement cores; 3) to utilize nanoindentation for the analysis of mineral fines in asphalt binder systems. Attempts have been made in many respects to analyze the properties of nanoscale asphalt schemes using molecular simulation technology [Greenfield 2007]. Greenfield studied two different asphaltene model structures and evaluated asphalt binder system properties such as the viscosity and compressibility (inverse of the bulk modulus). At the nanoscale levels, a high-frequency glass transition state above a temperature condition of 25 °C was observed. Greenfield further investigated that there existed a more pronounced transition and a greater bulk modulus for asphalt binder systems having a higher percentage of aromatic asphaltene than those with lower aromatic asphaltene fractions.

2.3 What Is a Nanoclay Particle?

Nanoclays has been defined as "clay that can be modified to make the clay complexes compatible with organic monomers and polymers" [Jahromi and Khodaii 2009]. In striking similarity with montmorillonite clay minerals, x-ray and microscopic detailing have indicated the presence of Si, O, Mg, Al and Fe elements in typical nanoclay materials, which are small in nature and have poor crystalline shapes (Ghille 2006). Ghille further reported that nanoclays do not normally have nitrogen elements in them. Based on plastic limit results between the range of 85.4 and 87.5%, Ghille added that nanoclays can be classified as among the family of expansive montmorrilonite.

2.4 Nanoclays in Asphalt Binders and Mixtures

Pavement researchers have suggested that nanoclays have interesting performance enhancing characteristics that could improve the mechanical strength properties of both asphalt binders and mixtures. In one of such notable studies, it has been shown that polymeric nanoclay composites can improve the physical, chemical and mechanical properties of asphalt binder systems when the nanoclays are evenly dispersed at the nanoscopic level within the binder system [Jahromi and Khodaii 2009]. Nanofil-15 and Cloiste-15A are two types of nanoclay materials that were used in their research. In terms of binder system properties like the stiffness, phase angle and aging resistance, *nanofil-15* and *cloisite-15A* have the potential to increase stiffness, decrease phase angle and reduce the aging characteristics of the asphalt binder.

Another research worth mentioning is work conducted by Ghille in the Netherlands. Ghille's work centered on the microscopic analysis on *nanofil* and *cloisite* nanoclays. Among the findings from the microscopic study is the scientific fact that nanoclays have very large aspect ratios [Ghille 2006]. Furthermore, nanoclays are usually of non-uniform size and arrangement. Between the nano cloisite and the nanofill particles, the former was found to be larger and less curly than the latter. At the rate of 6% input as additive in an asphalt binder, improved the permanent deformation or rutting behavior of the asphalt binder system and enhanced the resistance to both short and long-term aging (oxidation) of the asphalt binder system. For both the asphalt binder system and the designed asphalt paving mixture, the *nanofil* and nano cloisite were found to have lower fatigue resistance compared to unmodified asphalt binders and mixtures.

Nano-calcium carbonate material use in asphalt binder and mixtures has been explored in China [Liu et al, 2007; Ma et al. 2007]. Their work indicated strongly that using nano-$CaCO_3$ can reduce the occurrence of inelastic deformations which causes permanent deformation on an asphalt pavement. The basis for this occurrence is the formation of a uniform and steady system which improves the high temperature viscoelastic reaction of the system. Under Marshal Stability Test Method for the laboratory performance evaluation of nanoclay-modified asphalt mixtures, it has been determined that at 5% addition of the nanoclays, optimum mixture performance can be attained.

In asphalt paving technology, the use of styrene- butadiene-styrene (SBS) copolymer has been used as a modifier for producing high performing mixtures [Hanyu et al. 2005; Chen et al. 2006, Fu et al. 2007; Yildirim, 2007]. Nanoclays have been used as a secondary modifier to further enhance the performance properties of styrene-butadiene-styrene (SBS) copolymer modified asphalt (Yu et al. 2007b). In adding the sodium montmorillonite (Na-MMT) and organophilic montmorillonite (OMMT) nanoclays, it was found that: 1) the viscosity of the SBS-modified asphalt was increased; 2) the stiffness (complex modulus) was increased while the phase angle decreased. The research results are an indication that the Na-MMT and OMMT nanoclays have promising potential to reduce the permanent deformation or rutting of asphalt pavements. The Na-MMT and OMMT were detected to have form an intercalated and exfoliated structure within the SBS copolymer modified asphalt binder system (Yu et al. 2007a; Yu et al. 2007c).

The effect of nano-sized materials for asphalt mixtures under water or deicing solutions was studied recently (Goh, et al 2010). A total of 27 asphalt mixtures were prepared with various amount of nanoclay and/or carbon microfiber. The moisture susceptibility and deicer impacts were assessed by exposing the samples to water or deicing chemicals (NaCl, $MgCl_2$ and $CaCl_2$), and seven freeze-thaw cycles. It was found the addition of nanoclay and carbon microfiber would improve a mixture's moisture susceptibility in most cases (Goh, et al 2010).

3 Experimental Program

3.1 *Preparation of the Nanoclay Compound*

Surfactant-modified nanoclay A and B was first dispersed in isopropanol, a known organic solvent. The Nanoclay A and B were *Nanomer* and *Nanocloiste* nanoclay materials, respectively. Silane coupling agent was added to the surfactant-modified nanoclay and the mixture was continuously stirred in a water bath until all of the solvent evaporates. The use of a silane coupling agent has been known to contribute significantly towards the production of a homogenous uniformly dispersed nanoclay compound (Qian et al. 2007). At a temperature of about 80°C, the prepared products were dried in a vacuum, wetted and later filtered with a 280 mesh. Both the surfactant-modified nanoclay with the silane coupling agent was then dispersed in the control PG 64-28 asphalt binder using a high-shear mixer set at 2500 rpm for 3 hours. The PG 64-28 had been initially heated to 160°C and mixed with the surfactant-modified nanoclay which had the silane coupling agent. The control asphalt binder used for the research investigation was a designated Superpave™ Performance Grade (PG) 64-28. This asphalt binder was used by the Michigan Department of Transportation (MDOT) for its US-127 Highway Project. From this PG 64-28 asphalt binder, four types of nano-modified samples were prepared:

a) 2% Nanoclay A by weight of PG 64-28.
b) 4% Nanoclay A by weight of PG 64-28.
c) 2% Nanoclay B by weight of PG 64-28.
d) 4% Nanoclay B by weight of PG 64-28.

3.2 *Dispersion and Morphological Characterization of the Nanoclay-modified Asphalt*

3.2.1 Dispersion Characterization (Separation Tube and X-Ray Dispersion)

After the preparation of the *Nanomer* and *Cloisite*-modified PG 64-28 asphalt binders, two techniques are employed to evaluate the extent of dispersion of the *Nanomer* and *Nanocloisite* particles within the PG 64-28 asphalt binder system. The two techniques are the separation tube method and the X-ray dispersion method (XRD). The separation tube method involves the application of a thin-walled metal tube used to ascertain the level of uneven dispersion within an asphalt binder system. Fig. 1 shows the separation tube used during the test.

In using the separation tube, the hypothesis is that if the *Nanomers* and *Nanocloisite* particles are not sufficiently and uniformly dispersed within the asphalt binder matrix, vertical segregation will occur with higher nanoclay

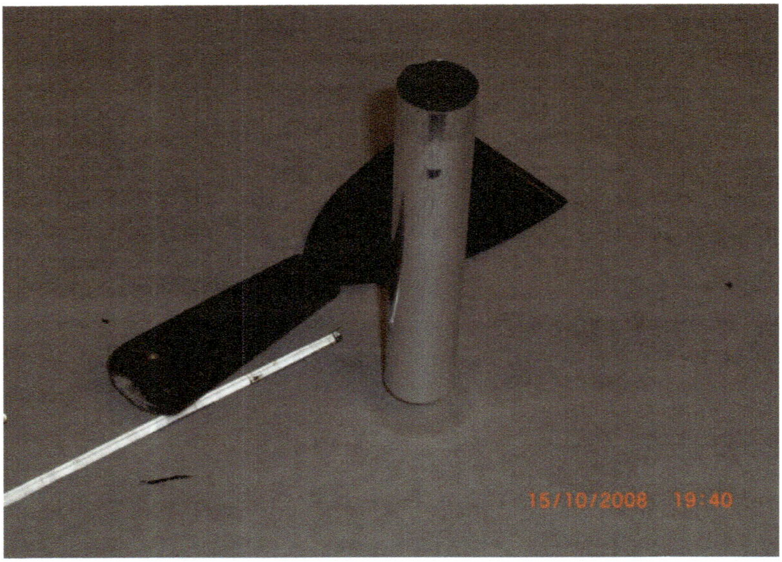

Fig. 1 The separation tube used in the dispersion analysis

concentrations at the bottom of tube than at the top. The *Nanomers* and *Nanocloisite*-modified PG 64-28 was heated to about 130 ^{0}C until fluid and this was poured into the separation tube. It was left to stand undisturbed and solidify at room temperature. The tube was later cut into three equal parts with the aid of a heated scraper. The three sections – top, middle and bottom – were tested individually using the dynamic shear rheometer (DSR) equipment. This stage of the procedure will confirm segregation or otherwise if there exists macroscopic changes in the rheological behavior or not.

In studying the morphology of the final nanoclay-modified PG 64-28 asphalt binder system, the X-ray diffraction (XRD) characterization technique was employed. The XRD technique has gained prominence in several scientific research fields for its ability to investigate the morphological and chemical features of complicated compounds including asphalt binders [Shirkoff et al. 1997; Siddiqui et al. 2002; Hesp et al. 2006]. Specifically, the XRD was used in this project specifically to give a quantitative analysis of the interlayer gallery spacing and the extent of dispersion of the nanoclays in the PG 64-28 asphalt binder system.

The XRD was conducted on the original or unmodified PG 64-28, Nanoclay A compound and finally the Nanoclay A modified PG 64-28 binder. From the XRD peaks, the research investigation sought to give an in-depth analysis of the extent of uniform mixing of the nanoclays within the asphalt. Therefore, the technique allowed for the evaluation of how the nanoclays with their electrostatic charges are attracting each neighbouring nanoclay particle in the blend.

3.3 *Characterization of the Nanoclay-Modified Asphalt Binder*

Based on the Superpave™ Standard Specifications for testing asphalt binders, the under-listed test procedures were conducted on both the control and the four nano-clay modified asphalt binders:

a) Rotational viscosity test (RV) - *Standard Method for Viscosity Determinations of Unfilled Asphalts Using the Thermosel Apparatus, ASTM 4402*).
b) Dynamic shear rheometer test (DSR) - *Rheological Properties of Asphalt Binder Using Dynamic Shear Rheometer AASHTO TP5*
c) Rolling thin film test (RTFO) - *Standard Method of Test for Effect of Heat and Air on a Moving Film of Asphalt, AASHTO T240*
d) Direct tensile test (DTT) - *Method for Determining the Fracture Properties of Asphalt Binder in Direct Tension, AASHTO TP3*

At temperatures of 80, 100, 130, 135, 150 and 175°C, the Brookfield Rotational Viscosity equipment was used to determine the viscous behavior of the control binder and the four types of prepared nanoclay-modified PG 64-28 binders. The speed for running the RV test was 100 rpm. In terms of the low and mid-temperature rheological properties, the DSR equipment was used at temperatures 13°C to 70°C and loading frequencies ranging from 0.01Hz to 25 Hz. The critical parameters sought here were the complex shear modulus (G*) and phase angle (δ) which are used to determine the rutting and fatigue cracking resistance behavior of the asphalt binder systems (Asphalt Institute 2003, Asphalt Institute). To determine the low temperature cracking resistance performance of the neat and nanoclay-modified asphalt binder system, the modified Superpave™ direct tensile test (DTT) procedure was employed. The DTT evaluated the low temperature ultimate tensile strain/stress at -18°C on both the neat and modified asphalt binders.

4 Discussion of Results

4.1 *Seperation Tube*

The separation tube tests were conducted on the nanomaterial at 2000 rpm at temperatures of 13 and 42 °C. The results for the 13 and 42 °C test conditions are provided in Fig 2 and Fig 3, respectively. The results reveal the DSR complex modulus (G*) trend for the nanoclay-modified binder from different segments of the separation tube. The different segments are the top, middle and bottom levels. As expected, the G* increases with increasing frequency for all the segments tested. This was in consonance with the fact that at lower frequencies, the shearing action had more pronounced impact on the asphalt binder and thus it resulted in less G* or stiffness condition. At higher or faster frequencies, the G* values are stiffer since the shearing action is less in time effect.

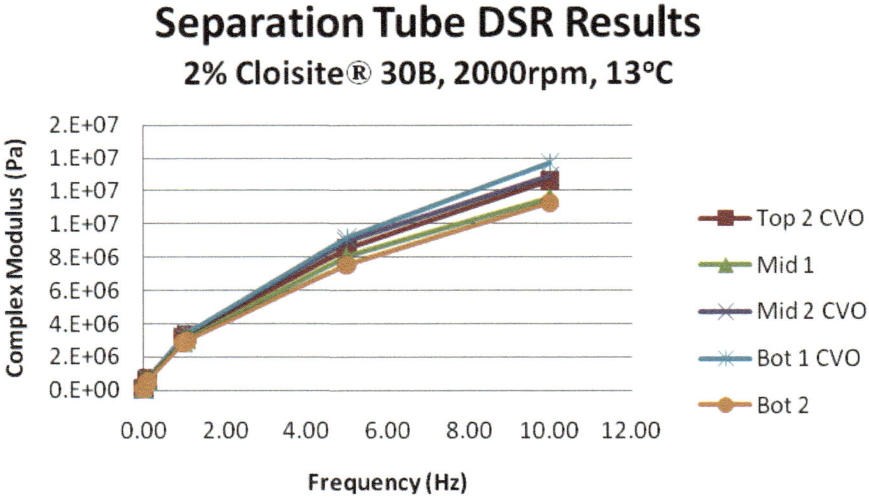

Fig. 2 DSR Results at 13 deg-C on Separation Tube Samples for 2% Nanomaterials

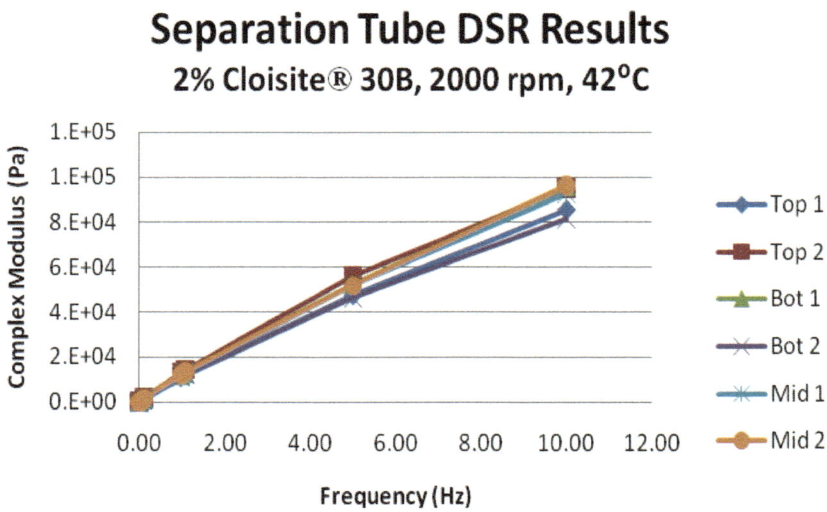

Fig. 3 DSR Results at 42 deg-C on Separation Tube Samples for 2% Nanomaterials

From Fig. 2, it is evident that between 0 and 10 Hz, the G* for the top, middle and bottom segments of the separation tube are close for the DSR tests at 13 ^{0}C. This qualitative assessment indicates that there is appreciable homogeneity within different segments of the nanoclay modified asphalt binder system. The trend seems similar to that observed in Fig. 3 which is the DSR results at 42 °C. From the two tests at both 13 and 42 °C, it was deducted that elevating the shear mixing speed from 2000 to 6000 rpm will improve the mixing condition and thus the uniform homogeinity of the asphalt binder and nanoclay medium. After this was the more accurate x-ray diffraction system used to confirm the analysis.

4.2 XRD Results

The nanoclays were shown to have been well distributed within the modified asphalt binder matrix. This condition is necessary for enhanced characterization testing of the mechanical and rheological properties. With the aid of the relative prominence of the nanoclays in the matrix compared to the unmodified asphalt binder system, the homogeneity of the system was determined. A comprehensive XRD results and data analysis is available in the paper by You et al. (2010).

4.3 Rotational Viscosity

The viscosity tests on the four different types of asphalt binder systems are shown in Fig. 4. From Fig. 4, it is evident that for temperatures ranging from 80, 100, 130, 135, 150 to 175°C, the increasing viscosity trend is:

Original < Nanoclay A-modified < RTFO < Nanoclay B-modified

In Table 1, the average viscosity changes are shown for all binder systems. The chemical reaction needs to be studied to understand how the chain-to-chain (bond-to-bond) formations between nanoclays and asphalt binders impact the viscosity; and if at all the increase in viscosity is as a result of new asphalt hybrid compound formations.

4.4 Dynamic Shear Rheometer (DSR)

In Fig. 5 and 6, two bar charts showing the DSR trends are presented here to establish the rheological characteristic of the original and nano-modified PG 64-28 asphalt binder systems. With 2% and 4% nanoclay addition to the PG 64-28 asphalt binder, the G* value increased by an average of 66% and 125%, respectively, from that of the original heat mixed PG 64-28 asphalt binder. When 2% of Nanoclay B was added, the trend showed an average G* increase of 184% while increasing the Nanoclay B additive to 4% generally increased the G* modulus by 196%.

The general increase can be due to the attraction of the polar ends of the nanoclays to the corresponding opposite polar ends of the asphalt molecules. Because of their relatively higher surface areas compared to other conventional asphalt modifiers, the nanoclays create a better interaction and reaction with the

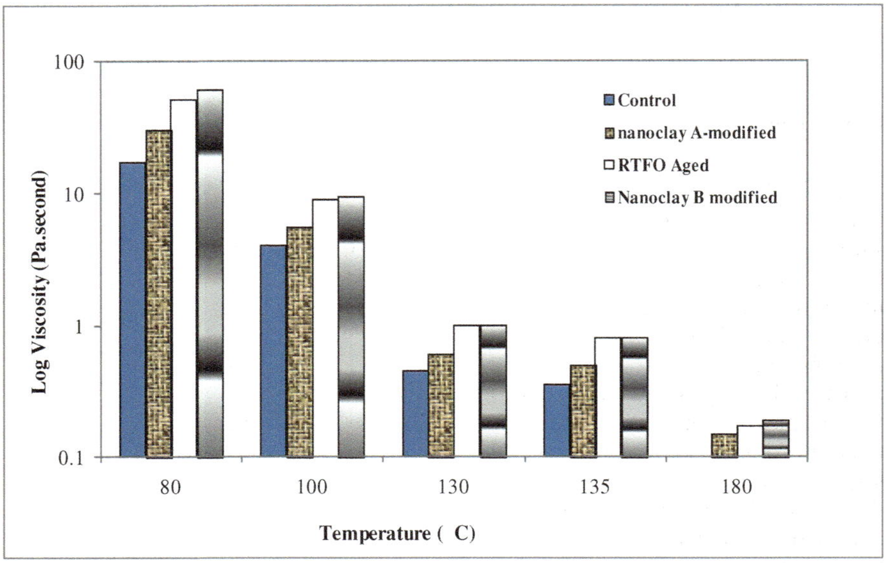

Fig. 4 Rotational viscosity plots for original binder, RTFO binder, Nanoclay A and Nanoclay B modified asphalt binder (after You at. el. 2010)

Table 1 Average viscosity changes between control, RTFO-aged and modified binders (after You at. el. 2010)

Average Viscosity Changes			
Log Original binder viscosity (Pa.s)	**2% Nanoclay A-modified**	**RTFO-aged**	**2% Nanoclay B-modified**
17.00	76%	194%	253%
4.00	38%	125%	138%
0.45	33%	122%	122%
0.35	43%	129%	129%
0.10	50%	70%	90%
Average Change	**48%**	**128%**	**146%**

asphalt binder. This interaction is believed to lead to the formation of new bonds which gives the asphalt binders system the reinforcing effect.

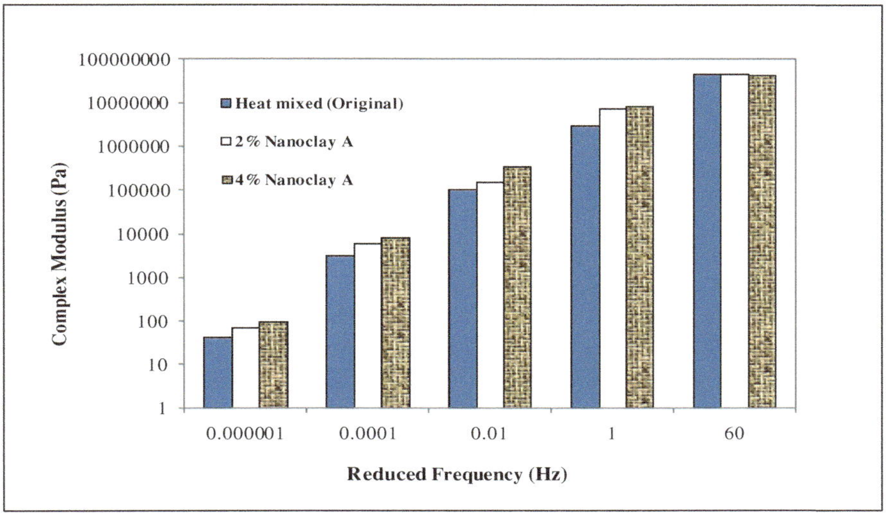

Fig. 5 DSR shear complex moduli (G*) plot on original asphalt, 2% Nanoclay A modified asphalt, and 4% Nanoclay A modified asphalt, after You at. el. 2010

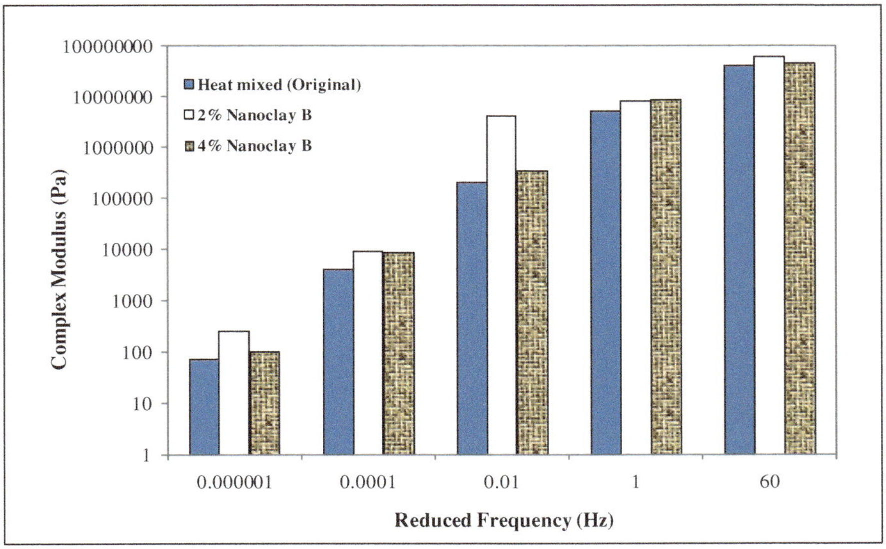

Fig. 6 DSR shear complex moduli (G*) plot on original, 2% Nanoclay B modified and 4% Nanoclay B modified asphalt binder, after You at. el. 2010

5 Conclusions

The research project initially involved the combination of a silane coupling agent and montmorillinite (MMT) surfactant-modified nanoclay to form an exfoliated nanoclay compound. The exfoliated MMT compound was then blended in a typical Michigan asphalt binder, PG 64-28, using a mechanical shear blending device operated at 2500 rpm. X-ray diffraction results prove that the shear mechanical blender at 2500 rpm was sufficient in creating a uniformly dispersed nanoclay particles within the PG 64-24 asphalt binder matrix. At 2 and 4% of Nanoclay A and B added to the PG 64-28 asphalt binder system, the viscosities and shear complex moduli all increased. From the Direct Tensile Test (DTT) results, it was observed that 2 and 4% additions of the nanoclays caused a reduction in the failure strain percent. The secant moduli from the DTT tests tend to increase with increasing nanoclay content.

From this preliminary research on the use of nanoclays in the asphalt binder, it can be inferred that nanoclays hold a potential in improving the mechanical properties of asphalt binders at both low and high temperatures. The nanoclays can be said to: 1) increase the viscosity of the asphalt binder system for resisting high temperature permanent deformation; 2) increase the shear complex modulus and thus improve the permanent deformation resistance at high temperature; 3) Increase the resistance to low temperature cracking if the nanoclays are added at more than 4% by weight of asphalt binder.

6 Future Work

This preliminary research has served as the guiding framework for future work on the use of nanoclays in asphalt binder systems. Future work will involve the use of nanoclays in asphalt binder at rates of more than 4% of the nanoclays by weight of the asphalt binder. Of course, it is not clear if the dispersion of the nano materials in asphalt is uniformly distributed. It is also a task to evaluate the performance of various nano materials blended in asphalt binder. One type of nano materials may not have similar effect as other nano materials.

It will be interesting also to determine the performance of asphalt paving mixtures that contains nanoclay modifiers or additives. With the known behavior of montmorillonite (clay) particles to attract water or moisture towards their lattice structures, it is pertinent to establish the moisture susceptibility of asphalt binder systems and mixtures that are subjected to moisture damage.

Acknowledgements. The research work was partially sponsored by the State of Michigan - Research Excellence Funds. The experimental work was completed in the Transportation Materials Research Center at Michigan Technological University, which maintains the AASHTO Materials Reference Laboratory (AMRL) accreditation on asphalt and asphalt mixtures. The contents of this article do not necessarily reflect the official views and policies of any institution or agency. The authors acknowledge the assistance of laboratory testing by Foley Justin and Kari Nasi.

References

Balaguru, P.N.: Nanotechnology and Concrete: Background, Opportunities and Challenges. In: Proceedings of the International Conference – Application of Technology in Concrete Design, Scotland, UK, pp. 113–122 (2005)

Chen, J.S., Huang, C.C.: Fundamental Characterization of SBS-Modified Asphalt Mixed with Sulfur. Journal of Applied Polymer Science 103(5), 2817–2825 (2006)

Fu, X., Chung, D.D.L.: Submicron Carbon Filament Cement-Matrix Composites For Electromagnetic Interference Shielding. Cement and Concrete Research 26(10), 1467–1472 (1996)

Fu, M.-F., Xiong, J.-G., Song, G.-Q.: Study and Application of Nano-materials in Concrete. Gongcheng Lixue/Engineering Mechanics 21(suppl.), 48–51 (2004)

Fu, H., Xie, L., Dou, D.L.: Storage Stability and Compatibility of Asphalt Binder Modified by SBS Graft Copolymer (styrene-butadiene-styrene), Construction and Building Materials

Furukawa, N., Nishio, K.: Applications of Electroactive Polymers. In: Scrosati, B. (ed.), ch. 5, pp. 150–181. Chapman & Hall, London (1993)

Ghile, D.B.: Effects of Nanoclay Modification on Rheology of Bitumen and on Performance of Asphalt mixtures. M.S. Thesis, Delft University of Technology, Delft, The Netherlands (2006)

Goh, S.W., Akin, M., You, Z., Shi, X.: Effect of deicing solutions on the tensile strength of micro or nano-modified asphalt mixture. Constr. Build. Mater. 25(1), 195–200 (2011), doi:10.1016/j.conbuildmat.2010.06.038

Greenfield, M.: Modeling for Nano-engineering: Molecular Simulation of Asphalt-like Materials. In: NSF Workshop on Nanomodification of Cementitious Materials at the University of Florida (2006)

Hanyu, A., Ueno, S., Kasahara, A., Saito, K.: Effect Of The Morphology of SBS Modified Asphalt on Mechanical Properties of Binder and Mixture. Journal of the Eastern Asia Society for Transportation Studies 6, 1153–1167 (2005)

Hesp, S.A.M., Serban, I., Shirokoff, J.W.: Reversible Aging in Asphalt Binders. Energy Fuels 21, 1112–1121 (2007)

Jahromi, S.G., Khodaii, A.: Effects of Nanoclay on Rheological Properties of Bitumen Binder. Construction and Building Materials 23(8), 2894–2904 (2009)

Li, H., Zhang, M., Ou, J.-p.: Flexural Fatigue Performance of Concrete Containing Nano-particles for Pavement (2006)

Li, Y.G., Ming Wang, P., Zhao, X.: Pressure-Sensitive Properties and Microstructure of Carbon Nanotube Reinforced Cement Composites. Cem. Concr. Compos. 29, 377–382 (2007)

Liu, D.-L., Bao, S.-Y.: Research of Improvement of SBS Modified Asphalt Pavement Performance by Organic Montmorillonite. Journal of Building Materials 10(4), 500–504 (2007)

Ma, F., Zhang, C., Fu, Z.: Performance and modification mechanism of nano-$CaCO_3$ modified asphal Wuhan Ligong Daxue Xuebao (Jiaotong Kexue Yu Gongcheng Ban. Journal of Wuhan University of Technology (Transportation Science and Engineering) 31(1), 88–91 (2007)

MacInnes Jr., D., Drury, M.A., Nigrey, P.J., Nairns, D.P., MacDiarmid, A.G., Heeger, A.J.: J. Chem. Soc., Chem. Comm., 317 (1981)

Mehta, P.K.: Concrete Structure, Properties and Materials. Prentice-Hall, Englewood Cliffs (1986)
Nishide, H., Iwasa, S., Pu, Y.J., Suga, T., Nakahara, K., Satoh, M.: Electrochim. Acta 50, 827 (2004)
Pauli, T., Beemer, A., Miller, J., Robertson, R.: Emerging Issues in Asphalt Binder Research and Application of Nanotechnology. In: NSF Workshop on Nanomodification of Cementitious Materials at the University of Florida (2006)
Shirkoff, J.W., Siddiqui, M.N., Ali, M.F.: Characterization of the Structure of Saudi Crude Asphaltenes by X-ray Diffraction. Energy & Fuels II, 561–565 (1997)
Siddiqui, M.N., Ali, M.F., Shirkoff, J.W.: Use of X-ray Diffraction to Assess the Aging Pattern of Asphalt Fractions. Fuel 18, 51–58 (2002)
Taylor, H.F.W.: Cement Chemistry, 2nd edn. Thomas Telford, London (1997)
Yildirim, Y.: Polymer Modified Asphalt Binders. Construction and Building Materials 21(1), 66–72 (2007)
You, Z., Mills-Beale, J., Justin, F., Roy, S., Odegard, G.M., Dai, Q., Goh, S.W.: Nanoclay-modified asphalt materials: Preparation and characterization. Constr. Build. Mater. 25(2), 1072–1078 (2011), doi:10.1016/j.conbuildmat.2010.06.070
Yu, J., Li, B., Zeng, X., Wang, X., Hu, H.: Preparation and Properties of Montmorillonite Modified Asphalts. Journal of Wuhan University of Technology 29(9), 65–67 (2007a)
Yu, J., Wang, L., Zeng, X., Wu, S., Li, B.: Effect of Montmorillonite on Properties of Styrene-Butadiene-Styrene Copolymer Modified Bitumen. Polymer Engineering & Science 47(9), 1289–1295 (2007b)
Yu, J., Zeng, X., Wu, S., Wang, L., Liu, G.: Preparation and Properties of Montmorillonite Modified Asphalts. Materials Science and Engineering: A 447(1-2), 233–238 (2007c)

About the Editors

Kasthurirangan (Rangan) Gopalakrishnan, Ph.D.

Prof. Kasthurirangan Gopalakrishnan is a Research Assistant Professor in the Department of Civil, Construction and Environmental Engineering at Iowa State University. He received his Ph.D. in civil engineering from the University of Illinois at Urbana-Champaign in 2004. His research interests include sustainable civil infrastructure, green engineering technology, bio-inspired computing, and Smart pavements. Dr. Gopalakrishnan has authored more than 125 publications and is also the lead editor of Springer's *Intelligent and Soft Computing in Infrastructure Systems Engineering: Recent Advances* and *Sustainable and Resilient Critical Infrastructure Systems: Simulation, Modeling, and Intelligent Engineering*.

Bjorn Birgisson, Ph.D.

Prof. Bjorn Birgisson is the Vice-President of Swedish Royal Institute of Technology (KTH) as well as a Professor of Road and Rail Technology in the Department of Civil and Architectural Engineering at KTH. He received his Ph.D. in civil engineering from the University of Minnesota in 1996. His research has covered a wide range frompavement design and materials development to operation and maintenance. Dr. Birgisson has authored more than 125 publications and is also the co-author of *American Roadmap for Nanotechnology-Based Concrete Research*. He serves as the Chairman of the *RILEM Technical Committee on Nanotechnology for Bituminous Materials* and *TRB Task Force on Nanotechnology-Based Concrete (AFN15T)*.

Peter C. Taylor, Ph.D., P.E.

Dr. Peter C. Taylor is the Associate Director of the National Concrete Pavement Technology Center (CP Tech Center) at Iowa State University. He received his Ph.D. in civil engineering from the University of Cape Town, South Africa in 1995. His prime area of interest is in concrete durability, including the development of test methods for assessing the potential durability of concrete mixtures. Dr Taylor is the author of more than 50 publications, and is actively involved in teaching applied materials technology to practicing engineers.He is an active member of TRB, ASTM and ACI with particular involvement in the committees relating to development of test methods for assessing potential durability of concrete, and in the use of slag and fly ash.

Nii O. Attoh-Okine, Ph.D.

Dr. Nii O. Attoh-Okine is an Associate Professor in the Department of Civil and Environmental Engineering at the University of Delaware. He received his Ph.D. in civil engineering from the University of Kansas in 1992. His research interests include computational intelligence and probabilistic reasoning in civil Infrastructure systems, MEMS application in civil Infrastructure systems and Hilbert-Huang Transform applications. Dr. Attoh-Okine is the author of more than 100 publications, and serves as an Associate Editor of *IEEE Transactions on Systems, Man, and Cybernetics, Part C: Applications and Reviews.*

Author Index

Batch number: 10614940

Printed by Printforce, the Netherlands